6995

Hazard Assessment of Chemicals

Advisory Board

Wesley J. Birge
University of Kentucky
Lexington, Kentucky

A. Wallace Hayes
R.J.R. Nabisco Inc.
Winston-Salem, North Carolina

Irwin H. Suffet
Drexel University
Philadelphia, Pennsylvania

Judith M. Hushon
Bolt Beranek and Newman Inc.
Arlington, Virginia

Donald MacKay
University of Toronto
Toronto, Ontario, Canada

Edmond J. LaVoie
Naylor Dana Institute
for Disease Prevention
Valhalla, New York

James E. Huff
National Toxicology Program
Research Triangle Park,
North Carolina

William L. Marcus
U.S. Environmental Protection Agency
Washington, D.C.

C. Richard Cothern
U.S. Environmental Protection Agency
Washington, D.C.

Topics Covered in the Series

Data Bases and Information Sources
Exposure Assessment
Monitoring and Analysis
Health Effects
Environmental Effects
Risk Assessment
Special Topics
Case Studies of Chemicals and Chemical Spills

Hazard Assessment of Chemicals

VOLUME 6

Edited by

JITENDRA SAXENA
Criteria and Standards Division
Office of Drinking Water
U.S. Environmental Protection Agency
Washington, D.C.

◑ HEMISPHERE PUBLISHING CORPORATION
A member of the Taylor & Francis Group
New York   Washington   Philadelphia   London

ŞT. PHILIP'S COLLEGE LIBRARY

HAZARD ASSESSMENT OF CHEMICALS: Volume 6

Copyright © 1989 by Hemisphere Publishing Corporation. All rights reserved.
Printed in the United States of America. Except as permitted under the United
States Copyright Act of 1976, no part of this publication may be reproduced or
distributed in any form or by any means, or stored in a data base or retrieval
system, without the prior permission of the publisher.

1 2 3 4 5 6 7 8 9 0   E B E B   8 9 8

This book was set in Times Roman by Hemisphere Publishing Corporation.
The editors were Linda A. Dziobek and Brenda Brienza, the production
supervisor was Peggy M. Rote, and the typesetter was Cindy Mynhier.
Edwards Brothers, Inc. was printer and binder.

Library of Congress Catalog card number: 82-640828

ISBN 0-89116-835-4
ISSN 0730-5427

ST. PHILIP'S COLLEGE LIBRARY

Contents

Sensitive Populations and Risk Assessment in Environmental Policymaking

JUDITH K. MARQUIS and GORDON C. SIEK

Factors Determining Target Doses: Extrapolation from a Shifting Basis

E. C. FOULKES

Approaches Used to Assess Chemically Induced Impairment of Host Resistance and Immune Function

PETER T. THOMAS

Detection of Mutagens and Carcinogens by Physiochemical Techniques

GEORGE BAKALE

Modeling of Combined Toxic Effects of Chemicals

ERIK R. CHRISTENSEN and CHUNG-YUAN CHEN

Principles and Applications of Surface Water Acidification Models

WILLIAM D. SCHECHER and CHARLES T. DRISCOLL

Fate and Transport of Sediment-Associated Contaminants

ALLEN J. MEDINE and STEVE C. McCUTCHEON

Theory and Practice of the Development
of a Practical Index of Hazardous Waste Incinerability

BARRY DELLINGER

Estimating Emissions from the Synthetic
Organic Chemical Manufacturing Industry:
An Overview

CLAY E. CARPENTER, CINDY R. LEWIS,
and WALTER C. CRENSHAW

Contributors

Numbers in parentheses indicate the pages on which the authors' contributions begin.

George Bakale (85), Case Western Reserve University, Radiology Department, Biochemical Oncology Division, Cleveland, Ohio 44106

Clay E. Carpenter (339), Versar, Inc., Springfield, Virginia 22151

Chung-Yuan Chen (125), School of Civil and Structural Engineering, Nanyang Technological Institute, Singapore

Erik R. Christensen (125), Department of Civil Engineering, University of Wisconsin-Milwaukee, Milwaukee, Wisconsin 53201

Walter C. Crenshaw (339), Versar, Inc., Springfield, Virginia 22151

Barry Dellinger (293), University of Dayton Research Institute, Dayton, Ohio 45469

Charles T. Driscoll (187), Department of Civil Engineering, Syracuse University, Syracuse, New York 13210

E. C. Foulkes (31), Department of Environmental Health and Physiology, University of Cincinnati, College of Medicine, Cincinnati, Ohio 45267

Cindy R. Lewis (339), Versar, Inc., Springfield, Virginia 22151

Judith K. Marquis (1), Department of Pharmacology and Experimental Therapeutics, Boston University School of Medicine, Boston, Massachusetts 02118

Steve C. McCutcheon (225), U.S. Environmental Protection Agency, Environmental Research Laboratory, Athens, Georgia 30613

Allen J. Medine (225), Water Science, Boulder, Colorado 80302

William D. Schecher (187), Department of Civil Engineering, Syracuse University, Syracuse, New York 13210

Gordon C. Siek (1), Department of Pharmacology and Experimental Therapeutics, Boston University School of Medicine, Boston, Massachusetts 02118

Peter T. Thomas (49), Life Sciences Department, IIT Research Institute, Chicago, Illinois 60616

Preface

Assessment of environmental and health hazards caused by chemicals requires a multidisciplinary approach. One needs to consider chemical economics, production, usage, environmental release, monitoring data, environmental behavior, and health and environmental effects. Prediction can often be made concerning environmental and health hazards based on the structure-activity relationship and the physicochemical characteristics.

A vast amount of new information about new pollutants, new effects, and new measures to deal with the problem of the increasing presence of chemicals in the environment is accumulating continually. This serial publication provides a single forum for comprehensive and authoritative reviews about new and significant developments within the broad field of chemical hazard assessment. Since the field is dynamic, the chapters of the various volumes in the series will change to reflect this dynamism. The topics covered in the series are data bases and information sources, exposure assessment, monitoring and analysis, health effects, environmental effects, risk assessment, special topics, and case studies of chemicals and chemical spills. However, no single topic is necessarily treated on an annual basis.

Volume 6 comprises nine reviews in the important areas of chemical hazard assessment. No case studies are covered in this volume. No attempt has been made to cover all areas of chemical hazard assessment in this or any other volume.

The need for information on the toxicological effects of environmental chemicals is clearly based on the need to project human health. Thus, four chapters in this volume are devoted to new and emerging areas of toxicology. Consideration of immunotoxicity for risk assessment has emerged, with the confirmation that many chemicals produce immunosuppression at dosages lower than those that alter other commonly used toxicological indices. The purpose of the chapter by Peter Thomas is to introduce readers to the concept of the immune system as a legitimate target organ for toxicity. The chapter presents a state-of-the-art review of the available approaches for evaluating chemically induced immunotoxicity and interlaboratory validation of these approaches. George Bakale has reviewed nonbiological physicochemical methods that are based on various structure-activity relationships, as well as chemical structures, for detection of chemical carcinogens and mutagens. The theoretical basis for these screening tests and their validation against the classical means of identifying chemical mutagens and carcinogens are also covered in this chapter.

Specific variables relating to the experimental animal, the chemical form, the environmental conditions during exposure, and other variables that may influence the outcome of toxicological studies and assessments of risk are reviewed by Ernest Foulkes. Special attention has been given in this chapter to the rate of exposure, as opposed to the total dose, and to the precise chemical nature of the toxicant. Setting environmental standards with an adequate margin of safety for protecting sensitive individuals in the population has been a difficult task and one of considerable controversy. To compensate for variability in susceptibility to the toxic effect of hazardous chemicals, a safety factor of 10 is often used in risk assessment and setting environmental standards. Judith Marquis's chapter examines the appropriateness of this safety factor in the context of the numerous factors that are capable of modifying an individual's response to a toxicant.

In the areas of environmental fate and exposure assessment, two important contributions are included in the volume. Allen Medine's chapter critically reviews the important processes that govern the association of contaminants and sediment and attempts to describe the mathematical formalism that can be used to describe the environmental behavior of sediment-associated contaminants. An understanding of the bioavailability of sediment-associated contaminants and their mobility is necessary to fully understand the effect of contaminants loading in aquatic systems. William Schecher's chapter, on the other hand, provides an overview of the major processes involved in the acidification of terrestrial and aquatic systems and the role of acidification models in predictive assessments. Although there is considerable consensus with regard to the environmental impact of acid deposition, the causes of acid deposition are a subject of much controversy. Therefore, an in-depth review on this subject is very much needed.

Erik Christensen's review deals with multiple toxicity, an area of aquatic toxicology in which additional information is urgently needed. He has provided an inclusive literature review of chemical mixtures dealing with aquatic life and a design for modeling the combined toxic effects of hazardous chemicals in the aquatic environment. The ultimate value of modeling the biological effect of multiple toxicants lies in the acquired predictive capability enabling estimation of the toxic effects from wastewater discharges and in the formulation of scientifically defendable water quality criteria.

The potential generation and release of partial combustion and pyrolysis effluent during thermal disposal of hazardous waste is an area of concern that needs to be addressed before this technology can be widely accepted. The complexity of thermal destruction processes and the differences in combustor and pyrolyzer design make prediction of partial combustion and pyrolysis effluents and absolute system performance essentially impossible. Barry Dellinger's chapter examines the concept of establishing guidelines for ranking the relative incinerability of hazardous waste and for predicting incomplete combustion products based on waste composition, system design, and operating parameters.

Critical and comprehensive reviews by leading authorities in the field in these areas of chemical hazard assessment should make Volume 6 a valuable addition to the series.

Jitendra Saxena

Hazard Assessment
of Chemicals

Sensitive Populations and Risk Assessment in Environmental Policymaking

Judith K. Marquis and Gordon C. Siek

Department of Pharmacology and Experimental Therapeutics
Boston University School of Medicine
Boston, Massachusetts

I. INTRODUCTION

Individuals vary in their susceptibility to the toxic effects of hazardous chemicals. A number of factors, including genetic predisposition, age, gender, nutritional status, preexisting disease, and concomitant exposure to other compounds, are all capable of modifying an individual's response to a toxicant (Fig. 1). Human populations are variable also with respect to diet, occupational and home environment, activity patterns, and other cultural factors. The following are factors that influence individual susceptibility to the toxic effects of exposure to xenobiotics:

- Genetic predisposition
- Age
- Gender
- Nutritional status
- Diet
- Body composition
- Concomitant disease
- Preexisting pathology
- Toxicant interactions
- Home environment

The authors are grateful to Mr. Richard Hamilton, Testall, Inc., for providing graphic illustrations of the concepts presented in this chapter. This work was supported in part by funds from the Laboratory of Analytical Toxicology, Boston University School of Medicine.

1

- Occupational factors
- Activity patterns
- Cultural factors

To compensate for this variability in sensitivity, a "safety factor" of 10 is often used in risk assessments and in setting exposure limits. In this chapter, we examine the role of these several factors in determining the sensitivity of individuals to toxicants, and we consider the appropriateness of using the "factor of 10" rule.

In a recent report, the GAO (1986) defined the concept of using safety factors to place quantitative limits on exposure to environmental chemicals, specifically pesticides. The safety factor was defined as "a number intended to provide a margin of safety and account for inherent uncertainty in projecting the results of animal toxicology studies to humans." The total safety factor for pesticides is usually 100-fold: 10-fold to account for the differences between humans and test animals, and 10-fold for the differences in sensitivity among different people. An experimentally derived no-observable-effect level (NOEL) can then be divided by this safety factor to yield a quantitative estimate of acceptable daily intake (ADI). For pesticides, ADI is defined as "a person's daily intake of a pesticide residue which, during a lifetime, is not expected to cause appreciable health risks on the basis of all facts known at the time" (GAO, 1986).

These definitions emphasize the importance of setting a safety factor that adequately accounts for the measured differences in susceptibility to toxicants within a given population (Fig. 2). The major alterations seen in the ability of different individuals to handle a given dose of toxicant are largely differences in toxicokinetics, i.e., "the rates of all metabolic processes related to the expres-

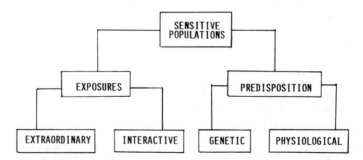

FIG. 1. Subgroups of sensitive populations. General summary of identified groups of people that exhibit altered susceptibility to the adverse effects of toxicants.

FIG. 2. Relative thresholds for toxic effects among various subgroups of a population. While the healthy worker overshadows the numerous sensitive populations identified in the text, there is increasing evidence to support a more conservative safety factor than presently utilized in setting acceptable exposure limits for the general population.

sion of toxicologic end points" (DiCarlo, 1982). Toxicokinetics are influenced by route of exposure and by many sources of variation within a species, including all of the numerous factors described and listed above. These variables especially influence the toxic effects of chronic interaction of the organism with a chemical and, thus are, of particular consequence in considering potential carcinogens.

II. GENETIC PREDISPOSITION

The influence of human genetic polymorphism on the effectiveness and toxicity of pharmacological agents has been extensively studied (Goldstein et al., 1974). Several well-known pharmacogenetic syndromes such as glucose-6-phosphate dehydrogenase deficiency and variations in *N*-acetyltransferase activity have also been implicated in various toxicant-related disorders. This has led to numerous investigations into the mechanisms underlying hereditary susceptibility to chemical toxicants (Omenn, 1983).

ST. PHILIP'S COLLEGE LIBRARY

A. Genetic Variation in Xenobiotic-Metabolizing Enzymes

Human populations show considerable variation in the metabolism of many xenobiotics. There are several cases where the underlying genetic polymorphism has been described and the possible effects on human sensitivity to toxins has been investigated.

1. N-Acetylation

N-Acetylation of aromatic amines in humans has been distinguished into fast and slow phenotypes. The slow phenotype is the result of a single recessive autosomal gene whose frequency varies widely in human populations. For example, in the United States the distribution of fast and slow acetylators is about 50:50, while among orientals, slow acetylators represent only about 10% of the population. The slow acetylator phenotype has been related to the appearance of a Lupus-like rheumatoid syndrome in patients given the antihypertensive agent hydralazine, the antiarrhythmic agent procainamide, and several other drugs (reviewed by Lunde et al., 1983). In 1978, Glowinski et al. published a study on the effects of N-acetylation phenotype on the metabolism of a number of carcinogenic arylamines including β-naphthylamine and benzidine. Using biopsied human liver tissue, they demonstrated that the fast and slow acetylator phenotypes exhibited a 5- to 13-fold difference in acetylation rate (depending on the substrate chosen). Since several of these arylamines had been previously associated with urinary bladder cancer among dye workers, a number of studies were initiated to examine the possible link between bladder cancer and N-acetylation phenotype. Several of these studies suggested that slow acetylators were at greater risk for bladder cancer (Lower et al., 1979; Cartwright et al., 1982).

However, Evans (1986) critically analyzed eight different studies on the relationship between N-acetyltransferase phenotype and bladder cancer. In his report, he pointed out deficiencies in several of the studies and reanalyzed the pooled data. He concluded that the association between phenotype and cancer was not significant and suggested that further study is needed to resolve this issue.

2. Debrisoquine Hydroxylation or C-center Drug Oxidation

Studies over the last decade have revealed that polymorphism of a single gene probably regulates the C-center cytochrome P-450 mediated oxidation of a wide number of xenobiotics, including the antihypertensives debrisoquine and guanoxan, the oxytocic sparteine, the local anesthetic encainide, the analgesic phenacetin, the antitussive dextromethorphan, several β-adrenergic blockers structurally related to propranolol, and several tricyclic antidepressants (Eichelbaum, 1984, 1986). Two phenotypes, poor metabolizers (Pm) and extensive metabolizers (Em), can be distinguished. The variation between the Pm and Em phenotypes is extraordinary. For example, Idle and Smith (1979) studied the metabolism of debrisoquine by 100 different individuals and found a 20-fold difference in the median metabolic ratios of debrisoquine/4-hydroxydebrisoquine between Pm and Em phenotypes. They then chose eight subjects, four of each phenotype, and found that the median metabolic rates for aromatic hydroxylation of the antihypertensive drug guanoxan varied 160-fold. The possible relationship between debrisoquine phenotype and several toxin-related cancers has been investigated, since oxidation may play a role in the activation of carcinogens.

Cartwright (1983) examined the debrisoquine phenotypes of bladder cancer patients which had been previously studied in connection with N-acetyltransferase phenotype and found no association. More recently, Ayesh et al. (1984) reported that the Em phenotype of debrisoquine hydroxylation is associated with the presence of bronchial carcinoma in smokers. They suggested that the gene that controls debrisoquine hydroxylation may also control metabolic oxidative activation of procarcinogens in cigarette smoke or may be a marker for other genes responsible for susceptibility to bronchial carcinoma.

3. Aryl Hydrocarbon Hydroxylase (AHH)

Polycyclic aromatic hydrocarbons are metabolized in part by certain cytochrome P-450 monooxygenases, collectively called aryl hydrocarbon hydroxylases (AHH). This often results in the formation of carcinogenic metabolites. A cytosolic receptor for aromatic hydrocarbons has been shown to regulate the induction of these enzymes through a specific gene (Ah). Evidence exists to suggest that the levels of inducible AHH activity exhibit genetic polymorphism in both animal models and humans (Nebert, 1978). Using cultured human lymphocytes, Kellerman et al. (1973) distinguished three "inducer" phenotypes— low, intermediate, and high. They also demonstrated that the "high inducer" phenotype could be associated with susceptibility to bronchial carcinoma.

Due to methodological problems, attempts to replicate this finding have been mixed. Kouri et al. (1983) identified conditions under which more reproducible results can be obtained. Their results confirm the original observation of Kellerman et al. (1973) linking lung cancer with the high inducer phenotype. Individuals of their control group who fell within the 95% confidence limits exhibited a 28-fold range of AHH activity.

In a comprehensive review of screening and monitoring procedures to identify both exposure and susceptibility to carcinogens, Warren and Beck (1987) discussed the use of AHH levels as a screening tool to detect cancer susceptibility. They conclude that studies on the induction of AHH in humans and its ability to predict lung cancer are really not definitive. Nevertheless, there remains a strong association between the inducibility of AHH and the presence of respiratory tract cancer in several subgroups of the population.

4. Epoxide Hydrolase

Epoxide hydrolase, a microsomal and cytosolic enzyme, hydrolyzes a number of epoxides to dihydrodiols (Kapitulnik et al., 1977). A number of epoxide metabolites of aromatic hydrocarbons have been implicated as carcinogens; thus, epoxide hydrolase may be important in detoxifying certain xenobiotics. Although this enzyme has not been shown to be expressed polymorphically, the wide range of activity seen in human populations suggests that it may influence the susceptibility of individuals to carcinogens. For example, Kapitulnik et al. (1977) demonstrated an eightfold variation in liver epoxide hydrolase activity, while Glatt and Oesch (1983) demonstrated a threefold variation in both human liver and blood. The relationship between differences in epoxide hydrolase activity and toxicant-induced disease has not yet been investigated.

5. Paraoxonase

Paraoxon, the toxic metabolite of parathion, can be hydrolyzed by a human serum enzyme paraoxonase. This activity has been characterized as an arylesterase with a preference for aromatic carboxylic acid esters, carbamates, and organophosphates and is expressed polymorphically in human populations. Under normal physiological conditions the high and low group vary in median activity by about threefold (LaDu and Eckerson, 1984). The level of activity is low in human blood, and it is unlikely to be an important route of detoxification following acute intoxication with large doses of organophosphates. However, there are in vitro data that suggest that variation of paraoxonase levels may be important during chronic, low-dose exposures to organophosphates (LaDu and Eckerson, 1984). This may have clinical relevance, since human responses to low doses of carbamates and organophosphates is quite variable (Levin et al., 1976; Stoller et al., 1965).

B. Genetic Variation in Target Tissues

In addition to genetic polymorphism of the metabolism of xenobiotics, the target tissues may also exhibit genetic variations in their sensitivity to toxicants. A number of examples are described below.

1. Glucose-6-Phosphate Dehydrogenase Deficiency

A well-studied example of gene polymorphism in humans is glucose-6-phosphate dehydrogenase (G6PD) deficiency. Many variants have been described, and the trait is expressed in about 10% of black males in the United

States and in many individuals of Mediterranean origin. G6PD helps maintain the supply of reduced glutathione which is important for protecting tissues from oxidative damage. Individuals with G6PD deficiency have activities that are about 10% of normal (Beutler, 1983). Individuals with G6PD deficiency are usually asymptomatic, unless they are exposed to chemicals that are strong oxidants.

Although many chemical compounds have been implicated in this mechanism of toxicity, only the following drugs and toxicants have been shown clinically to cause hemolytic anemia in G6PD deficient individuals, as reviewed by Beutler (1983):

- Acetanilid
- Methylene blue
- Nalidixic acid
- Naphthalene
- Niridazole
- Nitrofurantoin
- Phenylhydrazine
- Primaquine
- Pamaquine
- Pentaquine
- Sulfanilamide
- Sulfacetamide
- Sulfapyridine
- Sulfaphamethoxazole
- Thiazolesulfone
- Toluidine blue
- Trinitrotoluene
- Fava beans

2. Serum and Erythrocyte Cholinesterases

Organophosphate and carbamate pesticides exert their toxic effects by inhibiting acetylcholinesterase (AChE) at neuronal synapses. They are also capable of inhibiting erythrocyte AChE and serum cholinesterase (ChE). Although these particular ChE proteins are not, strictly speaking, the target for these toxicants, they can act as a "sink" for the compounds and, thus, can influence the extent of neuronal AChE inhibition (Russell and Overstreet, 1987).

Human populations exhibit considerable interindividual variation (approximately twofold to threefold) in the levels of erythrocyte AChE and serum ChE (Chu, 1985; Moses et al., 1986). This makes assessment of organophosphate and carbamate poisoning difficult, and further suggests that persons with low levels of blood ChE activities may be at greater risk for the toxic effects of ChE inhibitors at low-dose exposures. Since paraoxonase levels are also expressed polymorphically in humans and may influence individual responses to low doses of pesticides, as discussed above, the range of toxic doses for organophosphate pesticides is especially difficult to predict on a population basis.

C. Genetic Variation in DNA Repair and DNA Binding of Carcinogens

The sensitivity of different tissues to the action of individual carcinogens has been related to a number of factors including pharmacokinetic differences in tissue distribution and metabolism as well as differences in tissue activities of DNA repair mechanisms. For example, nervous tissue has been shown to be preferentially sensitive to the carcinogenic effects of nitrosoureas in comparison to liver tissue. This effect can be correlated with the half-life of the O^6-alkylguanine adducts formed by N-alkyl-N-nitrosoureas in each tissue, a reflection of the differences in the activity of O^6-alkylguanine transalkylase (Holbrook, 1982).

Studies such as these have led to investigations in the interindividual variation in both the tissue distribution of carcinogens and DNA repair mechanisms. Harris et al. (1983) reviewed a number of studies on human variation in both O^6-alkylguanine DNA transalkylase and uracil DNA glycosylase, two DNA repair enzymes. O^6-Alkylguanine alkyltransferase has been shown to vary from 2-fold in the esophagus to 42-fold in the small intestine, whereas uracil DNA glycosylase has been shown to vary from 3.1-fold in the liver to 65-fold in the small intestine (Harris et al., 1983; Myrnes et al., 1983). These results should be judged cautiously, however, since only small numbers of subjects (2–12) were studied. Furthermore, both enzyme activities may be induced by environmental factors such as diet. It remains to be demonstrated that these differences are true in larger groups of subjects and that they are of genetic origin.

Harris et al. (1983) also reviewed data on interindividual differences in the binding of carcinogens to DNA in cultured epithelial cells from a number of tissues. They report that the binding of benzo(a)pyrene, aflatoxin B_1, and N-nitrosodimethylamine varied by 68- to 150-fold depending on the tissues studied. The binding of these carcinogens to DNA in the cells studied represents a measure of a number of processes, such as a compound's ability to enter certain cell types, and the ability to those cells to metabolize the compound and repair the damage the compounds cause to DNA. The wide interindividual differences in different subjects suggests that all of the above processes vary in humans. However, the number of subjects studied was small, and it also remains to be shown that differences in in vitro binding to DNA relates to differences in susceptibility to carcinogens in vivo.

III. IMMUNOLOGICAL EFFECTS AND SENSITIVE POPULATIONS

Among the recent evidence for individual susceptibility to chemical toxicants are the numerous studies suggesting effects of toxicants on the immune system and altered response to toxicants in populations with diminished immune function. Activated or suppressed immune function is characteristic of hypersensitivity diseases, and the immune system is known to exhibit considerable variability even in a small group of people. Some of this variability is attributable to prior exposure or sensitization of particular individuals. However, immune alterations can be either a component or a consequence of the disease process and, thus, may be more difficult to quantitate than effects on other organ systems.

What are the markers for immunological changes? Serum immunoglobulins are frequently elevated in persons with hypersensitivity diseases. T-lymphocyte function is also depressed. Some hypersensitivity reactions result in complement activation followed by generation of chemotactic factors and release of proteins that enhance vascular permeability.

In animal studies a large number of tests can be used to characterize the capacity of chemicals to produce either immune suppression or enhancement of immune function. Rapid identification of affected subsets of immune function can be achieved readily by a relatively small battery of tests that measure host resistance, immune function, and macrophage function. Several of the tests presently available to examine immunotoxicity were described by Marquis (1986).

A. Alteration of Immune Function by Pesticides

Luster et al. (1982) summarized data for some of the environmental chemicals, including pesticides, that have been reported to affect the immune system. Immunosuppressive actions have been demonstrated for DDT, dieldrin, and organochlorines. Street and Sharma (1975) demonstrated alterations of induced cellular and humoral immune responses in rabbits by several pesticides, including the chlorinated hydrocarbons DDT and Aroclor 1254 and the carbamate anticholinesterase insecticide carbaryl. Their data illustrate the wide range of sensitivity to immunosuppression that is frequently observed both among the different tests for immune function and among the test animals in a study. The following are potential toxic effects measured in vivo in studies of the immune system (Dean et al., 1986):

- Complement activation
- Elevated serum immunoglobulins

* Increased susceptibility to infection
* Depression of cell-mediated immunity
* Depressed T-lymphocyte function
* Depressed levels of serum complement
* Altered responses to T- and B-cell mitogens
* Decreased spleen and thymus weight
* Altered natural killer cell cytotoxicity

Altered susceptibility to infection following exposure to pesticides provides another measure of chronic immunotoxicity. For example, Cranmer et al. (1982) demonstrated that chlordane may alter immune function in young adult mice following in utero exposure. Although considerable variability was evident in the data, a significant depression of cell-mediated immunity was observed in the exposed mice with no effects on the humoral immune system. This was accompanied by enhanced survival after influenza type A virus infection and increased production of antiviral antibodies in the exposed animals (Menna et al., 1985). These results are similar to those published earlier by Menna and co-workers (1980) following studies with percutaneous exposure of suckling mice to insecticide vehicles. Again, suppression of cell-mediated immunity in the exposed mice was correlated with enhanced survival following infection with influenza type A virus. The specificity of chlordane effects on the immune system requires further study, but it represents an area of concern for human exposure, as the compound has been a widely used termiticide for over 30 years.

Desi et al. (1978, 1980) have conducted several experiments to determine the effects of pesticides on immunological reactivity in laboratory animals. In a study with the chlorinated hydrocarbon, lindane, and two organophosphate insecticides, malathion and dichlorvos, they administered acute doses orally to rabbits and measured the development of antibodies to *Salmonella typhi*. All three pesticides were shown to cause a dose-related decrease in antibody titer, although, at the lowest dose, malathion did not significantly decrease brain or erythrocyte cholinesterase activity. It was later shown that oral dichlorvos compromises both the humoral immune response to *S. typhi* and cell-mediated immunity, measured by the tuberculin skin test, in rabbits given acute doses of dichlorvos. However, the dichlorvos was only 93% pure in the latter study, and the purity of the test material was not stated in the first study.

As several household insecticides contain organophosphates such as dichlorvos, a large number of people may be exposed to the compound. There have been suggestions in the literature that an increased incidence of blood dyscrasias may be associated with exposure of children to dichlorvos and methyl carbamates in household sprays (Reeves, 1982). There are also numerous epidemiological studies suggesting an increased incidence of hematological disorders among farmers (Blair et al., 1985). Whether these are directly related to chronic immunotoxicity of certain anticholinesterase insecticides or other pesticides remains to be documented but certainly represents an area of concern. Furthermore, these reports, while they are not definitive, at least point to certain groups of the population as especially susceptible to the toxic effects of pesticidal compounds.

Tiefenbach and Lange (1980) administered acute doses of the organophosphate dimethoate to mice and rats and demonstrated a depression of the immune system and release of glucocorticoids similar to that seen with prednisolone. Rodgers et al. (1987) reported immunotoxic effects of the organophosphate O,O,S-trimethylphosphorothioate, an impurity that may be formed during the manufacturing of some pesticides. Their data support the suggestion that organophosphate compounds can alter humoral and cellular immune responses in various species.

The immunotoxic properties of pesticides in humans received relatively little attention in the past, probably because immunology is still a new area of medicine. Street (1981) surveyed epidemiological data and suggested that immune system effects, including allergic dermatitis, may be the most frequent indications of toxic responses to pesticides in their normal uses. He reviewed the research and clinical literature, described a large number of case reports, and urged a more detailed appraisal of effects on immune processes during premarket testing of new chemicals, including pesticides. Current guidelines, published by the U.S. Environmental Protection Agency (U.S. EPA, 1984), recognize this need and request data on immunotoxicity of certain classes of compounds.

B. Interaction of the Immune System and the Nervous System

From the point of view of identifying individuals that may be particularly susceptible to immunotoxic effects, one may need to consider neurological factors as well as sensitization and allergy. There is considerable evidence for interaction between the immune system and the nervous system. Sanders and Munson (1985) described pharmacological mechanisms whereby sympathetic neurotransmitters can affect the magnitude of an immune response. Neurochemical modulation implies additional considerations of stress effects, diurnal effects, and neurological disease effects on immune function. Thus, two possibilities must be evaluated. First, chemical toxicants may be immunotoxic per se, and, second, the toxicant effects may be expressed indirectly through the immune system. Compounds that depress cholinergic (parasympathetic) function, for example, may produce a syndrome of sympathetic activation accompanied by altered immune function.

C. Susceptibility to Isocyanate Toxicity

Another group of compounds for which a clear pattern of increased reactivity has been demonstrated in susceptible individuals is the class of industrial isocyanates (Karol, 1986). These are highly reactive compounds with several different uses in organic chemistry. The compounds cause a variety of toxic effects, including tissue irritation, hypersensitivity reactions, and chronic pulmonary effects. Occupational exposure limits that include a consideration of the various factors necessary for a meaningful risk assessment have been established for the isocyanates.

The sensitization reactions to isocyanates are a good example of immunotoxicity and provide an opportunity to determine whether available data support a reduced exposure limit in consideration of susceptible individuals. Some data suggest a relationship between lymphocyte β-adrenergic receptor-mediated adenyl cyclase activity and isocyanate sensitization (Butcher et al., 1974). Whether a lymphocyte-receptor assay could be a useful tool for detecting isocyanate sensitization is undetermined. There is also evidence from studies to detect IgE antibodies that isocyanate sensitization is mediated by an immunological mechanism. Again, whether some individuals are more likely than others to evidence an immune response following isocyanate exposure is a question that requires further study.

Toluene diisocyanate (TDI) can also elicit a delayed clinical response that may evolve from bronchial hyperreactivity and/or cell-mediated mechanisms. Some patients with TDI-asthma exhibit increased lymphocyte reactivity. However, various studies are inconsistent in establishing a clear correlation between lymphocyte reactivity and isocyanate sensitivity. Other indicators of delayed sensitivity, including a skin response, are not consistently correlated with toxicant sensitivity.

It has also been suggested that hyperreactivity may be a prerequisite for the development of hypersensitivity to isocyanates (Chester et al., 1979). Hyperreactive conditions include upper respiratory tract infections and concomitant irritation of the airways which may produce heightened responsiveness to specific sensitizers. Clearly, there are no quantitative (i.e., dose-response) data available from these studies. However, susceptible groups are suggested—for example, smokers and workers with colds.

D. Altered Immunological Susceptibility in Aging

Several studies have reported a decline in immunocompetence among the elderly in a population. Whisler and Newhouse (1986) demonstrated abnormalities in T-cell proliferative responses and impaired B-cell functions in 50–80% of the elderly subjects examined. Murasko et al. (1986) studied a rather large group (260 subjects) of individuals with a mean age of 85 years and demonstrated decreased T-cell responses to phytohemagglutinin and concanavalin A. They also observed that the responses reached a minimum at about age 70 and decreased no further beyond that. B-Cell responses to pokeweed mitogen were not altered in this study group.

Bennett (1979) reviewed the effect of age on immune function and chemically induced cancer. Neonatal, fetal, and old animals are known to be particularly sensitive to chemical carcinogenesis. A number of reasons have been suggested for this sensitivity, including increased susceptibility of target organs or cells, altered hormone levels, and deficient immune functions. Basically, it is clear that susceptibility to cancer-causing chemicals is associated with altered function of the immune system. Thus, given a relationship of age to immune function, one is likely to observe an altered susceptibility to carcinogens with age. As with most of the current data on alterations of immune function, these findings are not quantitative, and even if changes in immune function are expressed in terms of a particular quantitative parameter, it may be difficult to extrapolate quantitatively to a corresponding toxic effect.

As is discussed below, several other factors can alter the susceptibility to carcinogens. For example, both humans and animals have exhibited increased risk for cancer if they are malnourished and exposed to a combination of infectious agents and environmental chemicals (Porter et al., 1984).

IV. CLINICAL ECOLOGY: THE SCIENCE OF SUSCEPTIBLE POPULATIONS?

One of the more controversial areas of environmental biology is undoubtedly the newly identified field of clinical ecology, a medical specialty that purports to study diseases related to xenobiotic exposures. Whether this is a meaningful subset of environmental or occupational medicine or just another label for psychosomatic illnesses is a debate that has provoked considerable discussion in both the medical and the lay press (Marshall, 1986). In fact, in terms of immunological responses, the cases presented in the literature of clinical ecology may be viewed as evidence that there exist sensitive groups within populations of individuals exposed to certain xenobiotics.

Brodsky (1983) reviewed the case histories of several individuals reportedly suffering from allergic responses to chemicals in the workplace. These patients were described as "environmentally ill" and were considered to be hypersensitive (although not necessarily immunologically allergic) to chemicals in their work environment. All were reviewed in consideration of workmen's disability compensation. Despite evidence that these patients were clearly "disabled," the apparent lack of physiological deficits as well as lack of a well-defined chemical exposure history led the author to conclude that the patients of the clinical ecologist are suffering primarily from psychosomatic illness. Terr (1986) reviewed an additional 50 cases of this type and came to the same conclusion despite some evidence in these cases of either acute or chronic exposures to environmental agents. The exposure history, however, was usually vague and undocumented. Furthermore, the patients showed no consistent alterations in any laboratory tests, including tests for immune function.

On the other hand, numerous case histories support the presence of both food and chemical susceptibility in certain individuals after some form of environmental chemical overexposure (Rea et al., 1978). King (1981) described a group of patients who exhibited evidence that exposure to environmental chemicals can provoke both psychological and somatic symptoms. McGovern et al. (1983) measured altered immune function in several people who exhibited an adverse clinical response to ingestion of certain foods or inhalation of certain chemicals. They concluded that these patients had developed a sensitivity to the food or chemicals. These individuals, similar to groups of patients who are sensitive to food colorings (Weiss, 1979), confirm the possibility that environmental illness may be a serious clinical diagnosis.

Furthermore, Levin et al. (1981) studied a large group of patients with food and chemical allergies and demonstrated a statistically significant increase in the levels of circulating immune complexes and prostaglandins. Their data support the suggestion of immune complex mediated intravascular inflammation in patients with food and chemical allergies, consistent with the patients' frequent reports of central nervous system symptoms such as migraine headaches. It is certainly conceivable that the victims of environmental illness represent a hitherto unidentified susceptible population and one for which no screening test exists at this time. Whether these patients have become sensitized because of preexisting biochemical or physiological conditions is an important question that requires considerable review.

V. PHYSIOLOGICAL PREDISPOSITION TO THE TOXIC EFFECTS OF XENOBIOTICS

Preexisting pathological conditions in individuals of all ages enhance their sensitivity to the toxic effects of a large number of chemicals including carcinogens. Certainly, the immunologically sensitive individuals described above, whether they are asthmatics or otherwise sensitized individuals, fall into this same category of physiologically predisposed populations. For the present discussion, however, we will examine the impact of preexisting disease conditions, including liver and kidney disease, on the threshold for experiencing the toxic effects of a given environmental chemical. In fact, toxicant responses in individuals with concomitant disease vary as a function of dose, duration of exposure, and severity of disease state as well as the many other physiological variables already described. A disease-state variable, such as renal failure or liver cirrhosis, increases the complexity of variability in human susceptibility to toxicants.

VI. KIDNEY AND LIVER DISEASE AND ALTERED SUSCEPTIBILITY TO TOXICANTS

Individuals with renal or hepatic disease are predisposed to the toxic effects of compounds, since the kidney and liver are the major organs for the elimination of xenobiotics. Renal disease is accompanied by a decreased ability to clear the blood of active toxicants and their active or inactive metabolites. Reduced renal function can result in the accumulation and correspondingly increased half-life of compounds that are eliminated by that route (as discussed in greater detail below with respect to age-related effects). Toxicants that are excreted by the kidney can thus be expected to have a greater effect for a longer duration in subjects with renal disease. These individuals can be identified easily by performing a creatinine clearance test (a measure of renal function), as the renal clearance of any compound can be related directly to the clearance of creatinine (Orme and Cutler, 1969).

Similarly, hepatic clearance or biotransformation of toxicants may be severely compromised in patients with liver disease. The effect of hepatic disease on the elimination of xenobiotics cleared by the liver depends on a number of factors, including the intrinsic clearance of the liver (i.e., the total activity of metabolic enzymes), blood flow, and plasma binding. Furthermore, the term hepatic disease includes a variety of pathological conditions such as cirrhosis and hepatitis.

Hepatic clearance is defined as

$$Cl_{hep} = Q \left(f_u \frac{Cl_{int}}{Q} + f_u Cl_{int} \right)$$

where Q = hepatic blood flow, Cl_{int} = intrinsic hepatic clearance (metabolic activity), and f_u = the fraction unbound by plasma proteins (Rowland, 1973). For compounds for which Cl_{int} is greater than Q, hepatic blood flow controls hepatic clearance, and hepatic disease is not expected to have a major effect on overall clearance. When Q is greater than Cl_{int}, the intrinsic clearance controls hepatic clearance, and hepatic disease is then expected to alter the overall clearance.

However, in studies of the effect of hepatic disease on clearance of several different drugs, it has been shown that liver disease, in fact, usually results in a decrease in total clearance whether the clearance is limited by hepatic blood flow or by intrinsic hepatic clearance (see review by Blaschke, 1983). There are a few exceptions, including tolbutamide, warfarin, and phenytoin. Overall, the experimental data indicate that individuals with impaired hepatic function can be expected to exhibit a reduced ability to detoxify and eliminate toxicants, and thereby exhibit increased risk for the toxic effects of chemicals.

It is important to realize also that plasma protein binding can modify the intrinsic clearance, since only unbound (free) xenobiotics can be metabolized. Hepatic disease can also be expected to modify plasma protein binding, since the liver is the major site of albumin synthesis.

A. Influence of Aging on the Chemical Effects and Distribution of Toxicants

Marked alterations in pharmacokinetic capabilities are well documented with increasing age, and these alterations will seriously influence the handling of a toxicant in the elderly. Changes in xenobiotic distribution, largely attributable to decreases in both hepatic and renal clearance, as well as in target organ sensitivity, are identified in older populations. These pharmacokinetic changes also interact with other factors including nutritional status and disease.

A number of factors contribute to the sensitivity of aged individuals to toxicants including changes in physiological functions, changes in body composition leading to pharmacokinetic alterations, the presence of preexisting disease, and the decreased ability of the elderly to repair tissue damage and respond to toxic insults.

Pharmacokinetic changes in the elderly have been studied extensively (for reviews, see Crooks et al., 1983, and Greenblatt et al., 1986). Some salient points of special interest to toxicologists will be discussed here. There are a number of age-related changes in body composition which result in changes in the distribution of xenobiotics. The proportion of body weight which is adipose tissue increases from 18 to 36% in men and from 33 to 48% in women (Novak, 1972), while the portion of lean body mass decreases (Forbes and Reina, 1970). Thus, highly lipid-soluble compounds are expected to accumulate with age. Indeed, Klotz et al. (1975) demonstrated that the longer plasma half-life of the drug diazepam in elderly patients is due to an increase in the volume of distribution and not to a change in total clearance. Plasma half-life $(t_{1/2}) = 0.693 \times$ volume of distribution (V_d)/total clearance (Cl_{tot}).

The total clearance of many xenobiotics is expected to change with age, since both renal and hepatic function, the major determinants of total clearance, change with age. Also, in humans, cardiac output falls 10% per decade between the ages of 20 and 80 years (Bender, 1965). Thus, perfusion of both the kidney and liver decrease with age.

It is well known that renal function decreases with age. The glomerular filtration rate falls approximately 35% between the ages of 20 and 90 years, and tubular secretion also decreases (reviewed by Crooks et al., 1983). Thus, xenobiotics that are cleared by renal mechanisms can be expected to have a longer half-life and higher steady-state concentration (concentration at steady state, C_{ss} = dosing rate/Cl_{tot}). This has been shown to be the case for a number of drugs, including digoxin, sulfamethizole, practolol, and kanamycin (Crooks et al., 1983).

The influence of age on the hepatic metabolism of xenobiotics is quite variable. Rikans and Notley (1982) demonstrated that the levels of cytochrome P-450, cytochrome b_5, and NADPH-cytochrome c reductase were all reduced in aged animals. However, age-related changes in specific enzyme activities was quite variable. The activity of benzphetamine N-demethylase was decreased by 63%, aniline hydroxylase by 26%, while the activity of nitrosole-O-demethylase was increased 250% in old rats when compared to young adult rats. Crooks et al. (1983) reviewed a number of studies that examined the effect of age on hepatic clearance of a number of drugs in humans. From the data in which the hepatic clearance was determined, it could be shown that the hepatic clearance of antipyrine and phenylbutazone decreases with age (25–28%), while the hepatic clearance of phenytoin increases by approximately 60%. There was no change in the hepatic clearance of paracetamol and diazepam as a function of age. These human studies support the previous animal studies which suggest that age has diverse effects on hepatic metabolism of xenobiotics, depending largely on the specific compounds studied and their physicochemical properties.

B. Changes in Tissue Sensitivity with Age

Aging results in a gradual decline in many physiological functions including cardiac output, vital capacity, glomerular filtration, and basal metabolic rate (Kanungo, 1980). Furthermore, the secretion of a number of hormones, especially those involved in anabolic processes, is also reduced (Meites et al., 1987). This suggests that aged individuals have a reduced capacity to respond to toxicants and that age-related changes may make specific tissues more susceptible to chemical insult. For example, cardiac responsiveness to catecholamines declines with age. Studies in rats demonstrate both a decline in β-adrenergic receptors and impaired postreceptor adenylate cyclase activity (Fan and Banerjee, 1985). In a review of age-related changes in the hepatotoxicity of allyl alcohol, galactosamine, and bromobenzene, Rikans (1984) demonstrated that the toxicity of allyl alcohol increased, galactosamine decreased, and bromobenzene toxicity was unchanged as a function of age. These differences were attributable to age-related changes in the ability of the liver to detoxify and metabolize xenobiotics.

C. Toxicant Sensitivity in Neonates and Young Children

Newborns and young children represent another sensitive population, largely because of their biochemical immaturity and differences in body composition. Many of these differences are well known, such as the reduced ability to form glucuronic acid conjugates and increased extracellular water as percentage of body weight. There are a number of other differences which are of interest to toxicologists (see review by Warner, 1986). Neonates exhibit enhanced percutaneous absorptive capacity which has resulted in several episodes of acute toxicity after administration of a number of therapeutic compounds, including hexachlorophene, water-soluble vitamin K, and boric acid. Levels of cytochrome P-450 and NADPH cytochrome c reductase are only about 50% of adult levels at birth, and renal function is also considerably lower at birth. These considerations suggest that the very young have a reduced capacity to metabolize and excrete toxicants which is further exacerbated by an enhanced ability to take up environmental xenobiotics.

VII. TOXICANT INTERACTIONS AND ENHANCED SUSCEPTIBILITY

Our environment provides a continual barrage of multiple chemical insults and, in many parts of the world, an uncertain supply of adequate food and water. Interactions among these factors probably represent a greater threat to human health than has actually been recognized, and should become an area of greater concern for environmental scientists and regulators. It is also an area of regulation that requires cooperative efforts both within and among nations. Perhaps a heightened awareness of the potentially predictable sources of interaction and their effects on sensitivity to toxicants can serve as a stimulus for more studies and, if necessary, greater control.

A. Dietary Effects on Pesticide Toxicity

Dietary and nutritional influences on pesticide toxicity are a good example of interacting factors and their effects on susceptibility to toxicants. While this is an area of research that is especially important for safe chemical use in parts of the world that suffer from chronic nutritional deficiencies, it is also important for understanding the enhanced susceptibility of specific subgroups in our own population.

Data from animals fed low protein diets demonstrate that the toxicity of several pesticides, including carbaryl, parathion, and the fungicide captan, is markedly increased (for an excellent review of these studies, see Shakman, 1974). The mechanisms of dietary influences on pesticide toxicity include liver microsomal enzyme induction, altered levels of enzyme cofactors, altered deposition of toxicants in fat depots, and mobilization of stored compounds under conditions of starvation. Conner and Newberne (1984) reviewed the various mechanisms of drug-nutrient interaction and provided evidence that high dietary protein intake protects animals from the toxic effects of many pesticides and herbicides, possibly by providing a stable amino acid pool for the synthesis of hepatic microsomal enzymes.

B. Effects of Dietary Factors on Xenobiotic Toxicity

While most available human data come from studies of therapeutic drugs, the same basic principles apply to dietary influences on the toxic responses to environmental chemicals. Dietary composition and patterns of eating can modify the absorption of toxicants as well as their metabolism and excretion. Hathcock (1985) reviewed the mechanisms of drug-nutrient interactions and pointed out several factors that are relevant influences on toxicant susceptibility, including: high-fat meals, which enhance the absorption of fat-soluble compounds and can also displace many compounds from binding sites on plasma proteins; high-protein diets, which increase the activity of mixed function oxidases and, thus, increase the metabolism of many compounds; acidic foods, which can increase the excretion of alkaline compounds; and basic foods, which can increase the excretion of acidic compounds.

Charcoal broiling has been shown to enhance the hepatic oxidative metabolism of many compounds. The polycyclic hydrocarbons formed on the surface of beef by charcoal broiling appear to be responsible for induction of mixed function oxidases in the liver as well as induction of gastrointestinal drug-metabolizing enzymes.

The relative proportion of carbohydrate and protein in the diet can also alter the rates of hepatic and gastrointestinal metabolism in humans. Fasting or starvation appears to affect primarily the elimination of highly protein-bound compounds. Since plasma free fatty acids rise dramatically after 12 hours of fasting, and, since the free fatty acids bind albumin, they are capable of displacing many highly bound xenobiotics, thereby accelerating their rate of elimination.

Certain kinds of foods also can alter xenobiotic metabolism. For example, the indoles present in vegetables such as brussel sprouts and cabbage are potent inducers of metabolic enzymes in the gastrointestinal tract. Studies of this effect indicate that such dietary effects can be sufficient to change an individual's response to a toxicant exposure.

C. Effects of Interaction between Diet and Aging on Toxicant Susceptibility

There are also important interactions between dietary deficiencies and aging. Special nutritional needs arise from metabolic disorders and chronic diseases that occur with increasing frequency in the elderly population. There are also some altered requirements for nutrients during normal growth and aging to keep up with the requirements for full functional capacity. It is, in fact, this loss of function that contributes to the development of degenerative disease. For example, poor absorption in postmenopausal women has been associated with the inability to maintain calcium equilibrium (Heaney and Recker, 1986), as evidenced in the onset of osteoporosis. Calcium deficiency has also been implicated as a factor in enhanced susceptibility to several toxicants, most notably aluminum (Yase, 1980). Whether this is an indication that postmenopausal women should follow more restrictive exposure limits is another concern in the process of assessing risk for susceptible populations.

D. Dietary Factors and Cancer Risk

In general, the risks for numerous diseases, including several forms of cancer, particularly cancer of the breast, colon, gall bladder, ovaries, prostate, and endometrium (American Cancer Society, 1986), are markedly increased in obesity. As already discussed above in relation to aging effects on susceptibility, the possibility that obesity or the body fat/lean body mass ratio (i.e., some measure of body composition) may be a relevant factor in defining individuals with increased susceptibility to the toxic effects of exposure to xenobiotics, especially carcinogens, is another concern for risk assessment.

Data linking nutritional factors with cancer risk arise from both laboratory studies with experimental animals and epidemiological investigations of various groups of people. While many effects and associations have been reported, it is not yet possible to quantitate the contribution of diet to the overall cancer risk or to determine the percentage reduction in risk that dietary modification may achieve. High-calorie diets have been linked to increased cancer risk in rodents. Several dietary constituents such as fiber, vitamins, and minerals have been investigated as possible inhibitors of carcinogenesis (for reviews of this subject, see Council for Agricultural Science and Technology, 1987, and National Academy of Sciences, 1982).

E. Other Interactive Factors

The overall effect of diet on carcinogenesis is not nearly as well established as the effect of the major factor, i.e., tobacco smoking. It is estimated that 30% of all cancer deaths in the United States are related to smoking (Council for Agricultural Science and Technology, 1987). Thus, the subgroup of smokers clearly represents a susceptible population with a well-defined increased risk for cancer.

A number of other interacting factors may have a significant effect on enhancing the risk for toxicants. These include alcohol consumption and consequent effects on liver function, solvent potentiation of toxicity by increasing absorption and overall bioavailability of toxicants, solvent induction of metabolic enzymes (for a review, see Andrews and Snyder, 1986), and several other factors that remain to be identified at this time.

VIII. SENSITIVE POPULATIONS AND RISK ASSESSMENT

A. Risk Assessment in Environmental Policymaking

The process of conducting a risk assessment generally includes four steps: hazard identification, dose-response evaluation, exposure determination, and risk characterization. The third step incorporates an awareness of sensitive populations, as it identifies populations exposed to toxicants, describes their composition and size, and examines the route, magnitude, frequency, and duration of the exposure.

Hazard depends on both human exposure to a chemical and the toxic effects of that chemical in humans. Exposure and response interact in a complex way to yield a toxic outcome that may be linear with dose or may depend on several modulating factors, including the fractionation of dose over exposure time, route of exposure, and synergistic exposures, as well as on the age, sex, race, and health status of the individual exposed.

Whether one should set priority levels for concern that take into account the biological characteristics of the exposed population is a difficult question for policymakers. The U.S. EPA, in particular, is acutely aware of the complexity of a risk assessment and the confounding elements related to individual variations in susceptibility to the adverse effects of a toxicant. In their recent series of guidelines for risk assessment, the U.S. EPA (1986a–1986e) described the problem but recognizes that it cannot be resolved in the absence of more specific and complete data. Recognizing that some individuals, for genetic or other reasons, are more susceptible than others, the U.S. EPA (1986a) pointed out that "Subpopulations with heightened susceptibility (either because of exposure or predisposition) should, when possible, be identified."

In a recent review Ames et al. (1987) defined, specifically for carcinogens, several variables or uncertainties that confound a risk assessment effort. These include quantitative factors related to the extrapolation from experimental data in laboratory animals, such as rodents, to real-life exposure situations in humans. Although these factors were discussed with specific reference to carcinogenicity, in a broad sense they are applicable to many other mechanisms of toxicity. These uncertainties arise largely from a lack of understanding of the mechanisms of toxicity, the relation of toxicity to age and life span, the timing and order of the steps in a toxic process, species differences in metabolism and pharmacokinetics, species differences among natural defenses against xenobiotics, and from the concerns we have addressed herein, namely, human heterogeneity. Evidence for human heterogeneity in response to carcinogens is abundant; for example, pigmentation is known to affect susceptibility to skin cancer from ultraviolet (UV) light. However, these differences are difficult to quantitate because so little empirical data are available to define these considerations in a rigorous fashion.

Wilson and Crouch (1987) indicated that different uncertainties may be factored into risk estimation in different ways. For example, the uncertainty related to variations among individuals in an exposed population is already considered in part by a safety factor, usually a factor of 10, as already discussed above, for the so-called health worker effect. Whether this is an adequate estimate is a question raised by the shadows in Fig. 2. The answers clearly depend on available data for the properties of a particular type of chemical and reports on clinical studies of individuals exposed to the chemical.

In the process of risk management policymakers must also consider the costs of reducing the risk for sensitive populations. Wilson and Crouch (1987) suggest that society is aware of the different costs that are "acceptable" for achieving protection against different hazards and is, in fact, willing to accept unusually large costs to reduce certain kinds of risks (e.g., risks from environmental exposures) and quite unwilling to support reduction of other kinds of risks (e.g., the cost of air bags in automobiles). Policymakers are asked to select an allowable range of hazardous choices and must, thus, determine appropriately conservative safety factors for setting exposure limits.

Thus, the discussion presented herein is intended to emphasize the responsibility of the scientific and medical community to provide complete and accurate scientific data to the decision-maker to help achieve an enlightened risk assessment policy. It is evident that before one can formulate policy to assess the variability in the risk of toxicant-induced disease that is attributable to variability among individuals or groups, one must be able to identify those particularly susceptible individuals or populations. Practically, medical histories may reveal preexisting disease conditions, and lab tests may be used to detect preexisting or genetically determined traits. For example, most risk assessments for developmental toxicants assume a threshold for developmental effects because of known repair capabilities in developing systems. Basic physiological data on the variations in repair capability in subgroups of the susceptible population, for example, pregnant females, would be useful for establishing a more precise factor to calculate safe exposure levels.

One approach to achieving this goal might be to prepare a comprehensive literature study, pooling data for toxicants that are reported to elicit a response in sensitive individuals that is not evident in the general population that is exposed to the compound. Similar efforts, under the auspices of the National Institute for Occupational Safety and Health (NIOSH), have proved to be very useful for identifying neurobehavioral toxicants and targeting specific areas of concern relative to neurotoxicity (Anger and Johnson, 1985). Furthermore, such a comprehensive project can help to identify hitherto unidentified sensitive populations for specific classes of chemical compounds. For example, are drug abusers a sensitive population? Are fetuses of undernourished pregnant females more sensitive to teratogenic toxicants? A comprehensive survey may also identify subgroups of the population that exhibit a unique sensitivity to chemicals.

For a comprehensive risk assessment, one should have some idea of the exposure histories of individuals as well as their particular sensitivity to a toxicant. However, Beck and Greaves (1988), in reviewing variations in susceptibility to inhaled pollutants, pointed out that a major fallacy associated with targeting individuals by their exposure alone is the assumption that exposure-dose and dose-response relationships are uniform in the population. Clearly, uniformity of response is not supported by the various data described above. Linearity of dose-response is also not readily supported by available data. Sensitization, for example, often occurs early in the course of exposure and is not always directly related to dose. In fact, determining the toxicity in a population by examining only the most exposed individuals may result in a review of individuals who are particularly resistant or tolerant. Thus, while exposure data are valuable, they are only one of the numerous pieces of data required for setting exposure limits.

B. Quantitative Models for Estimating Safety Factors in Risk Assessment

Traditionally, a safe level of exposure is defined as some arbitrary function of the dose level at which no effects are observed in a group of test animals (NOEL). That fraction includes so-called safety factors that take into consideration the extrapolation from a lab animal species to man and the variability inherent in a population, e.g., the difference between a minimum individual threshold for a toxic effect and the mean population threshold for that effect.

The quantitative aspect of risk assessment requires probability estimates derived from animal studies and, ideally, from epidemiology studies on the population of interest at the range of doses or exposures of interest. Epidemiological studies are fraught with difficulty, as experimental manipulation of the human population is not possible. It is nonetheless useful to consider what can be learned from each study and to review all available data, including both animal studies and epidemiological data, in calculating safety factors to establish exposure limits for environmental chemicals.

Although conservative assumptions have been utilized in risk assessment, there are certainly many cases where the true risk might be underestimated. The reasons for potential errors have already been discussed. First, people are not exposed to a single chemical, but rather to a number of chemicals. Even if they act independently, the risk will be the sum of the individual chemical risks. Second, the chemicals may interact and potentiate or dampen the effects of other chemicals. Third, some individuals may be particularly sensitive to individual chemicals, as illustrated schematically in Fig. 2.

While many different models have been utilized for quantitative risk assessment, each varying in its basic assumptions and methods of extrapolation to low-dose exposures, they are all designed primarily to calculate a NOEL to which the appropriate safety factor is then applied. The difficulties inherent in estimating a safety factor that adequately accounts for susceptible subgroups of the population have been emphasized throughout this chapter. Overall, it is certainly possible at this time to set priorities for concern, but it is really not possible to quantitate the level of concern for any given population that may be exposed to a xenobiotic.

IX. CONCLUSIONS

What then is the impact of sensitive populations on risk assessment and environmental policy? Does the present factor of 10 represent an adequate safety factor? In light of the many studies analyzed for the present review, we have concluded that a 10-fold adjustment in allowable exposure limits is not sufficiently conservative to account for the variable response to environmental toxicants that may be expressed among the individuals in a population. In fact, the factor of 10 may cover the range of responses observed in any one group of people or in response to any one class of toxicants. However, recognizing the extensive overlap among groups, e.g., elderly and immune suppressed, newborn and genetically predisposed, as well as the multiplicity of interactive exposures, one must conclude that a prudent safety factor of 50–100 will more likely account for the "healthy worker effect" and reduce the likelihood of overestimating safe exposure limits. Furthermore, even a casual review of the many sensitive groups described above reveals that the vast majority of the population in the United States falls into at least one of these categories, and that some level of susceptibility can be identified throughout the population.

REFERENCES

American Cancer Society. 1986. *Cancer Facts and Figures.* New York: ACS.

Ames, B. N., Magaw, R., and Gold, L. S. 1987. Ranking possible carcinogenic hazards. *Science* 236:271–280.

Andrews, L. S., and Snyder, R. 1986. Toxic effects of solvents and vapors. In *Casarett & Doull's Toxicology. The Basic Science of Poisons,* 3rd ed., eds. C. D. Klaassen, M. O. Amdur, and J. Doull, pp. 636–668. New York: Macmillan.

Anger, K., and Johnson, B. L. 1985. Chemicals affecting behavior. In *Neurotoxicity of Industrial and Commercial Chemicals,* vol. 1, ed. J. L. O'Donoghue, pp. 51–148. Boca Raton: CRC Press.

Ayesh, R., Idle, J. R., Ritchie, J. C., Crothers, M. J., and Hetzel, M. R. 1984. Metabolic oxidation phenotypes as markers for susceptibility to lung cancer. *Nature* 312:169–170.

Beck, B. D., and Greaves, I. 1988. Screening and monitoring for nonneoplastic lung disease. In *Variations in Susceptibility to Inhaled Pollutants: Identification, Mechanisms, and Policy Implications,* eds. J. D. Brain, B. D. Beck, A. J. Warren, and R. Shaikh, pp. 335–375. Baltimore: Johns Hopkins University Press.

Bender, A. D. 1965. Effect of increasing age on the distribution of peripheral blood flow in man. *J. Am. Geriatr. Soc.* 13:192–198.

Bennett, M. 1979. Effect of age on immune function in terms of chemically induced cancers. *Environ. Health Perspect.* 29:17–22.

Beutler, E. 1983. Glucose-6-phosphate dehydrogenase deficiency. In *The Metabolic Basis of Inherited Disease,* 10th ed., ed. J. B. Stanbury, J. B. Wyngaarden, D. S. Fredrickson, J. L. Goldstein, and M. S. Brown, pp. 1629–1653. New York: McGraw-Hill.

Blair, A. H., Malber, H., Cantor, K. P., Burmeister, L., and Wiklund, K. 1985. Cancer among farmers—A review. *Scand. J. Work Environ. Health* 11:397–407.

Blaschke, T. F. 1983. Protein binding and kinetics of drugs in liver disease. In *Handbook of Clinical Pharmacokinetics,* eds. M. Gibaldi and L. S. Prescott, pp. 126–139. New York: ADIS Health Science Press.

Brodsky, C. M. 1983. Allergic to everything: A medical subculture. *Psychosomatics* 24:731–742.

Butcher, B. T., Salvaggio, J. E., O'Neill, C. E., Weill, H., and Garg, O. 1977. Toluene diisocyanate pulmonary disease: Immunopharmacologic and mecholyl challenge studies. *J. Allergy Clin. Immunol.* 59:223–227.

Cartwright, R. A. 1983. Epidemiological studies on N-acetylation and C-center ring oxidation in neoplasia, pp. 359–361. In *Branbury Report 16.* New York: Cold Spring Harbor Laboratory.

Cartwright, R. A., Glashan, R. W., Rogers, H. J., Ahmad, R. A., Barham-Hall, D., Higgins, E., and Kahm, M. A. 1982. Role of n-acetyltransferase phenotypes in bladder carcinogenesis: A pharmacogenetic epidemiological approach to bladder cancer. *Lancet* ii:842–846.

Chester, E. H., Martinez-Catinchi, G. L., Schwartz, H. J., Horowitz, J., Fleming, G. M., Gerblich, A. A., McDonald, G. W., and Grethawer, R. 1979. Patterns of airway reactivity to asthma produced by exposure to toluene diisocyanate. *Chest* 75:229–231.

Chu, S. Y. 1985. Depression of serum cholinesterase activity as an indicator for insecticide exposure—consideration of the analytical and biological variations. *Clin. Biochem.* 18:323–326.

Conner, M. W., and Newberne, P. M. 1984. Drug-nutrient interactions and their implications for safety evaluations. *Fundament. Appl. Toxicol.* 4:S341–S356.

Council for Agricultural Science and Technology 1987. Diet and health, Report No. 111, CAST, Ames, IA.

Cranmer, J. M., Barnett, J. B., Avery, D. L., and Cranmer, M. F. 1982. Immunoteratology of chlordane. I. Cell-mediated and humoral immune responses in adult mice exposed in utero. *Toxicol. Appl. Pharmacol.* 62:402–408.

Crooks, J., O'Malley, K., and Stevenson, I. H. 1983. Pharmacokinetics in the elderly. In *Handbook of Clinical Pharmacokinetics,* eds. M. Gibaldi and L. Prescott, pp. 169–187. New York: ADIS Health Science Press.

Dean, J. H., Murray, M. J., and Ward, E. C. 1986. Toxic responses of the immune system. In *Casarett & Doull's Toxicology. The Basic Science of Poisons,* 3rd ed., chap. 9, eds. C. D. Klaassen, M. O. Amdur, and J. Doull. New York: Macmillan.

Desi, I., Varga, L., and Farkas, I. 1978. Studies on the immunosuppressive effect of organochlorine and organophosphoric pesticides in subacute experiments. *J. Hyg. Epidemiol. Microbiol. Immunol.* 22:115–122.

Desi, I., Varga, L., and Farkas, I. 1980. The effect of DDVP, and organophosphate pesticide on the humoral and cell-mediated immunity of rabbits. *Arch. Toxicol.* 4(Suppl):171–174.

DiCarlo, F. J. 1982. Metabolism, pharmacokinetics, and toxicokinetics defined. *Drug Metab. Rev.* 113:1–4.

Eichelbaum, M. 1984. Polymorphic drug oxidation in humans. *Fed. Proc.* 43:2298–2302.

Eichelbaum, M. 1986. Polymorphic oxidation of debrisoquine and sparteine. In *Ethnic Differences in Reactions to Drugs and Xenobiotics,* eds. W. Kalow, H. W. Goedde, and D. P. Agarwal, pp. 157–167. New York: Liss.

Evans, D. A. P. 1986. Acetylation. In *Ethnic Differences in Reactions to Drugs and Xenobiotics,* eds. W. Kalow, H. W. Goedde, and D. P. Agarwal, pp. 209–242. New York: Liss.

Fan, T. H. M., and Banerjee, S. P. 1985. Age-related reduction of beta-adrenergic sensitivity in rat heart occurs by multiple mechanisms. *Gerontology* 31:373–380.

Forbes, G. B., and Reina, J. C. 1970. Adult lean body mass declines with age: Some longitudinal observations. *Metabolism* 19:653–663.

GAO 1986. Pesticides—EPA's formidable task to assess and regulate their risks, Report to Congressional Requesters. Washington, D.C.: U.S. General Accounting Office.

Glatt, H., and Oesch, F. 1983. Variation in epoxide hydrolase activities in human liver and blood. In *Banbury Report 16.* New York: Cold Spring Harbor Laboratory.

Glowinski, J. R., Radtke, H. E., and Weber, W. W. 1978. Genetic variation in *N*-acetylation of carcinogenic arylamines by human and rat liver. *Mol. Pharmacol.* 14:940–949.

Goldstein, A., Aronow, L., and Kalman, S. M. 1974. In *Principles of Drug Action,* 2nd ed., chap. 6, Pharmacogenetics and drug idiosyncrasy, 437–487. New York: Wiley.

Greenblatt, D. J., Abernathy, D. R., and Shader, R. I. 1986. Pharmacokinetic aspects of drug therapy in the elderly. *Ther. Drug Monitor.* 8:249–255.

Harris, C. C., Autrup, H., Vahakangas, K., and Trump, B. G. 1983. Interindividual variation in carcinogen activation and DNA repair, pp. 145–153. In *Banbury Report 16.* New York: Cold Spring Harbor Laboratory.

Hathcock, J. H. 1985. Metabolic mechanisms of drug-nutrient interactions. *Fed. Proc.* 44:124–129.

Heaney, R. P., and Recker, R. R. 1986. Distribution of calcium absorption in middle-aged women. *Am. J. Clin. Nutr.* 43:299–305.

Holbrook, D. J. 1982. Chemical carcinogenesis. In *Introduction to Biochemical Toxicology,* 2nd ed., eds. E. Hodgson and F. E. Guthrie, pp. 310–329. New York: Elsevier.

Idle, J. R., and Smith, R. L. 1979. Polymorphisms of oxidation at carbon centers of drugs and their clinical significance. *Drug Metab. Rev.* 9:301–317.

Kanungo, M. S. 1980. *Biochemistry of Aging,* pp. 1–10. London: Academic Press.

Kapitulnik, J., Levin, W., Lu, A. Y. H., Morecki, R., Dansett, P. M., Jerina, D. M., and Conney, A. H. 1977. Hydration of arene and alkene oxides by epoxide hydrase in human liver microsomes. *Clin. Pharmacol. Ther.* 21:158–165.

Karol, M. H. 1986. Respiratory effects of inhaled isocyanates. *CRC Crit. Rev. Toxicol.* 16:349–379.

Kellerman, G., Shaw, C. R., and Luyter-Kellerman, M. 1973. Aryl hydrocarbon hydroxylase inducibility and bronchogenic carcinoma. *New Engl. J. Med.* 289:934–937.

King, D. S. 1981. Can allergic exposure provoke psychological symptoms? A double-blind test. *Biol. Psychiatry* 16:3–19.

Klotz, Y., Avant, G. R., Hoyumpa, A., Schenker, S., and Wilderson, G. A. 1975. The effects of age and liver disease on the disposition and elimination of diazepam in the adult man. *J. Clin. Invest.* 55:347–359.

Kouri, R. E., Levine, A. S., Edwards, B. K., McLemore, T. L., Vesell, E. S, and Nebert, D. W. 1983. Source of interindividual variations in aryl hydrocarbon hydroxylase in mitogen-activated human lymphocytes. In *Banbury Report 16.* New York: Cold Spring Harbor Laboratory.

LaDu, B., and Eckerson, H. 1984. The polymorphic paraoxonase/arylesterase isozymes of human serum. *Fed. Proc.* 43:2338–2341.

Levin, A. S., McGovern, J. J., Miller, J. B., Lecam, L., and Lazaroni, J. 1981. Immune complex mediated vascular inflammation in patients with food and chemical allergies. *Ann. Allergy* 47:138 (abstract).

Levin, H. S., Rodnitzky, R. L., and Mick, D. L. 1976. Anxiety associated with exposure to organophosphate compounds. *Arch. Gen. Psychiatry* 33:225–228.

Lower, G. M., Nilsson, T., Nelson, C. E., Wolf, H., Gamsky, T. E., and Bryan, G. T. 1979. N-acetyltransferase phenotype and risk in urinary bladder cancer: Approaches in molecular epidemiology. Preliminary results in Sweden and Denmark. *Environ. Health Perspect.* 29:71–79.

Lunde, P. K. M., Frislid, K., and Hansteen, V. 1983. Disease and acetylation polymorphism. In *Handbook of Clinical Pharmacokinetics,* eds. M. Gibaldi and L. Prescott, pp. 150–168. New York: ADIS Health Sciences Press.

Luster, M. I., Dean, J. H., and Moore, J. A. 1982. Evaluation of immune function in toxicology. In *Principles and Methods of Toxicology,* ed. W. Hayes, pp. 561–586. New York: Raven Press.

Marquis, J. K. 1986. *Contemporary Issues in Pesticide Toxicology and Pharmacology,* pp. 23–27. Basel: S. Karger.

Marshall, E. 1986. Immune system theories on trial. *Science* 234:1490–1492.

McGovern, J. J., Lazaroni, J. A., Hicks, M. F., Adler, J. C., and Cleary, P. 1983. Food and chemical sensitivity: Clinical and immunologic correlates. *Arch. Otolaryngol.* 109:292–297.

Meites, J., Goya, R., and Takahashi, S. 1987. Why the neuroendocrine system is important in the aging process. *Exp. Gerontol.* 22:1–15.

Menna, J. H., Moses, E. B., and Barron, A. L. 1980. Influenza type A virus infection of suckling mice pre-exposed to insecticide carrier. *Toxicol. Lett.* 6:357–363.

Menna, J. H., Barnett, J. B., and Soderberg, L. S. F. 1985. Influenza type A virus infection of mice exposed *in utero* to chlordane; survival and antibody studies. *Toxicol. Lett.* 24:45–52.

Moses, G. C., Tuckerman, J. F., and Henderson, A. R. 1986. Biological variation of cholinesterase and 5′-nucleotidase in serum of healthy persons. *Clin. Chem.* 32:175–177.

Murasko, D. M., Nelson, B. R., Silver, R., Matour, D., and Kaye, D. 1986. Immunologic response in an elderly population with a mean age of 85. *Am. J. Med.* 81:612–618.

Myrnes, B., Giercksky, K. E., and Krokan, H. 1983. Interindividual variation in the activity of O^6-methylguanine DNA methyltransferase and uracil-DNA glycosylase in human organs. *Carcinogenesis* 4:1565–1568.

National Academy of Sciences, National Research Council, Committee on Diet, Nutrition, and Cancer, Assembly of Life Sciences. 1982. *Diet, Nutrition & Cancer.* Washington, D.C.: National Academy Press.

Nebert, D. W. 1978. Genetic control of carcinogen metabolism leading to individual differences in cancer risk. *Biochimie* 60:1019–1029.

Novak, L. P. 1972. Aging, total body potassium, fat-free mass and cell mass in males and females between the ages of 18 and 85 years. *J. Gerontol.* 27:438–443.

Omenn, G. S. 1983. Risk assessment, pharmacogenetics, and ecogenetics. In *Banbury Report 16,* pp. 3–13. New York: Cold Spring Harbor Laboratory.

Orme, B. M., and Cutler, R. E. 1969. The relationship between kanamycin pharmacokinetics: Its distribution and renal function. *Clin. Pharmacol. Ther.* 10:543–550.

Porter, W. P., Hinsdill, R., Fairbrother, A., Olson, L. J., Jaeger, J., Yuill, T., Bisgaard, S., Hunter, W. G., and Nolan, K. 1984. Toxicant-disease-environment interactions associated with suppression of immune system, growth, and reproduction. *Science* 224:1014–1016.

Rea, R. J., Bell, I. R., Suits, C. W., and Smiley, R. E. 1978. Food and chemical susceptibility after environmental chemical overexposure: case histories. *Ann. Allergy* 41:101–110.

Reeves, J. D. 1982. Household insecticide-associated blood dyscrasias in children. *Am. J. Pediatr. Hematol. Oncol.* 4:438–439.

Rikans, L. E. 1984. Influence of aging on the susceptibility of rats to hepatotoxic injury. *Toxicol. Appl. Pharmacol.* 73:243–249.

Rikans, L. E., and Notley, B. A. 1982. Age-related changes in hepatic microsomal drug metabolism are substrate selective. *J. Pharmacol. Exp. Ther.* 220:574–578.

Rodgers, K. E., Imamura, T., and Devens, B. H. 1987. Investigations into the mechanism of immunosuppression caused by acute treatment with O,O,S-trimethylphosphorothioate: Generation of suppression macrophages from treated animals. *Toxicol. Appl. Pharmacol.* 88:270–281.

Rowland, M., Benet, L. Z., and Graham, G. G. 1973. Clearance concepts in pharmacokinetics. *J. Pharmacokin. Biopharmaceut.* 1:123–126.

Russell, R. W., and Overstreet, D. H. 1987. Mechanisms underlying sensitivity to organophosphorus anticholinesterase compounds. *Prog. Neurobiol.* 28:97–129.

Sanders, V. M., and Munson, A. E. 1985. Norepinephrine and the antibody response. *Pharmacol. Rev.* 37:229–248.

Shakman, R. A. 1974. Nutritional influences on the toxicity of environmental pollutants. *Arch. Environ. Health* 28:105–113.

Stoller, A., Krupinski, J., Christopher, A. J., and Blanks, G. K. 1965. Organophosphorus insecticides and major mental illness: An epidemiological investigation. *Lancet* i:1387–1388.

Street, J. C., and Sharma, R. P. 1975. Alterations of induced cellular and humoral immune responses by pesticides and chemicals of environmental concern: Quantitative studies of immunosuppression by DDT; aroclor 1254, carbaryl, carbofuran, and methylparathion. *Toxicol. Appl. Pharmacol.* 32:587–602.

Street, J. C. 1981. Pesticides and the immune system. In *Immunologic Considerations in Toxicology,* ed. R. P. Sharma, pp. 45–66. Boca Raton: CRC Press.

Terr, A. I. 1986. Environmental illness: a clinical review of 50 cases. *Arch. Intern. Med.* 146:145–149.

Tiefenbach, B., and Lange, P. 1980. Studies on the action of dimethoate on the immune system. *Arch. Toxicol.* 4(Suppl.):167–170.

U.S. EPA 1984. Data requirements for pesticide registration; final rule. *Fed. Reg.* 49:42856–42905.

U.S. EPA 1986a. Guidelines for carcinogen risk assessment. *Fed. Reg.* 51:33992–34003.

U.S. EPA 1986b. Guidelines for mutagenicity risk assessment. *Fed. Reg.* 51:34006–34012.

U.S. EPA 1986c. Guidelines for the health risk assessment of chemical mixtures. *Fed. Reg.* 51:34014–34025.

U.S. EPA 1986d. Guidelines for the health assessment of suspect developmental toxicants. *Fed. Reg.* 51:34028–34040.

U.S. EPA 1986e. Guidelines for exposure assessment. *Fed. Reg.* 51:34042–34054.

Warner, A. 1986. Drug use in the neonate: Interrelationships of pharmacokinetics, toxicity, and biochemical maturity. *Clin. Chem.* 32:721–727.

Warren, A. J., and Beck, B. D. 1988. Screening and monitoring for exposure and susceptibility to carcinogens. In *Variations in Susceptibility to Inhaled Pollutants: Identification, Mechanisms, and Policy Implications,* eds. J. D. Brain, B. D. Beck, J. Warren, and R. Shaikh, pp. 376–418. Baltimore: Johns Hopkins University Press.

Weiss, B. 1979. Food additives and hyperkinesis: Current evidence. In 87th Annual Meeting of the American Psychological Association, New York.

Whisler, R. L., and Newhouse, Y. G. 1986. Function of T cells from elderly humans: Reductions of membrane events and proliferative responses mediated via T3 determinants and diminished elaboration of soluble T-cell factors for B-cell growth. *Cell. Immunol.* 99:422–433.

Wilson, R., and Crouch, E. A. C. 1987. Risk assessment and comparisons: An introduction. *Science* 236:267–270.

Yase, Y. 1980. The role of aluminum in CNS degeneration with interaction of calcium. *Neurotoxicology* 1:101–109.

Factors Determining Target Doses: Extrapolation from a Shifting Basis

Departments of Environmental Health and Physiology/Biophysics
University of Cincinnati, College of Medicine, Cincinnati, Ohio

I. INTRODUCTION

Assessment of the risks associated with exposure of humans to toxic substances generally requires extrapolations from experiments involving exposure of animals to the same or similar compounds. Only in few instances, such as the study of victims of methyl mercury poisoning in Iraq (Clarkson and Marsh, 1976), has it been possible directly to assess the risks of human exposure. Where extrapolation is necessary, such as from animals to humans, or on the basis of the lowest-observed-adverse-effect level (LOAEL) of a toxicant rather than from a no-observed-adverse-effect level (NOAEL) or a no-observed-effect level (NOEL), allowance is usually made by introducing arbitrary safety or uncertainty factors, conventionally based on powers of 10. The guidelines of the National Academy of Sciences (1977) propose application of these uncertainty factors as follows:

> An uncertainty factor of 10 is used when good chronic or acute expo-
> sure data are available from humans, and they are supported by animal
> studies. An uncertainty factor of 100 is used when good chronic or acute
> toxicity data identifying a NOEL or NOAEL are available for animals but
> not humans. An uncertainty factor of 1000 is applied when limited or
> incomplete toxicity data only are available, or when they define a LOAEL
> rather than a NOEL or NOAEL.[1]

Additional smaller factors have also been applied, depending, for instance, on the quality of the toxicological data, the significance of any adverse effect observed, or on whether a toxicant was administered in food when attempts are

I am grateful for Dr. E. O'Flaherty's constant willingness to discuss the subject with me.
[1]Reprinted from *Drinking Water and Health*, Volume 1, 1977, with permission of the National Academy of Sciences, Washington, D.C.

made to assess the risk of its presence as a contaminant in drinking water. Unavoidably, application of such corrections always involves subjective judgments.

It is obvious that the derivation of an acceptable exposure limit should be based on data requiring a minimum number of uncertainty factors in extrapolating from experiment to regulation. As a consequence, the basis for the extrapolations needs to resemble the final exposure conditions as closely as possible. This conclusion implies, for instance, that no meaningful acute exposure limits can be derived from the results of chronic exposure studies. Such extrapolations have been made in an effort to calculate so-called 1-day or 10-day health advisories, which recommend limits for short-term exposures; the scientific basis for these calculations is sometimes lacking.

The first goal of the experimentalist, therefore, becomes the definition of the optimum exposure conditions from which to extrapolate to human exposures. This requires control over many factors, such as rate and route of exposure, the species, age, sex, and nutritional status of the experimental animals, the chemical form of the toxicant used, and the environmental conditions during exposure.

Such variables can critically influence the response of an animal to a toxic agent by many mechanisms. Consider, for instance, effects of diet on intestinal absorption of toxicants. There may be direct effects on the rate of absorption, as for instance the inhibition of cadmium absorption by calcium (Foulkes, 1980). Binding of a toxicant to nonabsorbed macromolecules or fibers might alter its absorption; changes in the microflora could influence the intestinal metabolism of a toxicant. Section III,B discusses some of these factors in greater detail.

This chapter considers some specific variables that may influence the outcome of toxicological studies. Many aspects of this field have been previously discussed, for instance by Doull (1975) in a useful chapter, Factors Influencing Toxicology, and in a recent short review, Route, Rate and Nature of Exposure as Determinants of Effects of Heavy Metals and Other Toxicants (Foulkes, 1986a). Special attention here focuses on the importance of the rate of exposure, as opposed to the total dose, and on the precise chemical nature of the toxicant. Empirical information collected without adequate control over such variables can provide at best only a shifting basis for risk extrapolations.

II. DOSE–EFFECT RELATIONSHIPS

The view that drugs or toxicants (i.e., agonists) exert their effect on target organs or enzymes by virtue of their reaction with specific receptors has implicitly been accepted for many years. Almost 100 years ago it provided the basis of Ehrlich's formulation of the lock-and-key analogy for the action of specific therapeutic agents. Quantitative formulations of the concept have been contributed by many authors, such as Clark (1926), in reference to the reaction of acetylcholine with muscle cells, and Briggs and Haldane (1925) and Michaelis and Menten (1913) in their work on enzyme inhibition.

The fundamental concept is that within certain dose ranges, the effect of an agonist depends on a stoichiometric reaction with specific receptors; such dose-effect relationships can readily be demonstrated, especially in isolated systems. Over limited dose ranges, the relationship may be linear; extrapolation over wider ranges is more difficult. The process may be complicated by such qualitatively different factors as the existence of a threshold dose below which no effect is produced, or a nonlinear relationship reflecting the fact that the reaction of the agonist with its receptor is a higher-order reaction, or the contribution of capacity-limited processes, as further discussed below.

Especially in the intact animal, interaction between agonist and receptor may become subject to dynamic factors such as the rate at which an agonist can reach the receptor, the rapidity with which the administered compound is metabolized either to an active form (Peters, 1951) or to an inactive product, and the efficiency with which the agonist is excreted. Each of these variables, in turn, may be influenced not only by the total dose of a toxicant but also by the rate at which it is administered or absorbed.

The mechanism of the dynamic interplay between these processes falls under the general heading of toxicodynamics (see, for instance, O'Flaherty, 1981). It follows that the internal dose of a toxicant may not be directly related to the total dose administered. The problem is further complicated by the fact that toxic interactions may neither be reversible nor necessarily very rapid in relation to the turnover of the toxicant. Ultimately, therefore, it is the time integral of that portion of the internal dose that is present at the target site in an available form (not tightly bound to plasma protein, for instance) that will determine the extent of a lesion. This integral, identified as the area under the curve describing effective agonist concentration at the target as a function of time, will here define the effective target dose of the agonist. To the extent that this target dose cannot be predicted from the total dose to which an experimental animal is exposed, it becomes difficult to extrapolate outside the experimental dose range, or from one exposure condition to another.

If a toxicant is not bound to plasma protein, and where the target is readily accessible to diffusion from plasma, the target dose of the toxicant can be approximated by the area under the curve describing the changes in its plasma concentration with time. However, such an approach would not always be applicable, for instance, for a neurotoxin with restricted ability to cross the blood-brain barrier, or for a nephrotoxin like cephaloridine (Tune et al., 1977), known to accumulate at its site of action by a capacity-limited mechanism. Any such saturable mechanism, be it involved in the absorption of the administered dose of a toxicant, its distribution in the body, its metabolism, or its excretion, will abolish the linear relationship between administered and target dose. This point was illustrated in a recent review by O'Flaherty (1986) of models in which capacity-limited elimination kinetics can account for a nonlinear response. Some additional processes that also may alter dose-effect relationships are considered below.

What clearly emerges is that it is not safe to assume that the effective target dose always bears a constant relationship to the administered dose. This difficulty may be illustrated by the attempt to compare the susceptibility of amino acid carriers to p-chloromercuribenzoate on brush border and basolateral cell membranes in the proximal tubule of the rabbit kidney (Foulkes, 1971). It was observed that an injected dose of the mercurial, sufficient to depress basolateral transport by 64%, exerted no apparent effect on the brush border. It would be incorrect, however, to conclude that the basolateral carrier systems are intrinsically more sensitive to the mercurial than are those on the brush border. The missing information, without which no such conclusion could be justified, includes the effective target doses of the drug at each membrane, the relationship of these doses to the respective number of sensitive carrier sites, and the degree to which the functional capacity of the transport systems on the two sides of the cells is saturated by its substrate under the conditions of the experiment.

A major source of uncertainty in evaluating dose-effect relationships thus arises from variations in the effective target dose of a drug or poison. A second source of uncertainty, even at a constant effective target dose, may be the intrinsic sensitivity of the target organ. There is, of course, no reason why similar targets in two different species should respond identically to a given effective target dose. In addition, the sensitivity of a target organ to toxic insult may vary also within a species. For instance, changes in zinc metabolism, such as may follow alterations in zinc intake, or as the result of endocrine factors, can alter the concentration of the metal-binding protein metallothionein and thus influence the toxic effects of cadmium on liver and kidney (Webb, 1979). Zinc and cadmium also interact directly at various sites. An example is the noncompetitive inhibition of cadmium uptake from the lumen of the rat jejunum by excess zinc (Foulkes, 1985). The response to an agonist can also be influenced by the level of the target organ's physiological activity: existence of a large safety margin of functional capacity (the redundancy factor) can increase the target dose of an agonist that is required critically to alter function. In summary, the physiological and nutritional state of an animal may greatly alter the toxicological response to a given dose of toxicant.

We next consider in greater detail the variables determining the pharmacodynamics of an administered toxicant and, therefore, also its effective target dose.

III. DETERMINANTS OF TARGET DOSES

The target dose of a toxicant has been defined above in terms of the time integral of its concentration at the target site. Many dynamic factors can influence this dose, and capacity-limited processes in particular can greatly alter the relationship between the dose of a toxicant administered and the effective dose at the target. This section considers some of these processes.

A. Route of Absorption

It is common knowledge that effects of a drug or toxic substance may depend on the route of its administration. For instance, deeper anesthesia is achieved by intravenous administration of a given amount of barbiturate than following its intramuscular or intraperitoneal injection. Depending especially on the rate of absorption from the injection site, widely differing plasma values of the drug may be obtained and, therefore, also significantly different effective target doses. (Section III,B considers the effect of the rate of absorption.)

There are many examples where the route of administration of a toxicant determines the toxic effects produced. Thus, pulmonary exposure to various carcinogens causes lung cancer and sarcomas may be produced at the site of injection of certain compounds. In both cases, it is presumably the dose at the site of contact that determines the local nature of the initial effect. In addition, the efficiency of absorption by different routes may vary considerably. For example, the efficiency of pulmonary absorption of Cd greatly exceeds that observed in the intestine (Lauwerys, 1982). As a result, the Cd concentration in the target organ (the renal cortex) after pulmonary exposure may exceed that seen following Cd ingestion.

In general, the route of absorption of a toxicant determines the organ through which it first passes, and which will therefore be exposed to higher concentrations than other organs. In the case of substances absorbed from the intestine, the first-pass effect primarily involves the liver. The liver may, therefore, become the primary site of the lesion. Alternatively, hepatic metabolism may prevent significant fractions of a toxicant absorbed at a low rate from reaching peripheral target sites. This effect, which depends on the metabolic capacity of the liver and on the rate of toxicant absorption, is further discussed in Section III,B.

The distribution of a toxicant in the body also depends on the rapidity with which the pool at the site of absorption or injection equilibrates with other body compartments. An interesting example is that of the slow equilibration of pulmonary Cr with other Cr pools in the body. This was reported by Baetjer et al. (1959) and is illustrated in Fig. 1, which compares tissue Cr levels following the intratracheal administration of 0.2 mg of hexavalent Cr to guinea pigs. Note

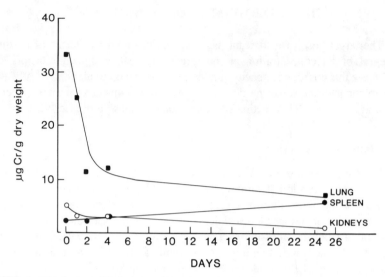

FIG. 1. Distribution of Cr in guinea pigs after intratracheal administration of 0.2 mg hexavalent Cr. (Based on work of Baetjer et al., 1959.)

that for at least 25 days the concentration of the metal in the lung remained significantly higher than that in, for example, the kidneys. A relatively slow turnover of pulmonary Cr was also observed by Yamaguchi et al. (1983): longer turnover times were reported for the lung than for kidneys and other tissues.

B. Rate of Toxicant Absorption

The rate at which a toxicant reaches a toxic target dose depends in the first place on the rate of its absorption into the body. Note that it is not the total dose administered that is emphasized here, but rather the absolute rate at which it is absorbed. Although the importance of absorption rate must be considered for any route of exposure, here it is discussed more in reference to toxicants taken up through the gastrointestinal tract, specifically the intestine.

Several processes contribute to making the rate of intestinal absorption one of the critical factors in establishing dose-effect relationships and in determining the ratio of target to total dose. The first of these is the intrinsic nature of intestinal absorption. This is known in many cases to represent apparently saturable processes, so that above a certain dosage, fractional solute absorption may decrease. In contrast, if an administered substance is largely bound to nonabsorbable macromolecules in the lumen until a dose is reached where the binding capacity of the luminal content is exceeded, fractional absorption may increase with dose. The characteristics of heavy-metal absorption provide an example of such factors at work. Thus, the presence of phytic acid in the intestinal lumen can prevent the absorption of Zn or Fe at low doses; at higher doses of the metals, a greater fraction may be present in diffusible and absorbable form. The reverse effect, a decreased fractional absorption or apparent saturation of the process, has been reported for certain heavy metals, and is illustrated in Fig. 2 for Cd (Foulkes, 1980). A first-order process, with a half time of about 25 min, removes Cd from a jejunal perfusate containing 20 μM CdCl$_2$; this half time is prolonged to over 100 min when the initial Cd concentration is raised to 200 μM. Note also that an excess of Ca depresses Cd uptake. The apparent saturability of Cd uptake and its sensitivity to inhibitors probably do not represent characteristics of specific carrier systems but rather result from nonspecific neutralization of fixed membrane charges (Foulkes, 1985). Whatever its mechanism, however, the consequence of the apparent saturability is the decreased fractional uptake of the metal at higher concentrations. Even below acutely toxic levels, the absolute amount of Cd taken up thus becomes relatively independent of concentration; this distorts any dose-effect relationship seen at lower doses.

The complexity of intestinal processes may render it difficult to predict an effect of dietary additives on toxicant absorption. Still in reference to heavy metals, Rose and Quarterman (1987) actually found that some metal-binding macromolecules may increase absorption indirectly, perhaps as a result of their influence on the magnitude of unstirred layers in the intestine, or on the

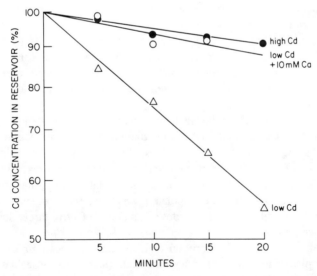

FIG. 2. Cd uptake from rat jejunum. Initial Cd concentrations in recirculating perfusate were 20 μM and 200 μM. (From Foulkes, 1980, with permission.)

production of gastric or enteric mucus. Influence of macromolecules on intestinal metabolism, mediated perhaps by an effect on the microflora, is indicated by the observation of deBethizy et al. (1983). These authors reported that addition of pectin to a purified rat diet increased severalfold covalent binding of ^3H from labeled 2,6-dinitrotoluene to hepatic macromolecules.

At higher absorption rates, processes other than absorption may also reach saturation. Thus (as further discussed in Section III,D), renal secretion or reabsorption of a toxicant may reach maximal values that are no longer dependent on concentration. As a result, the rate of turnover of the absorbed dose would be altered at higher plasma levels. The metabolism of absorbed toxicant also involves saturable processes. This has already been referred to in Section III,A in relation to the first-pass effect. It will be illustrated here by the toxic effects of ingested cyanide. This compound exists at physiological pH primarily in the form of undissociated HCN, in which it readily crosses cell membranes by passive diffusion. The driving force for its absorption, therefore, is not only the total dose but also the volume of intestinal contents determining the chemical concentration of HCN. The possibility of cyanide binding to food constituents is here neglected. Low concentrations of cyanide in portal venous blood can be cleared effectively by the liver, where the enzyme rhodanese leads to its detoxification (Himwich and Saunder, 1948). If, however, the rate of absorption is increased, either by raising the dose or by reducing the volume of the luminal contents, a situation may be reached where the enzymatic capacity of the liver is exceeded, and cyanide will reach other organs such as heart and brain. Clearly, no constant dose-effect relationship can be expected under these conditions.

Rate of absorption of a toxicant is also strongly influenced by the precise chemical nature of the compound. Thus, mercury in the form of methyl mercury compounds is much more readily absorbed than are inorganic compounds of the metal. The importance of such speciation is further considered in Section III,C.

In general, the rate of intestinal absorption of a solute in the lumen is controlled by a series of external and internal factors, as discussed for metals in detail in some recent reviews (Foulkes, 1984b, 1986b). They include the composition of the diet; physiological variables, such as age, sex, and lactation; internal factors, such as bile secretion and intestinal motility; and others. It is, therefore, somewhat risky to extrapolate from an external dose under one set of experimental conditions to internal doses, let alone effective target doses, under different conditions.

Further, if the rate of absorption of a toxicant exceeds, for example, the ability of plasma protein to react with the toxicant or even causes the solubility of the substance in plasma to be exceeded, major changes in body distribution and, therefore, in target doses would ensue. Whatever factors may actually be at work, the effect of dose on body distribution of a toxicant is well documented. Table 1 illustrates, as an example, the influence of the size of an oral dose of Pb on the body distribution of this metal in rats (Aungst et al., 1981).

Note that in Table 1 neither the areas under the curves (AUC) nor the tissue levels of Pb achieved 24 h after oral dosing were proportional to the dose. Several factors must here be involved. The drop in F, the bioavailability of Pb calculated from the ratio of AUC following administration of the particular dose by the oral and by the intravenous routes, presumably reflects the nonlinear absorption kinetics of the metal from the intestine. This, however, cannot explain the changes in the relative distribution of Pb between organs as the oral dose is increased. In addition to the change in the total dose administered, various capacity-limited processes, such as that of Pb absorption, may have been important variables in this study.

When exposure varies with time because of cyclic fluctuations in toxicant concentration, a time-weighted average is usually calculated for the exposure as a whole. However, even for a given time-weighted average, the effects of exposure may be significantly influenced by the nature of the fluctuations. As pointed out by Saltzman (1987), for "threshold models, higher concentrations produce disproportionately greater effects, and thus the mean concentration is not a good estimate for the effects calculation." On the other hand, if saturable processes are involved in toxicant absorption, etc., then the higher concentrations may cause disproportionately smaller effects. Saltzman and Fox (1986) illustrated the extent to which the rate of fluctuation of carbon monoxide in the atmosphere affects the CO saturation of rabbit blood. Figure 3 is based on their results and shows that when a rabbit is exposed to sine wave input of CO, with a maximum exposure level of 300 ppm and similar mean CO levels, then blood CO saturation shows significantly higher excursions from the mean

TABLE 1

Distribution of Pb after Oral Dosing in Rats

Dose (mg/kg)[a]	AUC (days × μg/ml)[b]	F (%)[c]	Brain	Liver (μg/g)[d]	Kidney	Kidney/ liver
1	829	42	0.15	1.01	2.97	2.9
10	2069	10	0.53	2.20	9.16	4.2
100	3450	1.7	0.38	4.18	43.04	10.3

Source: Based on data of Aungst et al., 1981.

[a]Single dose of Pb, given orally (p.o.) to fasted animals.

[b]AUC, area under the blood concentration curve.

[c]Apparent bioavailability, calculated from the expression F = 100 × AUC p.o./AUC i.v. (intravenously), normalized for a constant dose.

[d]As measured 24 h after dosing.

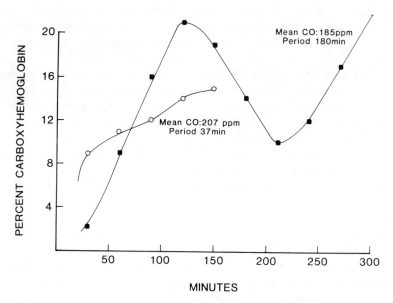

FIG. 3. CO saturation of blood during fluctuating exposure of rabbits to CO. (Based on results of Saltzman and Fox, 1986.) Results with short fluctuations were not available beyond 150 min.

fluctuation period of 180 than of 37 min. Clearly, in this manner the effective target dose may be significantly influenced by fluctuations in the exposure rate, even though the time-weighted average toxicant concentration is kept constant.

The rate of absorption of a toxicant also determines whether compensatory or repair processes can counteract any toxic effects produced. This consideration includes the intrinsic sensitivity of the target organ already referred to in Section II. At very low rates of absorption, even of a large total dose, the target dose may never reach critical levels unless the turnover of the toxicant is so slow that its action becomes in essence cumulative. Even at somewhat higher rates of absorption, compensatory reactions may prevent expression of toxicity. The compensation could involve homeostatic responses at the level of the whole organism, such as polycythemia in chronic CO exposure, or molecular mechanisms at the level of specific organs, such as the induction of metallothionein synthesis during chronic exposure to certain heavy metals (Webb, 1979). In other words, the system may become more resistant during low chronic exposure to a toxicant.

A diminished effect of an otherwise toxic dose of a compound, as stated above, may also result if the rate of administration is sufficiently slow so that tissue repair processes can maintain adequate function. Such a repair process is illustrated in Fig. 4, taken from the work of Nomiyama and Foulkes (1968) on the nephrotoxic action of uranyl ion in the rabbit. An intravenous dose of uranium equivalent to 0.3 mg/kg led to a maximal depression of creatinine clearance (glomerular function) and tubular capacity to secrete *p*-aminohippurate (PAH) after 5 days. Within 2 weeks, however, function had largely returned to normal. This repair is accompanied by histological evidence of cell regeneration. Higher doses of uranium caused irreversible renal shutdown.

C. Speciation

An essential precaution in extrapolating risks of exposure is to avoid, where possible, the use of different chemical forms of the toxicant for the experimental basis of the extrapolation and for the final exposure. This obvious restriction arises from the fact that even slight changes in the chemical form of a toxicant can greatly influence its uptake, distribution, and excretion, and

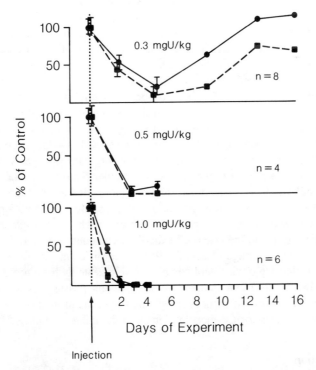

FIG. 4. Nephrotoxic action of uranium in rabbits. *Solid lines:* creatinine clearance (ml/min). *Dashed lines:* maximum tubular PAH secretion (μmol/min). Doses of U are expressed in mg/kg. (Modified from Nomiyama and Foulkes, 1968.)

therefore also its target dose and even the nature of the target. Numerous examples of such effects have been reported in the literature.

For instance, it is well known that compounds of trivalent Cr [Cr(III)] are generally less toxic than those of the hexavalent metal Cr(VI). The latter are usually more water soluble than the former, and thus more readily absorbed from the intestine. In addition, the greater absorption of Cr(VI) may also be related to its occurrence in complex anions, capable of being transported by nonspecific anion carrier mechanisms. In contrast, uptake of cationic Cr(III), when present in soluble ionized form, appears to involve nonspecific electrostatic binding, similar to what has been demonstrated for Cd and other heavy metals (Foulkes, 1984a); there is no evidence for specific carrier systems for most of these elements. Differences in solubility probably contribute to the great variations seen in the carcinogenicity of chromate salts following their intrapleural implantation in the rat (Hueper, 1961).

Another important variable is the lipid solubility of various species of a toxicant; this is well recognized to determine to a significant extent the ability of compounds to cross cell membranes. A classical example is given by the contrast between the toxic effects of inorganic mercury (Hg^{2+}) and of methyl mercury (Clarkson and Marsh, 1976). The organic mercurial crosses the intestinal barrier much more readily than does Hg^{2+}. It is, in addition, not appreciably excluded from the central nervous system by an impermeable blood-brain barrier, and its wider distribution in the body contributes to its longer retention. A greater fraction of ingested methyl mercury than of Hg^{2+} will therefore reach the brain; inorganic Hg preferentially accumulates in the kidney. As a result, the toxic effects of the two substances are quite distinct.

Similarly, the precise chemical form in which Cd is administered greatly influences its distribution in the body, and therefore its toxic effects. When Cd is injected into the renal artery as $CdCl_2$ it is neither filtered nor retained in the kidney (Foulkes, 1974). In contrast, thiol complexes of Cd are readily taken up by the kidney. Systemic administration of Cd leads to early renal damage only when the metal is given as a complex such as Cd-cysteine, or Cd-metallothionein; in the absence of the thiol compounds acutely injected Cd leads to testicular damage (Parizek, 1956). The different distributions of Cd given as the chloride or as a thiol complex have repeatedly been noted (Cherian et al., 1978; Johnson and Foulkes, 1980). Among possible explanations for these observations are differences in binding to plasma protein or the presence of specific receptors for Cd-metallothionein in the kidney (Foulkes, 1978).

D. Excretion

The second major factor that determines the effective target dose of a toxicant, i.e., the time integral of the effective toxicant concentration at the target site (see Section II), is the rate of toxicant clearance. The clearance concept is here used in the wider sense to include excretion from the body, removal from plasma by metabolic breakdown, or redistribution between different body compartments. The faster the clearance from a given volume of distribution, the shorter the half-life of the toxicant at the target site, and therefore the smaller the effective target dose. Obviously also, the half-life for a given rate of clearance will vary directly with the size of the distribution volume.

Few of the processes involved in clearance follow first-order kinetics. The nonlinear relationship between their rate and the concentration of a toxicant results in the concentration dependence of clearance. Involvement of saturable transport steps in the renal or hepatic clearance of many solutes is well established, as illustrated by the following well-documented examples. In the case of urinary excretion, fractional reabsorption of filtered solute may fall at higher concentrations, due to saturation of the tubular reabsorptive systems; this leads to a rise in clearance at higher concentrations. Inversely, for a solute actively secreted into urine or bile, saturation of the transport system may lead to a decreased clearance at higher concentrations. Clearance is also reduced by saturation of a metabolic pathway which normally would transform the xenobiotic into a more readily excreted form (e.g., conjugation reactions) or to breakdown of the toxicant. The important consequence of these facts is that clearance, and therefore the half-life of a toxicant at one concentration, cannot always be predicted from observations at another concentration.

Additional saturable processes may also come into play. The finite capacity of plasma protein to bind metals, for instance, will tend to reduce disproportionately their effective target dose at low metal concentrations. In general, transfer of a toxicant into a relatively inert but saturable compartment will decrease toxicity at low toxicant concentrations and interfere with extrapolation to higher concentrations.

E. Synergism and Antagonism

The complex mode of action of many toxicants may provide a multiplicity of points at which additional variables can influence the nature of magnitude of the final lesion. Reference has been made already to the fact that a diet high in Zn, for instance, may reduce the toxicity of Cd. The Zn may interfere with intestinal absorption of Cd (Foulkes, 1985), or it may induce synthesis of sufficient metallothionein to sequester Cd, or it may even compete directly with Cd at sensitive sites on enzymes. Even at a constant effective target dose of Cd, the target tissue may thus become less sensitive to the metal than it is in an animal on a smaller Zn intake.

A synergistic effect between unrelated environmental variables and toxicity of a test chemical may be seen, for instance, in the influence of atmospheric pressure on the action of CO. Because of the relative affinities of O_2 and CO for hemoglobin, the CO saturation of the protein increases with pressure, in spite of a constant ratio of partial pressures of the two gases. It follows that the maximum permissible CO concentration in air inspired at normal pressure is higher than that acceptable in the air supplied to a diver under pressure.

A third example is that of the influence of drugs stimulating the ability of an organ such as the liver to metabolize certain xenobiotics by the so-called cytochrome P-450 system. This is a multifunctional oxidase system intimately involved in the metabolism of many toxicants; it may lead to decreased toxicity if the active form of the toxicant is that reaching the liver, or it may increase toxicity if the active form is a metabolite of the administered substance. For instance, induction of the P-450 system decreases the sleeping time after barbiturate administration, i.e., it makes the animal more resistant to the drug. An instance of potentiation is the remarkable effect of isopropanol on the hepatotoxic action of carbon tetrachloride (Plaa et al., 1975). It is clear that such interactions can greatly alter the validity of risk extrapolations from one set of exposure conditions to another. This is especially important for nonexperimental exposure of humans, which often involves more than a single pure compound.

IV. IMPLICATIONS FOR RISK ASSESSMENT

Risk assessment must always include proper allowance for uncertainties by application of appropriate uncertainty or safety factors. This large and specialized field of study lies outside the scope of this chapter. Nevertheless, it is appropriate to reemphasize the need to reduce uncertainties to a minimum by careful attention to some of the factors discussed above. This requires that conditions of experimental exposure be kept as similar as possible to those of the human population potentially at risk. Route and rate of exposure, the matrix in which the toxicant is administered (e.g., food or water for intestinal exposure), the precise speciation of the toxicant, including its physicochemical properties (e.g., particle size for aerosols, valency for metals, and solubility), the state of nutrition of the animals, physiological factors such as age and sex, environmental conditions, etc., all need to be carefully controlled. Although this control has not always been achieved, useful values for acceptable limits for many chemical exposures have been obtained. In other cases, unnecessary uncertainty may have been introduced, for instance in attempts to calculate acceptable contaminant levels in drinking water on the basis of studies in which the toxicant had been administered admixed to food, or when acceptable levels for short-term exposures were to be derived from long-term exposure studies.

V. CONCLUSIONS

The risks attending human exposure to environmental toxicants are, in most instances, estimated on the basis of studies with experimental animals. Unavoidably, extrapolation from species to species, or from one set of exposure conditions to another, may involve some major uncertainties. In order to keep these at a minimum it is important to take into account a series of potentially important factors. The complexity of the necessary considerations is illustrated in this review by a discussion of some of these factors. Some determinants of dose-effect relationships are given in the following list. This listing is not exhaustive, and the categories of factors often overlap.

- Route of administration
- Rate of absorption and excretion
- Drug metabolism
- Constant vs. variable exposures
- Chemical form of administered drug
- Presence of synergists, antagonists
- Intrinsic sensitivity of target organ
- Sources of nonlinearity: threshold phenomena and capacity limitations
- Physiological variables (age, sex, etc.)

Careful selection of experimental studies to serve as the basis of risk assessment may require exclusion of a significant number of reports; this could greatly reduce the data base for extrapolation. Balanced against that is the lessened uncertainty involved in the extrapolation. The following suggestions represent some ideal guidelines for defining the optimum basis of risk extrapolations; in practice it may seldom be possible to follow them all.

1. Risks attending exposure by one route should not be deduced from studies in which the toxicant was administered by another route.
2. Standards for drinking water should not be based on studies in which the toxicant was administered in food, and vice versa.
3. Acute exposure studies cannot readily provide a basis for assessing risks attending chronic exposure, and vice versa.
4. To the extent possible, extrapolations of systemic effects of toxicants should be based not on the total dose administered, but rather on their effective concentration and turnover at the target site (i.e., the target dose), as approximated in some cases at least by the area under the curve relating blood concentration to time.

REFERENCES

Aungst, B. J., Doke, J. A., and Fung, H-L. 1981. The effect of dose on the disposition of lead in rats after intravenous and oral administration. *Toxicol. Appl. Pharmacol.* 61:48–57.

Baetjer, A. M., Damron, C., and Budacz, V. 1959. The distribution and retention of chromium in men and animals. *Arch. Indust. Health* 20:136–150.

Briggs, H. E., and Haldane, J. B. S. 1925. A note on the kinetics of enzyme action. *Biochem. J.* 19:338–339.

Cherian, M. G., Goyer, R. A., and Valberg, L. S. 1978. Gastrointestinal absorption and organ distribution of oral cadmium chloride and cadmium metallothionein in mice. *J. Toxicol. Environ. Health* 4:861–868.

Clark, A. J. 1926. The reaction between acetylcholine and muscle cells. *J. Physiol.* 61:530–546.

Clarkson, T. W., and Marsh, D. O. 1976. The toxicity of methyl mercury in man: Dose response relationships in adult populations. In *Effects and Dose Relationships of Toxic Metals,* ed. G. F. Nordberg, pp. 246–261. Amsterdam: Elsevier.

deBethizy, J. D., Sherrill, J. M., Ricket, D. E., and Hamm, T. E., Jr. 1983. Effects of pectin-containing diets on the hepatic macromolecular covalent binding of 2,6-dinitro-[^{3}H]toluene in Fischer-344 rats. *Toxicol. Appl. Pharmacol.* 69:369–376.

Doull, J. 1975. Factors influencing toxicology. In *Toxicology,* eds. L. J. Casarett and J. Doull, pp. 133–147. New York: Macmillan.

Foulkes, E. C. 1971. Effects of heavy metals on renal aspartate transport and the nature of solute movement in kidney cortex slices. *Biochim. Biophys. Acta* 241:815–822.

Foulkes, E. C. 1974. Excretion and retention of cadmium, zinc and mercury by rabbit kidney. *Am. J. Physiol.* 227:1356–1360.

Foulkes, E.C. 1978. Renal tubular transport of cadmium-metallothionein. *Toxicol. Appl. Pharmacol.* 45:505–512.

Foulkes, E. C. 1980. Some determinants of intestinal cadmium transport in the rat. *J. Environ. Pathol. Toxicol.* 3:471–481.

Foulkes, E. C. 1984a. Mechanism of chromium absorption. In *Trace Substances in Environmental Health,* vol. 18, ed. D. Hemphill, pp. 124–129. Columbia, Mo.: Univ. of Missouri.

Foulkes, E. C. 1984b. Intestinal absorption of heavy metals. In *Pharmacology of Intestinal Permeation,* ed. T. Z. Csaky, pp. 543–565. Berlin: Springer-Verlag.

Foulkes, E. C. 1985. Interactions between metals in rat jejunum: Implications on the nature of cadmium uptake. *Toxicology* 37:117–125.

Foulkes, E. C. 1986a. Route, rate and nature of exposure as determinants of effects of heavy metals and other toxicants. In *Trace Substances in Environmental Health,* vol. 20, ed. D. Hemphill, Columbia, Mo.: Univ. of Missouri, pp. 154–164.

Foulkes, E. C. 1986b. Cadmium absorption. In *Cadmium, Handbook of Experimental Pharmacology,* vol. 80, ed. E. C. Foulkes, pp. 75–100. Berlin: Springer-Verlag.

Himwich, W. A., and Saunder, J. P. 1948. Enzymatic conversion of cyanide to thiocyanate. *Am. J. Physiol.* 153:348–354.

Hueper, W. C. 1961. Environmental carcinogenesis and cancers. *Cancer Res.* 21:842–857.

Johnson, D. R., and Foulkes, E. C. 1980. On the proposed role of metallothionein in the transport of cadmium. *Environ. Res.* 21:360–365.

Lauwerys, R. 1982. The toxicology of cadmium. Environmental and quality of life, report EUR 7649 EN. Luxembourg: Commission of the European Communities.

Michaelis, L., and Menten, M. 1913. Die Kinetik der Invertinwirkung. *Biochem. Z.* 49:333–369.

National Academy of Sciences. 1977. Drinking water and health. U.S. Environmental Protection Agency, Washington, D.C., PB-269 519. Springfield, Va.: National Technical Information Service.

Nomiyama, K., and Foulkes, E. C. 1968. Some effects of uranyl acetate on proximal tubular functions in rabbit kidney. *Toxicol. Appl. Pharmacol.* 13:89–98.

O'Flaherty, E. J. 1981. *Toxicants and Drugs.* New York: Wiley-Interscience.

O'Flaherty, E. J. 1986. Dose dependent toxicity. *Comments Toxicol.* 1:23–34.

Parizek, J. 1956. Effect of cadmium salts on testicular tissue. *Nature* 177:1036–1037.

Peters, R. A. 1951. Lethal synthesis. *Proc. Roy. Soc. London* 139B:143–170.

Plaa, G. L., Traiger, G. J., Hanasono, G. K., and Witschi, H. P. 1975. Effect of alcohols on various forms of chemically induced liver injury. In *Alcoholic Liver Pathology,* eds. J. M. Khanna, Y. Israel, and H. Kalant, pp. 225–244. Toronto: Addiction Research Foundation.

Rose, H. E., and Quarterman, J. 1987. Dietary fibers and heavy metal retention in the rat. *Environ. Res.* 42:166–175.

Saltzman, B. E. 1987. Lognormal model for health risk assessment of fluctuating concentrations. *Am. Ind. Hyg. Assoc. J.,* 48:140–149.

Saltzman, B. E., and Fox, S. H. 1986. Biological significance of fluctuating concentrations of carbon monoxide. *Environ. Sci. Technol.* 20:916–923.

Tune, B. M., Wu, K. Y., and Kempson, R. L. 1977. Inhibition of transport and prevention of toxicity of cephaloridine in the kidney: Dose-responsiveness of the rabbit and the guinea pig to probenecid. *J. Pharmacol. Exp. Therap.* 202:466–471.

Webb, M. 1979. The metallothioneins. In *The Chemistry, Biochemistry and Biology of Cadmium,* ed. M. Webb, pp. 195–266. Amsterdam: Elsevier-North Holland.

Yamaguchi, S., Sano, K., and Shimojo, N. 1983. On the biological half-time of hexavalent chromium in rats. *Ind. Health* 21:25–34.

Approaches Used to Assess Chemically Induced Impairment of Host Resistance and Immune Function

Peter T. Thomas

Life Sciences Department
IIT Research Institute
Chicago, Illinois

I. INTRODUCTION

As knowledge of immune processes increases, it is becoming apparent that immunity plays a critical role not only in the resistance to infection but also in maintenance of homeostasis (Bellanti et al., 1981). A growing concern among toxicologists, health specialists, and regulatory agencies is that some pathological conditions observed in exposed human populations may be linked to alterations of immunological status caused by chemical xenobiotics. For example, in recent years, there have been reports documenting immune dysfunction in humans accidentally exposed to drug or chemical xenobiotics (Bekesi et al., 1978; Kalland, 1985; Kammuller, 1984; Biagini et al., 1985; Lee and Chang, 1985; Bekesi et al., 1987; Hoffman et al., 1986; Fiore et al., 1986). Such incidents are significant in light of the association between therapeutic use of immunosuppressives and increased incidence of infectious and neoplastic disease (Allen, 1976; Penn, 1985).

While it is relatively easy to demonstrate immunosuppression in the laboratory, one of the most critical issues is whether or not these xenobiotics present in the environment pose a health threat to the general population through inadvertent exposure. As a result, the study of chemical impairment of immune processes has evolved in recent years into a legitimate scientific discipline known as immunotoxicology. Several federal regulatory and scientific research agencies are now taking an active role both in the development of protocols for immune profile testing of chemical xenobiotics and of guidelines for risk assessment. This chapter introduces the concept of the immune system as a legiti-

I thank Pat Wagner for typing the manuscript and Dr. Robert Sherwood for his critical review.

mate target organ for toxicity. A review of the methodology currently in use to assess chemically induced impairment of immune function is also discussed. A summary follows of an interlaboratory validation exercise sponsored by the National Toxicology Program, of which our laboratory was a participant. The purpose of the exercise was to establish a tier testing approach for evaluating chemically induced immunotoxicity. A brief review of some environmentally relevant chemicals reported to alter the immune response then follows.

II. IMMUNE SYSTEM AS A TARGET ORGAN FOR TOXICITY

The immune system consists of a network of specialized cells that mount both specific and nonspecific responses in the presence of foreign substances known as antigens. The response ranges from production of soluble antibody (humoral-mediated immunity, HMI) to induction and modulation of cellular inflammatory responses (cell-mediated immunity, CMI; Roitt et al., 1985). Due to the complexity of the immune system it is necessary to appreciate the organizational structure in order to rationally evaluate the effects of a potential immunotoxicant. A simplified schematic diagram of the immune system is shown in Fig. 1. The bone marrow, a primary lymphoid organ, is the site of origin of pluripotent stem cells from which mature cells subsequently develop. Stem cells that subsequently migrate to and mature in the thymus, another primary lymphoid organ, are referred to as thymus-derived lymphocytes or T-cells.

The development of monoclonal antibodies has made possible the identification, in different species, of T-cell subpopulations which help (T helper/inducer) or suppress (T suppressor/cytotoxic) the immune response. T-cells can be further characterized into subsets that mediate delayed hypersensitivity, kill foreign cells in allogeneic tissue grafts, or mediate cell-mediated immune responses against certain intracellular bacterial pathogens such as *Listeria monocytogenes* or *Mycobacterium tuberculosis*. These lymphocytes are important in immunoregulation because they can augment or suppress immune function either directly or indirectly through the production of biologically active molecules known as lymphokines. These lymphokines can induce a chemotactic response (i.e., cause other cell types to migrate to the site of inflammation), suppress the function of other cells, or mediate tumor specific immune reactions, among other functions. Immature, uncommitted lymphocytes can also differentiate into B-cells in the bursa of Fabricius (in avian species) or in the bone marrow, Peyer's patches or other gut-associated lymphoid tissue in mammals. Upon antigenic stimulation, these cells differentiate into antibody-producing plasma cells in lymph nodes, spleen, or other secondary lymphoid organs.

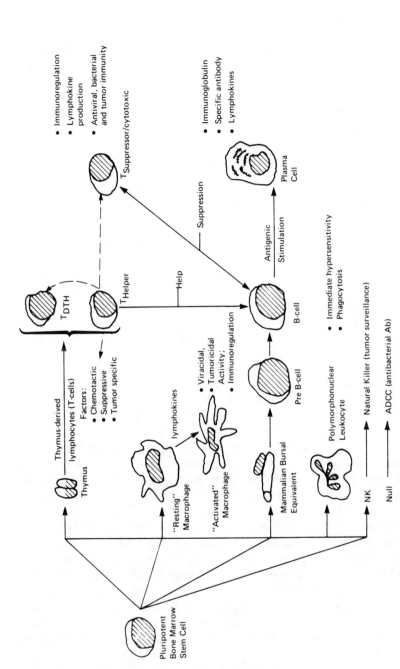

FIG. 1. Simplified schematic diagram of immune cell differentiation.

Other lymphocytic cells active in the immune response to not share cell surface markers characteristic of T lymphocytes or macrophages. These include natural killer (NK) cells and so-called null cells. NK cells were described initially by Herberman (1982). Evidence indicates that these cells play a major role in innate host resistance against neoplasia. These cells do not require prior antigeneic stimulation to exert their effector function (Kiessling and Wigzell, 1979; Koo et al., 1980). In addition, evidence suggests that NK cells are important in host resistance to certain bacterial (Holmberg et al., 1981) and viral infectious processes (Stein-Streilein and Guffee, 1986). NK cells are a definite lymphocyte lineage separate from T-cells because there are strains of mice with a normal T-cell compartment that exhibit deficient NK cell activity (Roder and Duwe, 1979).

The null cell lacks most but not all cell surface markers common to NKs, T-, or B-cells (Balch et al., 1980). This particular cell is active in antibody-dependent cell-mediated cytotoxicity (Perlmann et al., 1972). Null cells have been reported to mediate killing of target cells that have been coated with opsonizing antibody via IgG Fc receptor binding.

The pluripotent bone marrow stem cell pool also serves as the source of progenitors for cells of the monocyte/macrophage lineage. Macrophages are actively phagocytic and mediate a wide variety of immune processes (Roitt et al., 1985). The macrophage is present in a variety of tissues, including blood (blood-borne monocytes), liver (Kupffer cells), lung (alveolar macrophages), and spleen. Once the macrophage has phagocytized antigen, the material is presented to T-lymphocytes when both T-cells and macrophages are in close contact. In order for the immune response to occur, the antigen must be recognized by the T-cell in the presence of other cellular recognition structures present on the macrophage. Antigen presentation and T-cell activation are central for both the induction of antibody responses to so-called T-cell dependent antigens and specific activation of T-lymphocytes. Once macrophages are activated, they exhibit augmented physiological responses including increased protein synthesis, production of intracellular enzymes, and expressed several cell surface receptors (reviewed by Adams and Hamilton, 1984). Activation of macrophages from a resting state is accomplished by lymphokines, endotoxins, and other mediators. Once activated, the macrophage can act in a nonspecific fashion to phagocytize foreign bacteria, act against viral or bacterial pathogens, and also increase immunoregulatory activity through the release of such cytokines as interleukin-1. Therefore, the system-wide distribution of macrophages serves to not only localize antigen regionally but also modulate the immune response at those local sites.

The polymorphonuclear leukocyte (PMN) plays an initial first role in phagocytosis, inflammation, and immunity to pathogenic organisms. For PMNs to exert their full biological activity, they must be capable of mobilizing to the area of need, phagocytizing and killing the offending organism. Intrinsic cellular abnormalities of mobility are seen in Chediak-Higashi (CH) syndrome, diabetes mellitus, "lazy" leukocyte syndrome, and other diseases. Abnormalities of killing are seen in chronic granulomatous disease, myeloperoxidase deficiency, and CH syndrome (Baehner, 1974; Miller, 1975). Abnormal PMN function in humans is associated with certain bacterial infections.

The immune response also includes another major arm responsible for host protection. This manifests itself through the production of soluble, biologically active proteins including antibodies, immunoglobulins, and the complement components. As previously mentioned, B-lymphocytes, once stimulated with antigen, go on to differentiate into plasma cells and secrete in large quantities, specific proteins known as antibodies. At present there are five well-characterized classes of antibodies. These include immunoglobulins (Ig) G, A, M, D, and E. Each Ig isotype plays a specific biological role in the immune response. One important antibody isotype whose function is well understood is the large molecular weight pentameric protein IgM. This molecule is produced in large amounts early in an infectious process. It is very efficient in agglutinating offending organisms which increases their rate of removal by the reticuloendothelial system. Another antibody whose function is relatively well understood is the 150,000 molecular weight monomer IgG. This antibody isotype appears in large amounts in secondary immune responses. It plays a major role in phagocytosis (opsonization) and cell lysis in conjunction with complement. Once free antigen is removed from the host, these antibodies serve an immunoregulator role by suppressing their further production. However, in certain situations, antibody may act alone or in conjunction with antigen and damage certain host tissues leading to autoimmune disease.

The complement system consists of at least 20 interacting serum proteins, many of which are enzymes. These serum proteins mediate a series of reactions ultimately leading to cell lysis. Activation of the complement system can occur through a variety of mechanisms. In the classical pathway for activation, antigen bound to the cell surface with complexes of antibody initiates the enzyme cascade. In the alternative activation pathway initial steps in the cascade phenomenon can occur following exposure to several different biological products such as bacterial lipopolysaccharides. In either case, by-products of the enzymatic cleavage process lead to the production of many biologically active protein fragments which have the capability to induce chemotaxis, inhibit cell migration, cause changes in vascular permeability, and induce smooth muscle contraction.

The specific site of action of chemical xenobiotics or drugs can determine the nature and severity of immune modulation. If the target organ is the bone marrow, severe and broad-ranging deficiencies in immunity can occur. For example, exposure to benzene, a widely used chemical solvent, causes severe depletion of all bone marrow derived cellular elements (Irons et al., 1979). Death in experimental animals usually results from overwhelming infection. If the target is at the immune effector cell level, the defects may be more specific. For example, cyclophosphamide (Cy) is an immunosuppressive agent which, when administered at or near the time of experimental immunization, severely depresses the production of humoral antibody in a dose-dependent fashion. In addition, Cy also can inhibit the function of certain T-cell subsets such as T-suppressor cells. This was suggested by experiments where increased antibody and cell-mediated immunity was noted after subchronic in vivo low-dose Cy pretreatment (Rollinghoff et al., 1977; Turk and Parker, 1982).

In summary, as more is learned about the complex interaction of the cells, tissues, and soluble products involved in an immune response, it becomes increasingly difficult to functionally separate the cellular and humoral immune components. Both compartments are highly interrelated and exquisitely regulated to provide protection of the host from a wide variety of environmental insults under many different conditions.

III. METHODOLOGIES USED TO MEASURE CHEMICALLY INDUCED IMMUNOTOXICITY

The complex interactions among the various cells and tissues of the immune system (Fig. 1) are advantageous to the host but confound studies assessing the impact of chemical compounds on the immune system. Exposure to a chemical xenobiotic may alter the response or reserve capacity in one area of the immune system. Yet measurement of the host's ability to respond to an experimentally administered antigen or survive infectious challenge may be unchanged, leading to false negative interpretations.

To maximize the accuracy of the test and of the ability to make reasonable risk assessment decisions, one must evaluate the immune system at different levels. These levels can be categorized. They include assessment of immunopathology; antibody- and cell-mediated immunity; innate, nonspecific immunity; and lastly, host resistance to infectious or tumor challenge.

The animal model most commonly used for these studies is the mouse. This species is used primarily because a great deal is already known about the structure and function of its immune system. Furthermore, it is generally accepted that the human and murine immune systems are similar with respect to ontogeny, regulation and effector cell maturation, differentiation, and function. Brief descriptions of the methodologies and their advantages and disadvantages when used to assess chemical for potential to alter the immune response of rodents are discussed in the following section. Tables 1 and 2 summarize immune function

TABLE 1

Commonly Used Immune Function Assays to Detect Chemically Induced Immunotoxicity

Assay system	Endpoint measured	In vivo immune correlate
Plaque forming cell (PFC) response to sheep red blood cells (SRBCs)[a]	Quantitation of antibody producing cells in the spleen of previously immunized mice	AMI,[b] macrophage function
Lymphocyte proliferation to B-cell mitogens[a]	Incorporation of radiolabeled nucleic acid precursors into DNA of dividing B-lymphocytes	AMI; (nonspecific lymphocyte blastogenesis)
Delayed type hypersensitivity response to keyhole limpet hemocyanin (KLH)[c]	Measurement of influx of ^{125}IUdR labeled monocyte precursors into the site of elicitation of response to antigen	CMI,[d] hypersensitivity
Mixed lymphocyte culture[a]	Incorporation of radiolabeled nucleic acid precursors into DNA of dividing T-lymphocytes	CMI; (antigen recognition; cell proliferation)
Cytotoxic T-lymphocyte activity[c]	^{51}Cr-release from labeled target cells following incubation with sensitized T-lymphocytes	CMI; (antigen recognition, cell proliferation, effector function)
Natural killer (NK) cell function[a]	Percent cytotoxicity of spleen cells vs. ^{51}Cr-labeled YAC-1 lymphoma target cells	Natural or innate immunity
Macrophage phagocytosis[c]	Phagocytosis of ^{51}Cr-labeled CRBCs	Macrophage function
Luciferin, luciferase chemiluminescence	Macrophage ATP levels	Macrophage function
Determination of macrophage ectoenzyme activity	Leucine aminopeptidase, 5′-nucleotidase, alkaline phosphodiesterase	Status of macrophage activation and differentiation
Alveolar macrophage bactericidal activity	Pulmonary clearance of inhaled ^{35}S-labeled *Klebsiella pneumoniae*	Macrophage bactericidal activity

[a]National Toxicology Program (NTP) Tier I test (Luster et al., 1988).
[b]Antibody-mediated immunity.
[c]NTP Tier II test.
[d]Cell-mediated immunity.

55

TABLE 2

Commonly Used Host Resistance Assays to Detect Chemically
Induced Immunotoxicity

Host resistance model	Endpoint measured	Protective mechanism
Listeria monocytogenes	Cumulative percent mortality, mean survival time	CMI,[a] macrophage function
Streptococcus sp.	Cumulative percent mortality, survival time	AMI,[b] polymorphonuclear leukocytes, nonspecific immunity
PYB6 fibrosarcoma	Tumor latency, size following implantation	CMI, NK
B16-F10 melanoma tumor	Lung tumor burden	CMI, macrophage function, NK
Plasmodium berghei	Mouse malaria	AMI, CMI, macrophage function
Trichinella spiralis	Worm expulsion rates	CMI, AMI?
Influenza A2/Taiwan virus	Cumulative percent mortality, survival time	AMI, interferon, NK
Herpes simplex virus	Cumulative percent mortality, survival time	CMI, macrophage function

[a]Cell-mediated immunity.
[b]Antibody-mediated immunity.
Note: With the exception of the herpes virus and *Trichinella* models, all are classified as NTP Tier II tests (Luster et al., 1988).

and host resistance assays most frequently used to assess chemically induced immunotoxicity.

A. Immunopathology

Routine clinico- and histopathology currently are among the few tests routinely performed in an animal toxicology study that can detect chemically induced immunotoxicity. These include histology of the primary and secondary lymphoid organs; spleen and thymus weights and cellularity; and complete blood count (CBC), differential, and cellularity. Because of the structural division of the spleen and lymph nodes into thymus-dependent (T-cell) and thymus-independent (B-cell) areas, careful microscopic examination and the use of specific staining techniques may indicate preferential chemical effects on T- or B-cell compartmentalization and numbers. Bone marrow differential and cellularity determinations can aid in immunotoxicity assessment and can be performed on tissue sections, marrow smears, or single cell suspensions. Performing these determinations on single cell suspensions also allows the measurement of both cell numbers and viability, important parameters if one wishes to elucidate mechanisms of a functional immune defect.

The advantages of these clinico- and pathological tests are that they are quite sensitive and are relatively inexpensive to perform. The disadvantage is that they often provide no indication of functional defects in the immune response. If the data from the above studies suggest an immunological defect, one can utilize specific immunohistochemical methods to determine if functional changes in immunity can be explained by alterations in the number and location of specific T- or B-lymphocyte or monocyte subpopulations. Specific antibody reagents can be used for qualitative and quantitative analysis of B-cells, T-cells, stem cells and macrophages by flow cytometry (Hudson et al., 1985).

B. Humoral-Mediated Immunity

1. Antibody Responses to Experimentally Introduced Antigens

Among the many tests to measure integrity of the humoral immune response, the antibody plaque-forming-cell (PFC) assay (Cunningham and Szenberg, 1968) allows for quantitation of antibody-producing cells following immunization with an antigen. The most common antigen employed is the sheep erythrocyte (SRBC). The response to SRBCs requires the cooperation of a number of cell populations, including B-cells, T-helper cells, and macrophages. T-cells aid in antigen recognition and presentation via surface membrane glycoproteins, as well as by production of mediators required for cell proliferation (e.g., B-cell growth and stimulatory factors) and differentiation (e.g., B-cell differentiation factor). The macrophage is involved in antigen processing as well as production of interleukin-1, a soluble-growth factor. Assessment of T-cell independent responses can be assessed with the use of antigens such as TNP-ficoll which do not require T-cell presentation to activate B-lymphocytes to differentiate into specific antibody-producing plasma cells. Results most commonly are reported as PFC/spleen or PFC/10^6 nucleated viable cells in order to correct for chemically induced changes in splenic cellularity or viability.

2. In Vitro Lymphoproliferative Responses

Blastogenesis and proliferation represent sequential steps in the development of humoral immune responses. This phase of the immune response can be measured by determining the degree to which B-lymphocytes can be polyclonally stimulated to proliferate in the presence of a variety of bacterial cell well constituents, such as bacterial lipopolysaccharide (LPS). The magnitude of the response is quantitated by the incorporation of [^{3}H]thymidine into DNA.

One great advantage to using this assay is that microtiter plate adaptation coupled with computer-aided data analysis makes possible the simultaneous measurement, in large groups of animals, of B-lymphocyte mitogenic responses in the presence of several mitogen concentrations. However, suppression of humoral immune responses as measured by the anti-SRBC PFC response and B-lymphocyte mitogenesis do not always correlate (Luster et al., 1988). Lymphocyte blastogenesis to mitogens measures only a "window" in cell maturation and, unlike the antigen-induced maturation measured in the PFC assay, it does not measure early events in activation or terminal B-cell differentiation into antibody-secreting cells (Klaus and Hawrylowicz, 1984).

C. Cell-Mediated Immunity

Assessment of the steps necessary for any acquired cell-mediated immune reaction, namely antigen recognition and processing; blastogenesis and clonal proliferation of immune effector cells; migration of memory T-cells to the site of challenge and production and release of inflammatory mediators and lymphokines can be measured by the assays described below.

1. Delayed-Type Hypersensitivity (DTH) Responses

Hypersensitivity responses can be categorized into four general types (Coombs and Gell, 1975). Of the four types, classical delayed-type hypersensitivity (DTH) responses (type IV) are widely used in assessment of CMI. In a previously sensitized individual, this response is characterized as a slowly evolving inflammatory lesion, usually not appearing for at least 10 h and not peaking in inflammation and in duration until 24–72 hours following a subcutaneous injection of antigen. Histological examination of the lesion reveals a primarily mononuclear cell infiltrate with few polymorphonuclear leukocytes present (Davis et al., 1973). The DTH response is distinguished from an immediate type I response (Arthus) by the longer response kinetics, differing appearance of the lesion, and by the mononuclear cell infiltrate. The DTH reaction can be quantitated by measuring footpad or ear swelling (Luster et al., 1982b; Lefford, 1974) or by popliteal lymph node enlargement (Noble and Norbury, 1986). One commonly used antigen is Keyhole limpet hemocyamin (KLH). KLH is a large molecular weight protein capable of eliciting a T-dependent DTH response without the risk of subsequent infection or contact sensitization (Luster et al., 1982b). Radioisotopic procedures have been developed to measure either vascular permeability or monocyte influx (Holsapple et al., 1984). The magnitude of DTH can be reported as a stimulation index, which is calculated for each sensitized animal as a ratio of the cellular influx at the challenge site to that at a comparable unchallenged site. Although DTH is not as sensitive as lymphoproliferation or cytotoxic T-lymphocyte (CTL) responses, it represents a good overall assessment for CMI which correlates with immunodeficiency and decreased resistance to infectious disease (MacLean, 1979).

2. Mixed Leukocyte Response (MLR)

The MLR assesses the ability of T-lymphocytes to recognize antigens present on the surface of foreign cells and undergo proliferation. From a clinical viewpoint, the MLR measures the same mediators involved in graft vs. host reactions and graft rejection, and has been shown to be predictive of general immunocompetence and survival in cancer patients (Golub et al., 1974). As with mitogen-induced blastogenesis, stimulation by allogeneic leukocytes is quantitated by [^{3}H]thymidine uptake. For many xenobiotics, the response of spleen cells from chemically exposed mice to allogeneic cells is a very sensitive indicator of CMI which correlated with T-cell mediated host resistance models (Luster et al.,1988).

3. Generation of Cytotoxic T-Lymphocytes

A direct follow-up to the MLR is the evaluation of cytotoxic T-cell effector function. Interaction of T-cells with tumor or nonself antigens of the major histocompatibility complex results in the acquisition of IL-2 reactivity. In the presence of IL-2, these cells develop into fully cytotoxic T-effector cells capable of lysing syngeneic or allogeneic tumor cells. The ability of splenocytes to generate CTL can be assessed by a previously described method using P815 tumor target cells (Murray et al., 1985).

D. Nonspecific Immunity

1. Natural Killer Cell Activity

NK cells possess innate cytotoxicity against a variety of neoplasms, inhibiting both growth and metastatic dissemination of tumor cells in addition to various infectious agents (reviewed by Herberman, 1985). Splenic NK cell activity is quantitated in individual mice using an in vitro assay in which YAC-1 tumor cells are used as the target cell (Reynolds and Herberman, 1981). This assay shares the same advantages as lymphocyte blastogenesis, namely that it can be adapted to microtiter plates.

2. Macrophage Function

Assays to evaluate macrophage function are important because these cells are key constituents in maintaining host defenses (Skamene and Gros, 1983). Their importance as accessory cells to specific humoral- and cell-mediated immune processes and their contribution of various cytokines involved in lymphocyte activation and regulation also is recognized. Determination of resident peritoneal cell number and differential count is a rapid and simple way to assess the effects of nonparenterally administered xenobiotics on macrophage function. Immunotoxic chemicals, as well as known immunomodulators, have been shown to change the differential of resident peritoneal cell counts (Lewis and Adams, 1985).

The phagocytosis of pathogens and other foreign material by macrophages and PMNs represents a major first line defense mechanism in nonspecific immunity. A rapid means of measuring phagocytic capabilities has been to add fluorescent covaspheres or ^{51}Cr-labeled chicken erythrocytes to cultures containing adherent peritoneal exudate cells (PECs). The degree of phagocytosis then can be measured by fluorescent microscopy, flow cytometry (Duke et al., 1985) or by radiometric means (Thomas et al., 1985b). Quantitation of macrophage ATP levels by chemiluminescence is a sensitive assay to measure macrophage bactericidal activity (Walker et al., 1981) and activation (Muto et al., 1985). Changes in ectoenzyme activities have also been associated with activation, acquisition of tumoricidal activity, or change in resident macrophage status (Morahan et al., 1980). The advantage of these endpoints is that they are quite sensitive and lend themselves to more mechanistic studies (i.e., cell mixing, metabolic activation experiments). However, the high labor intensive nature of these assays precludes their use in a "screening mode." Furthermore, experience has shown that these macrophage endpoints don't always correlate with chemically induced changes in host resistance (Luster et al., 1988).

E. Host Resistance Assays

Increased susceptibility to infection or tumor cell challenge can be an in vivo manifestation of chemically induced immunotoxicity. The host resistance models discussed below were chosen because the mechanisms of resistance are relatively well understood and generally do not overlap. Furthermore, the models are reproducible and are technically easy to establish (Morahan et al., 1984; Dean et al., 1982b; Fugmann et al., 1983). In general, to document impaired host defenses, a concentration of infectious agent or tumor cells that will kill or produce a tumor incidence of 10–30% in vehicle-treated control animals is employed. The disadvantage of these test systems is that they require large numbers of animals to detect statistically significant differences in mortality or survival. Furthermore, they lack some of the sensitivity of the in vitro endpoints discussed above. However, the great advantage of host resistance models in immunotoxicity assessment is that they provide some measure of overall immune competence critical for meaningful risk assessment. Several commonly used bacterial, tumor, parasitic, and viral models are described below.

1. Listeria monocytogenes

The pathogenesis and immune mechanisms involved in recovery of animals from this infection have been studied extensively. Recovery from infection depends on specific sensitization of T-cells which subsequently activate macrophages for enhanced nonspecific bactericidal activity (North, 1973; Newborg and North, 1980). Initially, fixed macrophages are involved in inactivation and clearance of the bacterium. Development of specific T-cell mediated immunity and subsequent activation of macrophages occurs within 2–4 days following systemic infection. Resistance to the bacterium can be assessed by mortality or by quantitation of bacteria in the liver and spleen, which are the major sites for replication.

2. Streptococcus *sp.*

The pathogenesis of and immune response to streptococcal infections have been well characterized. In general, recovery from infection depends on the induction of opsonizing antibody, resulting in enhanced phagocytosis and intracellular destruction of the organism (Bradley and Morahan, 1982). Serum complement also plays a role, particularly during the early phase of the infection (Winkelstein, 1981). The streptococcal pneumonia infectivity model has been used for a number of years and proven to be a sensitive indicator of air pollutant-induced pulmonary injury (Ehrlich, 1980).

3. PYB6 Tumor

The PYB6 fibrosarcoma tumor was originally induced with polyoma virus in C57BL/6 mice and was developed as a screening model in the B6C3F1 mouse (Dean et al., 1981). Resistance to tumor growth, which reflects CMI and NK cell function is measured by differences in tumor latency and size following subcutaneous challenge into the thigh with 1×10^3 to 5×10^3 viable tumor cells (approximate TD_{10-30}). Tumor cells for challenge are best obtained from an animal bearing a 10- to 12-mm tumor rather than from tissue culture. However, the PYB6 cell line can be maintained for fairly long periods in culture or frozen prior to in vivo passage in a mouse for animal challenge studies.

4. B16F10 Metastatic Pulmonary Tumor

The B16F10 tumor has weak tumor-specific antigens and is well characterized with regard to metastasis and the response to immunotherapeutic and chemotherapeutic agents (Fidler et al., 1978). NK cells and macrophages are probably the major immune mechanisms responsible for clearance and growth inhibition of pulmonary tumors. Because of the multiple immunological mechanisms believed responsible for limiting the growth and metastasis of this tumor, it has been used in studies designed to evaluate holistic immune responses following exposure to dimethylnitrosamine (Thomas et al., 1985a), diethylstilbestrol (DES) (Fugmann et al., 1983), and benzo[*a*]pyrene (Dean et al., 1983a), among others. Furthermore, alterations in tumor resistance have been correlated with changes in in vitro immune parameters reflective of natural and adoptive immune responses (Dean et al., 1983b; Luster et al., 1988).

5. Plasmodium berghei/Trinchinella spiralis *parasites*

P. berghei is an intracellular sporozoan parasite causing a fatal malaria in mice. Host protection from infection depends on proper functioning and interactions between B- and T-lymphocytes and macrophages (Loose, 1982). Antiparasitic antibody and specifically sensitized T-lymphocytes are produced during the course of infection. *Trinchinella spiralis* infection requires intact T-cell mediated immunity for efficient adult worm expulsion (Larsh et al., 1974). Furthermore, a role for humoral immunity for expulsion has been indicated in studies with athymic mice (Ruitenberg et al., 1977). DES, a potent estrogenic compound and immunosuppressant, decreases host resistance to this organism at doses that alter in vitro immunocompetence in mice (Fugmann et al., 1983).

6. Influenza Virus

Influenza virus, an important respiratory virus in the young, aged, and infirm, has been used as a viral challenge model in mice to evaluate both T- and non-T-cell-mediated host defense mechanisms. Cellular factors (Leung and Ada, 1980), antibody (Vireligier, 1975), and interferon (Hoshino et al., 1983) play major roles in mediating resistance to this virus. Resistance to infection in the mouse is most easily assessed by mortality, which occurs within 14 days following aerosol or intratracheal challenge (Thomas et al., 1985a). Resistance to infection with this virus correlates well with certain measures of humoral immune function (Luster, et al., 1988).

7. Herpes Simplex Virus (HSV)

HSV is a widespread human pathogen which can cause disease with similar symptoms in mice. Persistence of the virus in the latent state with the potential for reactivation can take place. Much evidence suggests that cell-mediated immunity is of major importance in the regulation of herpes virus infections (Morahan, 1983; Peterman et al., 1985). In mice, depression of cell-mediated immunity by such agents as DES (Fugmann et al., 1983) or cyclophosphamide (Cy) (P. T. Thomas, unpublished observations) results in increased mortality after infection.

IV. TIER TESTING APPROACH TO EVALUATE
CHEMICALS FOR POTENTIAL IMMUNOTOXICITY

The previous section reviewed the general categories and specific methodologies commonly used to assess chemically induced immunotoxicity. It is clear that no single test can ascertain whether or not the immune system is a target organ for toxicity. In order to rationally evaluate a compound for potential immunotoxicity, a tiered approach incorporating some of these various tests has been validated in the B6C3F1 mouse (Luster et al., 1988). The goal of these developmental studies was to select a battery of tests among a large number of functional and host resistance assays which could be organized into a two-tiered testing scheme and incorporated, on an as-needed basis, into more traditional toxicology tests conducted by the National Toxicology Program.

The testing approach was developed using a number of immunomodulators having well-known effects on multiple immune cell types. These compounds included DES, Cy, cadmium, dimethylnitrosamine, and benzo[a]pyrene. Tier I includes assays for both HMI and CMI (Table 1) as well as routine clinico- and histopathology. The immune function and host resistance assays organized into Tier II (Tables 1 and 2) provide information on the mechanism(s) of immunotoxicity and effects on the holistic immune response, respectively.

Considerable effort was made in the tier development studies to document intra- and interlaboratory assay reproducibility, develop a historical control data base of optimal responses, define dose-response relationships, and establish correlations between functional and host resistance assays. The study results showed that as the degree of methodology standardization among laboratories improved, so did the interlaboratory reproducibility of the test results. As a result, a large historical control data base has been established for many tests. The sensitivity of the immune assays relative to other toxicological endpoints was, in part, established by documenting whether or not significant dose-response relationships could be produced. The results suggest that the functional immune endpoints were, in many instances, more sensitive for detecting subtle biological changes than were more routine toxicological parameters such as body weight change, organ weights, or histopathology (Luster et al., 1988). In any case, gross pathology and histopathology of tissues relevant to the immune system examined in Tier I are critically important in decisions to proceed to more detailed testing. Equally important is documentation of dose-related effects at nonovertly toxic exposure levels.

These validation studies also helped to further our understanding of the relationship between chemically induced suppression of specific immune functions and altered host resistance. As with most organ systems studied in toxicology, certain functional changes in immunity are presumed to be predictive of significant health effects. Furthermore, the degree of immune modulation is critical since there is significant immunological reserve capacity. Thus, depression of a particular function (e.g., lymphoproliferation or antibody plaque-forming cell response) beyond a critical point must occur before altered host resistance is seen. Similarly, pleiotropic immune responses occur following infection or tumor challenge. Therefore, mild loss of more than one specific immune function may result in increased susceptibility. This complexity is exemplified in Table 2 where various immunological mechanisms involved in resistance to infectious agents or transplantable tumor cells are summarized.

Table 3 summarizes the relationship between host susceptibility to challenge and immune suppression with data from mice exposed to various chemical compounds (Luster et al., 1988). Loss of the ability to generate PFCs in response to SRBCs significantly correlated with increased mortality following challenge with mouse malaria (*Plasmodium yoelli*) as well as with mouse-adapted influenza virus. On the other hand, susceptibility to challenge with PYB6 sarcoma or B16F10 tumor cells correlated significantly with a reduction in NK cell cytotoxicity. Recognition of alloantigens as measured by the MLR was also correlated significantly with increased susceptibility to PYB6 tumor challenge. Finally, susceptibility to challenge with *Listeria* correlated significantly with the loss of MLR and with the lymphoproliferative responses to the T-cell mitogen PHA. There was no correlation observed between loss of NK cell activity and challenge with the infectious agents tested.

It was also important to establish correlations among the various host resistance models in order to prevent needless duplication. Representative results

TABLE 3

Correlation between Host Susceptibility to Infection or Tumor and Depressed Immune Function

Challenge model	Natural killer cell function	Mixed lymphocyte response	Lymphocyte blastogenesis PHA	Lymphocyte blastogenesis LPS	Antibody PFC response	Delayed type hypersensitivity
PYB6 sarcoma	$0.45^{a,b}$	0.46^{b}	0.20	0.02	0.22	0.61^{b}
B16F10 melanoma	0.54^{b}	0.02	0.15	0.16	0.15	ND
Listeria	0.01	0.47^{c}	0.37^{b}	0.08^{b}	0.01	0.19
Influenza	0.11	0.78^{b}	0.03	0.70^{b}	0.83^{c}	ND^{d}
Plasmodium	0.24	0.59	0.67^{b}	0.64^{b}	0.78^{c}	ND

[a]Correlation coefficient as determined by Spearman's rank correlation test (rho values).
[b]Significant correlation at $p < 0.05$.
[c]Significant correlation at $p < 0.01$.
[d]ND = not done.
Source: From Luster et al., 1988.

TABLE 4

Degrees of Correlation among Various Infectious Host Resistance Models

	Correlation by challenge organism			
	Influenza virus	HSV-2	HSV-1	*Streptococcus*
Listeria monocytogenes	-0.40^a	0.63^b	0.09	-0.36
Influenza virus		-0.35	0.14	-0.02
HSV-2			0.23	-0.58
HSV-1				-0.67

[a]Correlation coefficient.
[b]Significant correlation at $p < 0.05$.

obtained from comparing changes in host resistance following exposure to a number of chemicals (i.e., $CdCl_2$, DES, dimethylvinyl chloride, Cy, dexamethasone) are shown in Table 4. These studies suggest that resistance to *Listeria monocytogenes* and *Herpes simplex* virus type 2 (HSV-2) are mediated by similar immunological mechanisms since significant correlations were seen. This is not surprising since T-cell and macrophage function are thought to play a major role in resistance to both *Listeria* and HSV infections (Newborg and North, 1980; Morahan, 1983). Surprisingly, resistance to HSV-1 did not correlate with HSV-2 or *Listeria,* suggesting that a slightly different mechanism of host protection may be involved in resistance to this agent. Although not statistically significant, there were degrees of negative correlation between the infectious models where resistance is mediated by antibody, interferon, or complement (influenza virus, *Streptococcus* sp.) and those models where resistance is primarily of the cell-mediated variety (*Listeria,* herpes infections).

In summary, a tier testing approach has been developed and is being used by the National Toxicology Program in special studies for immunotoxicity assessment. For the most part, the assays incorporated in the testing scheme have been validated using known immunomodulators and a historical control data base established in the mouse model. Although alternative testing approaches and species have been proposed (Spiers et al., 1978; Koller and Exon, 1985), other government agencies are considering this approach as a prototype for future testing guidelines (U.S. EPA,1982).

It should be stressed that the phased testing scheme outlined above represents only one approach in one animal species. The functional heterogeneity of the immune response requires a flexible approach to immune profile development of unknown compounds. In some cases, this approach must be tailored to provide meaningful data for risk assessment. For example, health effects determinations can best be made using data from the whole animal infectious or tumor challenge models. In this case, the Tier I initial screen should be modified to include relevant model(s) of host resistance. Regardless of the overall approach or the specific test used, it is important to remember that the immune system is a legitimate target organ and that these data are as important to the risk assessment process as are other toxicology data.

V. ENVIRONMENTALLY IMPORTANT CHEMICALS REPORTED TO AFFECT THE IMMUNE SYSTEM

While the actual number of environmental toxicants and chemicals found to affect the immune system is not large when compared to those causing other toxic symptoms, the immunotoxic effects of many have been known for years. Depending on the nature of the xenobiotic (e.g., protein, air pollutant, or low molecular weight chemical) or route of exposure (e.g., dermal, inhalation; local or systemic) a variety of possible adverse effects could occur. It is beyond the scope of this chapter to exhaustively review the literature on chemically induced immunotoxicity. The reader is directed to several comprehensive reviews and books on the subject (reviewed by Vos, 1977; Dean et al., 1982a; and Descotes, 1986). Table 5 summarizes a partial list of environmentally important chemicals with well-known immunotoxic effects in laboratory animals. Examples with each chemical class are discussed below.

A. Metals

Many heavy metals are toxic to mammals following deliberate or inadvertent exposure. The immune system has been implicated as a target organ for toxicity. However, the mechanisms of immunotoxicity of heavy metals are still largely unknown. One of the most intensively studied metals is cadmium, largely because of its widespread use and great potential for environmental contamination. One of the more consistent immunotoxic findings is suppression of humoral immune response following both in vivo and in vitro exposure. This effect, however, varies with animal strain, exposure route, dose, and endpoint determination relative to exposure. The effects on CMI and host resistance are less well understood.

B. Organochlorine Compounds

Organochlorine pesticides are another group of environmentally important chemicals that affect the immune system. Considering their widespread use and tendency to accumulate in the environment and in the food chain, surprisingly little is known about their potential for immune modulation. DDT is probably

TABLE 5

Some Environmentally Important Chemicals Reported to Affect the Immune System[a]

Class	Chemical	Clinical, histo-pathology	HMI	CMI	Host resistance models	Non-specific defenses	Aller-genicity	Reference
						Effects[b]		
Heavy metals	Arsenic	—	↓	↓	↓	—	—	Blakley et al., 1980
	Cadmium chloride	↕	↓↑	↓	↓?	↓	—	Thomas et al., 1985b
								Ohsawa et al., 1983
								Balter et al., 1982
								Koller and Roan, 1980a
								Bozelka and Burkholder, 1979
								Koller et al., 1979
								Barnes and Munson, 1978
								Koller and Brauner, 1977
								Koller and Roan, 1977
								Koller, 1973
	Lead acetate	↓	↓	↓↑	↓	↓↑	↑	Laschi and Tachon, 1982
								Lawrence, 1981
								Blakley and Archer, 1981
								Faith et al., 1979
								Cook et al., 1975
	Lead nitrate	—	—	↓	↓	—	—	Descotes et al., 1984
								Hemphill et al., 1971
	Mercuric chloride	↓	↓↑	↓	↕	—	↑?	Dieter et al., 1983
								Magnusson and Kligman, 1969
								Thaxton and Parkhurst, 1973
	Methylmercury	—	↓	↕	—	—	—	Koller and Roan, 1980b
								Koller et al., 1977
	Dioctyltin dichloride	↓	↓	↓	—	↕	—	Seinen et al., 1977b
								Seinen et al., 1977a

See footnotes on p. 70.

67

TABLE 5 (Continued)

Class	Chemical	Effects[b]						Reference
		Clinical, histopathology	HMI	CMI	Host resistance models	Non-specific defenses	Allergenicity	
Organochlorine insecticides	DDT	→	↑↓	↑↓	↓?	↑↓	—	Andre et al., 1983 Subba Rao and Glick, 1977 Crocker et al., 1976 Lukic et al., 1973 Wasserman et al., 1969 Bernier et al., 1987
	Dieldrin	—	→	→	→	—	—	Friend and Trainer, 1974 Wasserman et al., 1972
	Methylparathion	—	↑↓	↑↓	—	—	—	Street and Sharma, 1975
	Parathion	→	↑↓	—	—	—	—	Casale et al., 1983
	Malathion (oral, in vitro exposure)	→	↑↓	→	—	—	—	Rodgers et al., 1986 Casale et al., 1983
Carbamate insecticides	Carbaryl (oral, in vitro exposure)	—	←	→	—	—	—	Desi et al., 1978 Rodgers et al., 1986 Andre et al., 1983
	Carbofuran	—	↑↓	→	↓	—	—	Street and Sharma, 1975
	Aldicarb	↑↓	↓	↓	↓	—	—	Streete and Sharma, 1975 Thomas et al., 1987 Fiore et al., 1986
Polychlorinated phenols	Hexachlorobenzene (oral, in utero exposure)	↑↓	↑↓	↑↓	↑↓	↑↓	↕	Vos et al., 1979 Loose et al., 1979 Exon and Koller, 1983 Forsell et al., 1981
	Pentachlorophenol	↑↓	↑↓	↓?	—	←	—	McConnell et al., 1980
Monocyclic and polycyclic	Benzene (in vitro exposure)	—	—	↑↓	—	—	—	Moszczynski and Lisiewicz, 1983 Thurman et al., 1978

68

Chemical class	Agent (exposure)						References
aromatic hydrocarbons	7,12-Dimethylbenz[a]-anthracene (sc exposure)	—	—	→	→	—	Ward et al., 1985
	Methylcholanthrene	—	→	→	—	—	Brai et al., 1977; Baroni et al., 1970; Stjernsward, 1966
Polyhalogenated biphenyls	Aroclor 1248	—	→	↕	→	—	Thomas and Hinsdill, 1978; Lubet et al., 1986
	Aroclor 1254	—	→	↕	→	—	Bekesi et al., 1978; Bekesi et al., 1987
	Firemaster FF-1 (oral, dermal exposure)	↑↓	↑↓	→	↕	—	Fraker, 1980; Loose et al., 1981; Luster et al., 1980b
Isocyanates	Toluene diisocyanate (in vitro exposure)	—	→	→	—	↑	Karol, 1985; Rao et al., 1981
Dibenzo-p-dioxins, dibenzofurans	2,3,7,8-Tetrachlorodibenzo-p-dioxin (oral, in vitro, ip exposure)	↑↓	↑↓	↑↓	↑↓	—	Thurman et al., 1978; Vecchi et al., 1987; Greenlee, et al., 1985; Clark et al., 1981; Luster et al., 1980a; Vecchi et al., 1980; Thomas and Hinsdill, 1979; Vos et al., 1978; Faith and Moore, 1977; Thigpen et al., 1975; Kerkvliet and Brauner, 1987
	1,2,3,4,6,7,8-Heptachlorodibenzo-p-dioxin	—	→	→	—	—	Bandiera et al., 1984
	Tetrachlorodibenzofuran	→	→	→	—	—	Luster et al., 1979; Vecchi, 1987
Aliphatic hydrocarbons	Ethyl carbamate (ip exposure)	→	→	→	→	→	Luster et al., 1982a
	Methyl carbamate	↑	↕	↕	↕	↕	Luster et al., 1982a; Johnson et al., 1987
	Dimethylnitrosamine	—	→	→	→	→	Munson et al., 1987; Thomas et al., 1985a

See footnotes on p. 70.

TABLE 5 (Continued)

| Class | Chemical | Effects[b] | | | | | | Reference |
		Clinical, histo-pathology	HMI	CMI	Host resistance models	Non-specific defenses	Aller-genicity	
Aliphatic hydrocarbons	Vinyl chloride (inhalation, in vitro exposure)	↑	—	↑	—	—	—	Sharma et al., 1980; Sharma and Gehring, 1979
	1,3-Butadiene (inhalation)	—	↑↓	↑↓	—	—	—	Thurmond et al., 1986
Gaseous air pollutants (inhalation exposures)	Ozone	—	—	↓	→	—	—	Ehrlich, 1980
	Nitrogen dioxide	—	—	—	→	—	—	Peterson et al., 1981; Ehrlich, 1980
	O₃, SO₂, (NH₄)₂SO₄ mixtures	—	↔	↑	→	↑↓	—	Aranyi et al., 1983
Miscellaneous chemicals	Trimellitic anhydride (inhalation exposures)	—	—	—	—	↑	↑	Leach et al., 1987
	Gallium arsenide	—	—	—	↑↓	→	—	Lysy et al., 1987
	Vinylcyclohexene diepoxide (dermal exposure)	↔	→	→	↑↓	→	—	Thomas et al., 1985c

[a]Unless otherwise indicated, oral exposure routes were used. Symbols and abbreviations used: ↑↓, varying results across studies; —, not measured; ↓, significant suppression (p ≤ 0.05); ↑, significant stimulation, cellular hyperplasia or allergic response (p ≤ 0.05); ?, no significant trend; ↔, no significant effect; sc, subcutaneous injection; ip, intraperitoneal injection.

[b]Clinical and histopathology includes lymphoid organ weights, WBC counts, differential, histology. Humoral-mediated immunity (HMI) includes T-dependent antibody response, B-cell proliferation, B-cell rosettes, immunoglobulin levels. Cell-mediated (CMI) includes delayed type hypersensitivity, T-cell mitogenesis, mixed lymphocyte culture. Nonspecific defenses include NK cell activity, macrophage phagocytosis, carbon clearance. Animal species include mouse, rat, rabbit, cow, human, subhuman primate. Host resistance models include pseudorabies, encephalomyocarditis, influenza, herpes viruses; bacterial endotoxin challenge; Bacillus Calmette-Guerin; Salmonella sp.; Plasmodium Berghei; Klebsiella pneumoniae; Streptococcus sp.; Listeria monocytogenes; PYB6 sarcoma; B16F10 melanoma.

the most intensively studied in this class of compounds. Studies in laboratory animals have demonstrated significant modulation of both humoral and cell-mediated immunity. Furthermore, increased incidences of infection coupled with impaired neutrophil function have been seen in recent epidemiological studies in humans (Hermanowicz et al., 1982).

C. Carbamates

The carbamate compounds are another class of pesticides whose mode of action is reversible inhibition of acetylcholinesterase activity. To date, there is no conclusive evidence to implicate this class of compounds as immunotoxicants. One compound that has received considerable attention recently is aldicarb, one of the most acutely toxic carbamate insecticides. Mice exposed to aldicarb levels comparable to those found in contaminated well water exhibited significantly reduced antibody PFC responses (Olson et al., 1987). The response exhibited an inverse dose-response relative to increasing exposure levels. These results, however, were not substantiated in other studies (Thomas et al., 1987). Meanwhile, alterations in T4+ (helper/inducer) to T8+ (suppressor/cytotoxic) cell ratios were noted in humans exposed to aldicarb contaminated well water (Fiore et al., 1986). Although not conclusive, these results underscore the need for further research.

D. Polychlorinated Phenols

Of the polychlorinated phenols, one of the most intensively studied is hexachlorobenzene (HCB). HCB is of particular importance because most of the 4.3 million pounds synthesized annually are unwanted by-products from various industrial processes (Courtney, 1979); hence there is the potential for inadvertent occupational and environmental exposure. HCB, depending on animal strain or exposure route used, has different effects on similar measures of the immune response. Rats exposed in utero exhibited significant increases in T-dependent antibody responses to tetanus toxoid (Vos et al., 1979). Conversely, decreases in anti-SRBC plaque responses were seen in mice exposed orally for 6 weeks to HCB (Loose et al., 1979). Unpublished observations from the author's laboratory indicate that pulmonary macrophage bactericidal activity of mice exposed to HCB aerosols is suppressed following four exposures. However, after 16 exposures, compensation by as yet unknown mechanisms results in normal antibacterial activity.

E. Monocyclic and Polycyclic Aromatic Hydrocarbons

Benzene is a widely used industrial solvent and carcinogen. Workers exposed to benzene exhibited reduced numbers of total circulating white blood cells and T-lymphocytes. This did not correlate, however, with reduced delayed-type hypersensitivity or T-lymphocyte blastogenic transformation (Moszczynski, 1982; Moszczynski and Lisiewicz, 1983). In addition, reduced human T-cell blastogenesis to mitogens, following in vitro exposure to benzene, although at relatively high concentrations, has been reported (Thurman et al., 1978).

Among the polycyclic aromatic hydrocarbons (PAH), 7,12-dimethylbenz[a]anthracene (DMBA), a known carcinogen, has been extensively studied. Like other PAHs, DMBA is metabolized via the cytochrome P450 mixed function oxidases into reactive diol-epoxides. These intermediates potentially can form adducts with cellular macromolecules. Suppression of humoral, cellular, and natural immunity and host resistance to bacterial or tumor challenge has been reported in adult mice given single or multiple subcutaneous injections of DMBA (Ward et al., 1985). Although profoundly immunosuppressive as a class, only those PAHs which are carcinogenic appear to share this property. For example, benzo[a]pyrene, a known carcinogen, is immunosuppressive. The noncarcinogenic congener benzo[e]pyrene is not (Ward et al., 1985).

F. Polyhalogenated Biphenyls

The polychlorinated and polybrominated biphenyls (PCBs, PBBs, respectively) represent two distinct classes of biphenyl compounds. Each consists of many isomers having different degrees of halogen substitution on the biphenyl ring. Due to their chemical stability, PCBs were widely used as dielectric fluids, in printing inks, as heat exchangers, and as pesticide extenders. Although banned for use in the United States, significant levels of PCBs still remain in the environment. Both mixtures and individual PCB isomers have been the subject of intensive study in laboratory animals (reviewed by Kerkvliet, 1984). In general, toxicity, immune modulation, and bioaccumulation (i.e., resistance to biodegradation) increase with increasing chlorine substitution.

There have been two major episodes of mass human exposure to PCB through the consumption of contaminated rice bran oil. The first, termed "Yusho," occurred in Japan in 1968. The second (Yu-Cheng) occurred in Taiwan in 1979. Patients from both incidents exhibited immune suppression manifested as persistent skin infections and respiratory bronchitis (Lee and Chang, 1985).

Compared with the PCBs, relatively little is known about the immunosuppressive properties of PBBs. Accidental contamination of a cattle feed supplement with PBBs resulted in exposure of dairy farmers and consumers who ingested contaminated beef or poultry. Compared to a cohort of nonexposed Wisconsin farmers, farmers in Michigan exhibited chronic reductions in T-lymphocyte blastogenic responses to mitogens and alloantigens (Bekesi et al., 1978).

G. Isocyanates

Isocyanate compounds have a marked capability to induce hypersensitivity. The widespread use of both aliphatic and aromatic isocyanates in insulation and in paints and coatings makes human exposure a health issue. The most widely used of these is toluene diisocyanate (TDI). It has been widely reported that TDI induces contact and respiratory sensitivity responses in experimental animals (reviewed by Karol, 1985). In addition, isocyanate-specific antibody has been produced following intradermal injection of the material in guinea pigs. Furthermore, antibody responses in workers exposed to TDI was dependent on the amount of TDI exposure; antibody was produced only following exposure to large amounts of material. Recently, the presence of TDI reactive lymphocytes in the blood of guinea pigs immunized with TDI and adjuvant implicated cellular as well as humoral factors in the development of TDI sensitivity (Karol et al., 1986).

H. Dioxins

Polychlorinated dibenzo-p-dioxins (PCDD) and polychlorinated dibenzofurans (PCDF) are two series of cyclic aromatic hydrocarbons which exhibit similar chemical and biological properties. Up to 75 PCDD and 135 PCDF isomers are possible, depending on the number and location of chlorine atom substitutions. Both the PCDD and PCDFs are unwanted by-products of a large number of industrial chemical manufacturing processes. The immunotoxicity of tetrachlorodibenzo-p-dioxin (TCDD) is probably the best understood. Both humoral and cell-mediated responses are altered following TCDD exposure. Sensitivity to immune modulation is somewhat dependent on the species tested and the stage of immune ontogeny, relative to TCDD exposure (Thomas and Hinsdill, 1979). From the extensive data reported in the literature, profound thymic atrophy is the most consistent finding. This effect segregates in mice with the Ah locus, which regulates the expression of a cytosolic receptor having high affinity for TCDD. Immunotoxicity is more marked in strains expressing this receptor, compared with those with undetectable receptor levels (reviewed by Vecchi, 1987).

Compared with PCDDs, there are few studies designed to document immune alterations by PCDFs in laboratory animals. As with the PCDDs, thymic atrophy and immune suppression following PCDF exposure correlates with Ah receptor induction (Vecchi, 1987). Furthermore, both PCDD and PCDF contamination have been implicated in the immunotoxicity seen in experimental animals exposed to technical grade pentachlorophenol (Kerkvliet et al., 1985) and other chlorinated phenols.

I. Aliphatic Hydrocarbons

Urethan, or ethylcarbamate is a known carcinogen with immunotoxic properties. Unpublished studies from the author's laboratory and others (Luster et al., 1982a) have shown that this compound is immunosuppressive. Conversely, methylcarbamate, a noncarcinogenic related compound has no significant immunotoxic activity (Luster et al., 1982a).

Dimethylnitrosamine (DMN), a commonly encountered N-nitroso compound and carcinogen has documented effects on immune responsiveness. Significant impairment of HMI, CMI, host resistance to infection and innate, nonspecific immune responses have been observed (Johnson et al., 1987; reviewed by Munson et al., 1987; Thomas et al., 1985a). Recent studies suggest that the B-lymphocyte may be a major target for DMN-induced immunosuppression (Johnson et al., 1987).

J. Gaseous Air Pollutants

Some of the earliest studies to document chemically induced immune impairment were conducted by Ehrlich (1966), who noted increased susceptibility to experimental respiratory infection in mice exposed to nitrogen dioxide. Subsequent studies with ozone and other widespread gaseous and particulate air pollutants (Aranyi et al., 1983) have indicated that the aerosol infectious challenge model is a sensitive system which reflects the exacerbation of bacterial respiratory infection by inhalation of air pollutants.

K. Miscellaneous Chemicals

Trimellitic anhydride (TMA) is a widely used chemical intermediate in the paint and plastics industry. TMA induces four separate syndromes in humans, three of which are immunologically based. Recent studies in rats exposed to varying concentrations of TMA aerosols demonstrated dose-dependent increases in lung weights, hemorrhagic lung foci, accumulation of alveolar macrophages, alveolar hemorrhage and pneumonitis, and lung and mediastinal lymph node IgG and C3. Rats exposed and rested recovered from most effects. Those rested and rechallenged exhibited lesions similar to those seen after the first series of exposures (Leach et al., 1987). The timing and nature of symptoms and lesions are consistent with some of the syndromes seen in TMA-exposed workers.

Gallium arsenide (GaAs) is used in microcircuitry production for the semiconductor industry. Occupational exposure during the manufacturing process can result from inhalation of GaAs particles. Mice receiving a single intratracheal injection of GaAs exhibited increased resistance to challenge with *Listeria monocytogenes*, a facultative intracellular parasite requiring intact macrophage and T-cell function for resistance; increased susceptibility to B16F10 melanoma tumor, suggesting impairment of natural killer cell function; and no change in susceptibility to *Streptococcus pneumoniae* infection, suggesting that complement and neutrophil factors are intact (Lysy et al., 1987).

Vinylcyclohexene diepoxide is a member of a family of compounds used commercially as reactive intermediates in the manufacture of epoxy resins. Recent studies in our laboratory (Thomas et al., 1985c) have suggested that dermal exposure to this compound in mice impairs host resistance to *Listeria monocytogenes* infection. This may be explained in part by a dose-related suppression of macrophage phagocytosis, since exacerbation of infection is macrophage dependent.

VI. CONCLUSIONS

This chapter introduced the concept of the immune system as a legitimate and important target organ for toxicity, reviewed commonly used methodologies for assessing immunotoxicity, and discussed how these methods can be incorporated into a tier testing scheme. Examples of environmentally important chemicals reported to modulate the immune system were presented.

It is important to understand that alterations in immune function can be either in a positive or negative direction. Extremes in both cases can be detrimental to the host. For example, enhancement of immune potential initially could lead to increased host resistance to infection or tumor but, in the extreme, may result in hypersensitivity or autoimmune disease. Conversely, decreased immune function could lead to anergy (reduced responsiveness to skin test antigens, for example) and ultimately to increased frequencies of infection or cancers. The immune system is constantly bombarded with foreign substances from the environment. As a result of these insults, the status of immune function in most individuals varies slightly on either side of what is considered "normal." The science of clinical immunology is concerned with the diagnosis and treatment of diseases resulting from extreme modulations of the immune response. The science of "epidemiological immunology" is concerned with more subtle alterations in immune function caused by environmental agents, including chemicals. The challenge to immunotoxicologists and epidemiologists working in this field is to determine whether or not these changes will be detrimental to the host in the long term.

REFERENCES

Adams, D., and Hamilton, T. 1984. Cell biology of macrophage activation. *Ann. Rev. Immunol.* 2:283–318.

Allen, J. C. 1976. Infection compromising neoplastic disease and cytotoxic therapy. In *Infection and the Compromised Host*, ed. J. C. Allen, pp. 151–171. Baltimore: Williams & Wilkins.

Andre, F., Gillon, F., Andre, C., et al. 1983. Pesticide–containing diets augments anti-sheep red blood cell nonreaginic antibody responses in mice but may prolong murine infection with *Giardia muris*. *Environ. Res.* 32:145–150.

Aranyi, C., Vana, S., Thomas, P., et al. 1983. Effect of subchronic exposure to a mixture of O_3, SO_2, and $(NH_4)_2SO_4$ on host defenses of mice. *J. Toxicol. Environ. Health* 12:55–71.

Baehner, R. L. 1974. Molecular basis for functional disorders of phagocytes. *J. Pediatr.* 84:317–327.

Balch, C. M., Ades, E. W., Loken, M. R., et al. 1980. Human "null" cells mediating antibody-dependent cellular cytotoxicity express T-lymphocyte differentiation antigens. *J. Immunol.* 124:1845–1851.

Balter, N., Kawecki, M. E., Gingold, B., et al. 1982. Modification of skin graft rejection and acceptance by low concentrations of cadmium in drinking water of mice. *Int. J. Immunopharmacol.* 4:324.

Bandiera, S., Sawyer, T., Romkes, M., et al. 1984. Polychlorinated dibenzofurans (PFCFs): Effects of structure on binding to the 2,3,7,8-TCDD cytosolic receptor protein, AHH induction and toxicity. *Toxicology* 32:131–144.

Barnes, D. W., and Munson, A. E. 1978. Cadmium-induced suppression of cellular immunity in mice. *Toxicol. Appl. Pharmacol.* 45:350.

Baroni, C. D., Bertoli, G., Pesando, P., et al. 1970. Delayed hypersensitivity in Balb/c and Charles River mice injected at birth with a single dose of 7,12-dimethylbenz[a]anthracene (DMBA). *Experientia* 26:899–900.

Bekesi, J. G., Holland, J. F., Anderson, H. A., et al. 1978. Lymphocyte function of Michigan dairy farmers exposed to polybrominated biphenyls. *Science* 199:1207–1209.

Bekesi, J. G., Roboz, J., Fischbein, A. S., et al. 1987. Clinical immunology studies in individuals exposed to environmental chemicals. In *Immunotoxicology*, eds. A. Berlin, J. Dean, M. Draper, E. Smith, and F. Sperafico, pp. 347–361. Boston: Martinus Nishoff.

Bellanti, J. A., Balter, N. J., and Gray, I. 1981. A unifying model for immunotoxicology: A summation presentation. In *Biological Relevance of Immune Suppression*, eds. J. H. Dean, and M. Padarathsingh, chap. 20. New York: Van Nostrand Reinhold.

Bernier, J., Hugo, P., Krzystyniak, K., and Fournier, M. 1987. Suppression of humoral immunity by dieldrin. *Toxicol. Lett.* 35:231–240.

Biagini, R. E., Bernstein, I. L., Gallagher, J. S., et al. 1985. The diversity of reaginic immune responses to platinum and palladium metallic salts. *J. Allergy Clin. Immunol.* 76:794–802.

Blakley, B. R., and Archer, D. L. 1981. The effect of lead acetate on the immune response in mice. *Toxicol. Appl. Pharmacol.* 61:18.

Blakley, B. R., Sisodia, C. S., and Mukkur, T. K. 1980. The effects of methylmercury, tetraethyl lead and sodium arsenite on the humoral immune response in mice. *Toxicol. Appl. Pharmacol.* 52:245–254.

Bozelka, B. E., and Burkholder, P. M. 1979. Increased mortality of cadmium-intoxicated mice infected with the BCG strain of *Mycobacterium bovis. J. Reticuloendothel. Soc.* 26:229–237.

Bradley, S. G., and Morahan, P.S. 1982. Approaches to assessing host resistance. *Environ. Health. Persp.* 43:61–69.

Brai, M., Patrucco, A., and Osler, A. G. 1977. Chemical carcinogenesis and humoral antibody synthesis. *J. Immunol.* 109:317–323.

Casale, G. P., Cohen, S. D., and Dicapua, R. A. 1983. The effects of organophosphate-induced stimulation on the antibody response to sheep erythrocytes in inbred mice. *Toxicol. Appl. Pharmacol.* 68:198–205.

Clark, D., Gauldie, I., Szewcyuk, M., et al. 1981. Enhanced suppressor cell activity as a mechanism of immuno-suppression by 2,3,7,8-tetrachlorodibenzo-p-dioxin. *Proc. Soc. Exp. Biol. Med.* 168:290–299.

Cook, J. A., Hoffmann, E. O., and DiLuzio, N. R. 1975. Influence of lead and cadmium on the susceptibility of rats to bacterial challenge. *Proc. Soc. Exp. Biol. Med.* 150:741–747.

Coombs, R. R. A., and Gell, P. G. H. 1975. Classification of allergic reactions responsible for clinical hypersensitivity and disease. In *Clinical Aspects of Immunology,* eds. P. G. H. Gell, R. R. A. Coombs, and P. J. Lachman, p. 761. Oxford: Blackwell Scientific Publications; Philadelphia: J. B. Lippincott.

Courtney, K. 1979. Hexachlorobenzene (HCB): A review. *Environ. Res.* 20:225–266.

Crocker, J. F. S., Ozere, R. L., Safe, S. H. et al. 1976. Lethal interaction of ubiquitous insecticide carriers with virus. *Science* 192:1351–1353.

Cunningham, A. J., and Szenberg, A. 1968. Further improvements in the plaque technique for detecting single antibody-forming cells. *Immunology* 14:599–601.

Davis, B., Dulbecco, R., Eisen, H., et al. 1973. Cell-mediated hypersensitivity and immunity. In *Microbiology,* 2nd ed. Hagerstown: Harper & Row.

Dean, J., Luster, M., Boorman, G., et al. 1981. Host resistance models as endpoints for assessing immune alterations following chemical exposure: Studies with diethylstilbestrol, cyclophosphamide and 2,3,7,8-tetrachlorodibenzo-p-dioxin. In *The Biological Relevance of Immune Suppression Induced by Therapeutic and Environmental Agents,* eds. J. Dean, M. Padarathsingh, pp. 233–255. New York: Van Nostrand Reinhold.

Dean, J. H., Luster, M. I., and Boorman, G. A. 1982a. Immunotoxicology. In *Immunopharmacology,* eds. P. Sirois, and M. Rola-Pleszczynski, pp. 349–397. New York: Elsevier Biomedical Press.

Dean, J. H., Luster, M. I., Boorman, G. A., et al. 1982b. Application of tumor, bacterial and parasite susceptibility assays to study immune alterations induced by environmental chemicals. *Environ. Health Perspect.* 43:81–87.

Dean, J. H., Luster, M. I., Boorman, G. A., et al. 1983a. Selection immunosuppression resulting from the exposure to the carcinogenic congener of benzopyrene in B6C3F1 mice. *Clin. Exp. Immunol.* 52:199–206.

Dean, J. H., Luster, M. I., Boorman, G. A., et al. 1983b. Immunotoxicity of tumor promoting environmental chemicals and phorbol diesters. In *Advances in Pharmacology 2,* ed. J. W. Hadden, pp. 23–31. Oxford: Pergamon.

Descotes, J., Evreux, J. C., Laschi-Locquerie, A., et al. 1984. Comparative effects of various lead salts on delayed hypersensitivity in mice. *J. Appl. Toxicol.* 4:265–266.

Descotes, J. 1986. *Immunotoxicology of Drugs and Chemicals.* Amsterdam: Elsevier Science.

Desi, I., Varga, L., and Farkas, I. 1978. Studies on the immunosuppressive effect of organochlorine and organophosphoric pesticides in subacute experiments. *J. Hyg. Epidemiol. Microbiol. Immunol.* 22:115–122.

Dieter, M. P., Luster, M. I., Boorman, G. A., et al. 1983. Immunological and biochemical responses in mice treated with mercuric chloride. *Toxicol. Appl. Pharmacol.* 68:218–228.

Duke, S. S., Schook, L. B., and Holsapple, M. P. 1985. Effects of N-nitrosodimethylamine on tumor susceptibility. *J. Leuk. Biol.* 37:383–394.

Ehrlich, R. 1966. Effect of nitrogen dioxide on resistance to respiratory infection. *Bacterial Rev.* 30:604–613.

Ehrlich, R. 1980. Interactions between environmental pollutants and respiratory infections. *Environ. Health Perspect.* 35:89–100.

Exon, J. H., and Koller, L. D. 1983. Effects of chlorinated phenols on immunity in rats. *Int. J. Immunopharmacol.* 5:131–136.

Faith, R. E., and Moore, J. A. 1977. Impairment of thymus-dependent immune functions by exposure to the developing immune system by 2,3,7,8-tetrachlorodibenzo-p-dioxin (TCDD). *J. Toxicol. Environ. Health* 3:451–464.

Faith, R. E., Luster, M. I., and Kimmel, C. A. 1979. Effect of chronic developmental lead exposure on cell-mediated immune functions. *Clin. Exp. Immunol.* 35:413–420.

Fidler, I. J., Gersten, D. M., and Hart, I. R. 1978. The biology of cancer invasion and metastasis. *Adv. Cancer Res.* 28:149–250.

Fiore, M., Anderson, H., Hong, R., et al. 1986. Chronic exposure to aldicarb-contaminated groundwater and human immune function. *Environ. Res.* 41:633–645.

Forsell, J. H., Shull, L. R., and Kateley, J. R. 1981. Subchronic administration of technical pentachlorophenol to lactating dairy cattle: Immunotoxicologic evaluation. *J. Toxicol. Environ. Health* 8:543–558.

Fraker, P. J. 1980. The antibody-mediated and delayed-type hypersensitivity response of mice exposed to polybrominated biphenyls. *Toxicol. Appl. Pharmacol.* 53:1–7.

Friend, M., and Trainer, D. O. 1974. Experimental dieldrin–duck hepatitis virus interaction studies. *J. Wildl. Manag.* 38:896–902.

Fugmann, R. A., Aranyi, C., Barbera, P. W., et al. 1983. The effects of DES as measured by host resistance and tumor susceptibility assays in mice. *J. Toxicol. Environ. Health* 11:827–841.

Golub, S. H., O'Connell, T. X., and Morton, D. L. 1974. Correlation of in vivo and in vitro assays of immunocompetence in cancer patients. *Cancer Res.* 34:1833–1837.

Greenlee, W., Dold, K., Irons, R., et al., 1985. Evidence for direct action of 2,3,7,8-tetrachlorodibenzo-p-dioxin (TCDD) on thymic function. *Toxicol. Appl. Pharmacol.* 79:112–120.

Hemphill, F. E., Kaeberle, M. I., and Buck, W. B. 1971. Lead suppression of mouse resistance to *Salmonella typhimurium*. *Science* 172:1031–1032.

Herberman, R. B. (ed.) 1982. *NK Cells and Other Natural Effector Cells.* New York: Academic Press.

Herberman, R. B. 1985. Immunologic mechanisms of host resistance to tumors. In *Immunotoxicology and Immunopharmacology,* eds. J. H. Dean, M. I. Luster, A. E. Munson, and H. Amos, pp. 69–77. New York: Raven Press.

Hermanowicz, A., Nawarska, Z., Borys, D., et al. 1982. The neutrophil function and infectious diseases in workers occupationally exposed to organochlorine insecticides. *Int. Arch. Occup. Environ. Health* 50:329.

Hoffman, R. E., Stehr-Green, P. A., Webb, K. B., et al. 1986. Health effects of long-term exposure to 2,3,7,8-tetrachlorodibenzo-p-dioxin. *J. Am. Med. Assoc.* 255:2031–2038.

Holmberg, L. A., Springer, T. A., and Ault, K. A. 1981. Natural killer cell activity in the peritoneal exudates of mice infected with *Listeria monocytogenes:* Characterization of the natural killer cells by using a monoclonal rat anti-murine macrophage antibody (M1/70). *J. Immunol.* 127:1792–1799.

Holsapple, M. P., Page, D. G., Shopp, G. M., et al. 1984. Characterization of the delayed hypersensitivity response to a protein antigen in the mouse. *Int. J. Immunopharmacol.* 6:399–405.

Hoshino, A., Takenaka, H., Mizukoshi, P., et al. 1983. Effect of anti-interferon serum on influenza virus infection in mice. *Antiviral Res.* 3:59–68.

Hudson, J. L., Duque, R., and Lovett, E. 1985. Applications of flow cytometry in immunotoxicology. In *Immunotoxicology and Immunopharmacology,* eds. J. H. Dean, M. I. Luster, A. Munson, and H. Amos, pp. 159–178. New York: Raven Press.

Irons, R., Heck, H., Moore, B., et al. 1979. Effects of short-term benzene administration on bone marrow cell cycle kinetics in the rat. *Toxicol. Appl. Pharmacol.* 51:399–409.

Johnson, K., Munson, A., and Holsapple, M. 1987. Primary cellular target responsible for dimethylnitrosamine-induced immunosuppression in the mouse. *Immunopharmacology* 13:47–60.

Kalland, T. 1985. Immunotoxicology of diethylstilbestrol in man. In *Immunotoxicology and Immunopharmacology,* eds. J. H. Dean, M. I. Luster, A. Munson, and H. Amos, pp. 407–414. New York: Raven Press.

Kammuller, M. E., Penninks, A. H., and Seinen, W. 1984. Spanish toxic oil syndrome is a chemically induced GVHD-like epidemic. *Lancet* i:1174–1175.

Karol, M. 1985. Hypersensitivity to isocyanates. In *Immunotoxicology and Immunopharmacology,* eds. J. H. Dean, M. I. Luster, A. Munson, and H. Amos, pp. 475–488. New York: Raven Press.

Karol, M., Koros, A., Magreni, C., et al. 1986. Detection of TDI-specific lymphocytes in animals sensitized with toluene diisocyanate (TDI). *Toxicologist* 6:15.

Kerkvliet, N., and Brauner, J. 1987. Mechanisms of 1,2,3,4,6,7,8-heptachlorodibenzo-*p*-dioxin (HpCDD)-induced humoral immune suppression: Evidence of primary defect in T-cell regulation. *Toxicol. Appl. Pharmacol.* 87:18–31.

Kerkvliet, N. 1984. Halogenated aromatic hydrocarbons (HAH) as immunotoxicants. In *Chemical Regulation of Immunity in Veterinary Medicine,* pp. 369–387. New York: A. R. Liss.

Kerkvliet, N., Brauner, J., and Matlock, J. 1985. Humoral immunotoxicity of polychlorinated diphenyl ethers, phenoxyphenols, dioxins and furans present as contaminants of technical grade pentachlorophenol. *Toxicology* 36:307–324.

Kiessling, R., and Wigzell, H. 1979. An analysis of the murine NK cell as to structure, function and biological relevance. *Immunol. Rev.* 44:165–208.

Klaus, G. G. B., and Hawrylowicz, C. M. 1984. Cell-cycle control in lymphocyte stimulation. *Immunol. Today* 5:15–19.

Koller, L. D. 1973. Immunosuppression produced by lead, cadmium and mercury. *Am. J. Vet. Res.* 34:1457–1458.

Koller, L. D., and Brauner, J. D. 1977. Decreased B-lymphocyte response after exposure to lead and cadmium. *Toxicol. Appl. Pharmacol.* 42:621–624.

Koller, L. D., and Exon, J. 1985. The rat as a model for immunotoxicity assessment. In *Immunotoxicology and Immunopharmacology,* eds. J. H. Dean, M. I. Luster, A. Munson, and H. Amos, pp. 99–112. New York: Raven Press.

Koller, L. D., and Roan, J. G. 1977. Effects of lead and cadmium on mouse peritoneal macrophages. *J. Reticuloendothel. Soc.* 21:7–12.

Koller, L. D., and Roan, J. G. 1980a. Effects of lead, cadmium and methylmercury on immunological memory. *J. Environ. Pathol. Toxicol.* 4:47.

Koller, L. D., and Roan, J. G. 1980b. Response of lymphocytes from lead, cadmium and methylmercury-exposed mice in the mixed lymphocyte culture. *J. Environ. Pathol. Toxicol.* 4:393.

Koller, L.D., Exon, J. H., and Arbogast, B. 1977. Methylmercury: Effect on serum enzymes and humoral antibody. *J. Toxicol. Environ. Health* 2:1115–1123.

Koller, L.D., Roan, J. G., and Kerkvliet, N. I. 1979. Mitogen stimulation of lymphocytes in CBA mice exposed to lead and cadmium. *Environ. Res.* 19:177–188.

Koo, G. C., Jacobson, J. B., Hammerling, G. J., et al. 1980. Antigenic profile of murine natural killer cells. *J. Immunol.* 125:1003–1006.

Larsh, J. E., Race, G. J., Martin, J., et al. 1974. Studies on delayed (cellular) hypersensitivity in mice treated with *Trinchinella spiralis*. VIII. Serologic and histopathologic responses of recipients injected with spleen cells of donors suppressed with ATS. *J. Parasitol.* 60:99–106.

Laschi, A., and Tachon, P. 1982. Influence of lead and nickel on passive cutaneous anaphylaxis in the rat. *Toxicol. Lett.* 13:185–188.

Lawrence, D. A. 1981. In vivo and in vitro effects of lead on humoral and cell-mediated immunity. *Infect. Immunity* 31:136–143.

Leach, C., Hatoum, N., Ratajczak, H., et al. 1987. The pathologic and immunologic responses to inhaled trimellitic anhydride in rats. *Toxicol. Appl. Pharmacol.* 87:67–80.

Lee, T., and Chang, K. 1985. Health effects of polychlorinated biphenyls. In *Immunotoxicology and Immunopharmacology*, eds. J. H. Dean, M. Luster, A. Munson, and H. Amos, pp. 415–422. New York: Raven Press.

Lefford, M. J. 1974. The measurement of tuberculin hypersensitivity in rats. *Int. Arch. Allergy Appl. Immunol.* 47:570–585.

Leung, K., and Ada, G. 1980. Two T-cell populations mediated delayed-type hypersensitivity to murine influenza virus infection. *Scand. J. Immunol.* 12:481–487.

Lewis, J. G., and Adams, D. O. 1985. The mononuclear phagocyte system and its interaction with xenobiotics. In *Immunotoxicology and Immunopharmacology*, eds. J. H. Dean, M. I. Luster, A. Munson, and H. Amos, pp. 23–44. New York: Raven Press.

Loose, L. D. 1982. Macrophage induction of T-suppressor cells in pesticide-exposed and protozoan-infected mice. *Environ. Health Perspect.* 43:89–97.

Loose, L. D., Silkworth, J. B., Mudzinski, S. P., et al. 1979. Modification of the immune response by organochlorine xenobiotics. *Drug Chem. Toxicol.* 2:111–132.

Loose, L. D., Mudzinski, S. P., and Silkworth, J. B. 1981. Influence of dietary polybrominated biphenyl on antibody and host defense responses in mice. *Toxicol. Appl. Pharmacol.* 59:25–39.

Lubet, R., Lemaire, B., Avery, D., et al. 1986. Induction of immunotoxicity in mice by polyhalogenated biphenyls. *Arch. Toxicol.* 59:71–77.

Lukic, M. L., Popeskovic, L., and Jankovic, B. D. 1973. Potentiation of immune responsiveness in rats treated with DDT. *Fed. Proc.* 32:1037.

Luster, M. I., Faith, R. E., and Lawson, L. D. 1979. Effect of 2,3,7,8-tetrachlorodibenzofuran (TCDF) on the immune system of guinea pigs. *Drug Chem. Toxicol.* 2:49–60.

Luster, M. I., Boorman, G. A., Dean, J. H., et al. 1980a. Examination of bone marrow, immunologic parameters and host susceptibility following pre- and post-natal exposure to 2,3,7,8-tetrachlorodibenzo-*p*-dioxin (TCDD). *Int. J. Immunopharmacol.* 2:301–310.

Luster, M. I., Boorman, G. A., Harris, M. W. et al. 1980b. Laboratory studies on polybrominated biphenyl-induced immune alterations following low-level chronic and pre-/post-natal exposure. *Int. J. Immunopharmacol.* 2:69–80.

Luster, M. I., Dean, J. H., Boorman, G. A., et al. 1982a. Immune functions in methyl and ethyl carbamate treated mice. *Clin. Exp. Immunol.* 50:223–230.

Luster, M. I., Dean, J. H., and Moore, J. A. 1982b. Evaluation of immune functions in toxicology. In *Principles and Methods of Toxicology*, ed. W. W. Hayes, pp. 561–586. New York: Raven Press.

Luster, M. I., Munson, A. E., Thomas, P., et al. 1988. Development of a testing battery to assess chemical-induced immunotoxicity: National Toxicology Program's Guidelines for Immunotoxicity Assessment. *Fund. Appl. Toxicol.* 10:2–19.

Lysy, H., McCay, J., Snyder, N., et al. (1987). Modulation of host resistance following intratracheal instillation of gallium arsenide (GaAs). *The Toxicologist* 7:225.

Maclean, L. E. 1979. Host resistance in surgical patients. *J. Trauma* 19:297–304.

Magnusson, B., and Kligman, A. M. 1969. The identification of contact allergens by animal assay. The guinea-pig maximization test. *J. Invest. Dermatol.* 52:268–276.

McConnell, E. E., Moore, J. A., Gupta, B. N., et al. 1980. The chronic toxicity of technical and analytical pentachlorophenol in cattle. I. Clinicopathology. *Toxicol. Appl. Pharmacol.* 52:468–490.

Miller, M. E. 1975. Pathology of chemotaxis and random mobility. *Semin. Hematol.* 12:59–82.

Morahan, P. S., Edelson, P. J. and Gass, K. 1980. Changes in macrophage ectoenzymes associated with anti-tumor activity. *J. Immunol.* 125:1312–1321.

Morahan, P. S. 1983. Interactions of herpes viruses with mononuclear phagocytes. In *Immunology of Herpes Simplex Virus Infection,* eds. B. Rouse and C. Lopez, pp. 71–84. Boca Raton: CRC Press.

Morahan, P. S., Bradley, S. G., Munson, A. E., et al. 1984. Immunotoxic effects of DES on host resistance. Comparison with cyclophosphamides. *J. Leukocyte Biol.* 35:329–338.

Moszczynski, P. 1982. Effect of organic solvents in the course of professional exposure on the E rosette test and skin reactions against streptase and tuberculin. *Folia Haematol.* 109:224.

Moszczynski, P., and Lisiewicz, J. 1983. Nitrotetrazolium blue reduction and the time of occupational exposure to benzene and its homologues. *Haematologica* 68:689.

Munson, A. E., Holsapple, M., and Duke, S. 1987. Effects of dimethylnitrosamine. In *Immunotoxicology,* eds. A. Berlin, J. Dean, M. Draper, E. Smith, and F. Sperafico, pp. 333–346. Boston: Martinus Nishoff Publishers.

Murray, M. J., Lauer, L. D., Luster, M. I., et al. 1985. Correlation of murine susceptibility to tumor, parasite and bacterial challenge with altered cell-mediated immunity following systemic exposure to the tumor promoter phorbol myristate acetate. *Int. J. Immunopharmacol.* 7:491–500.

Muto, S., Yamasaki, A., Yamamoto, et al. 1985. Rapid effect of lymphokines on macrophage activities determined by chemiluminescence. *Chem. Pharm. Bull.* 33:4041–4044.

Newborg, M., and North, R. 1980. On the mechanisms of T-cell independent anti-*Listeria* resistance in nude mice. *J. Immunol.* 124:571–576.

Noble, C., and Norbury, K. C. 1986. Use of the popliteal lymph node enlargement assay to measure rat T-cell function in immunotoxicologic testing. *Int. J. Immunopharmacol.* 8:449–453.

North, R. 1973. Importance of thymus-derived lymphocytes in cell-mediated immunity to infection. *Cell Immunol.* 7:166–176.

Ohsawa, M., Sato, K., Takahashi, K., et al. 1983. Modified distribution of lymphocyte subpopulations in blood and spleen from mice exposed to cadmium. *Toxicol. Lett.* 19:29–35.

Olson, L. J., Erickson, B., Hinsdill, R., et al. 1987. Aldicarb immunomodulation in mice: An inverse dose-response to parts per billion levels in drinking water. *Arch. Envir. Contam. Toxicol.* 16:433–439.

Penn, I. 1985. Neoplastic consequences of immunosuppression. In *Immunotoxicology and Immunopharmacology,* eds. J. H. Dean, M. I. Luster, A. E. Munson, and H. Amos, pp. 79–89. New York: Raven Press.

Perlmann, P., Perlmann, H., and Wigzell, H. 1972. Lymphocyte mediated cytotoxicity in vitro. Induction and inhibition by humoral antibody and natural effector cells. *Transplant. Rev.* 13:91–114.

Peterman, G., Altman, L., and Corey, L. 1985. Cell-mediated immunity in human herpes virus infection: Analysis by monoclonal antibodies. *Clin. Immunol. Immunopathol.* 37:83–92.

Peterson, M., Smialowicz, R., Harder, S., et al. 1981. The effect of controlled ozone exposure on human lymphocyte function. *Envir. Res.* 24:299.

Rao, K. S., Betso, J. E., and Olson, K. J. 1981. A collection of guinea-pig sensitization test results—grouped by chemical class. *Drug Chem. Toxicol.* 4:331–351.

Reynolds, C. W., and Herberman, R. B. 1981. In vitro augmentation of rat natural killer (NK) cell activity. *J. Immunol.* 126:1581–1585.

Roder, J., and Duwe, A. 1979. The beige mutation in the mouse selectively impairs natural killer activity. *Nature* 278:451–453.

Rodgers, K., Leung, N., Imamura, T., et al. 1986. Rapid in vitro screening assay for immunotoxic effects of organophosphorus and carbamate insecticides on the generation of cytotoxic T-lymphocyte responses. *Pest. Biochem. Physiol.* 26:292–301.

Roitt, J., Brostoff, J., and Male, D. 1985. *Immunology.* St. Louis, Mo.: C. V. Mosby.

Rollinghoff, M., Starzinski-Powitzz, A., Pfizenmaier, K., et al. 1977. Cyclophosphamide-sensitive T-lymphocytes suppress the in vivo generation of antigen-specific cytotoxic T-lymphocytes. *J. Exp. Med.* 145:455–459.

Ruitenberg, E., Elgersma, A., Kruizinga, W., et al. 1977. *Trichinella spiralis* infection in congenitally athymic (nude) mice. Parasitological studies with observations on intestinal pathology. *Immunology* 33:581–587.

Seinen, W., Vos, J. G., Van Spanje, I., et al. 1977a. Toxicity of organotin compounds. II. Comparative in vivo and in vitro studies with various organotin and organolead compounds in different animal species with special emphasis on lymphocyte cytotoxicity. *Toxicol. Appl. Pharmacol.* 42:197–212.

Seinen, W., Vos, J. G., Van Krieken, P., et al. (1977b). Toxicity of organotin compounds. III. Suppression of thymus-dependent immunity in rats by di-*n*-butyltindichloride and di-*n*-octyltindichloride. *Toxicol. Appl. Pharmacol.* 42:213–224.

Sharma, R. P., and Gehring, P. J. 1979. Immunologic effects of vinyl chloride in mice. *Ann. N.Y. Acad. Sci.* 320:551–563.

Sharma, R. P., Yakel, H. O., and Gehring, P. J. 1980. Immunotoxicologic studies with vinyl chloride in rabbits and mice. *Int. J. Immunopharmacol.* 2:295–299.

Skamene, E., and Gros, P. 1983. Role of macrophages in resistance against infectious disease. *Clin. Immunol. Allerg.* 3:539.

Spiers, R., Benson, R. W., and Schiffman, G. 1978. Models for assessing the effect of toxicants on immunocompetence in mice. The effect of diphtheria, pertussis and tetanus vaccine on antibody response to Type II pneumococcal polysaccharide. *J. Environ. Pathol. Toxicol.* 1:689–699.

Stein-Streilein, J., and Guffee, J. 1986. In vivo treatment of mice and hamsters with antibodies to asialo GM-1 increases morbidity to pulmonary influenza infection. *J. Immunol.* 136:1435–1441.

Stjernsward, J. 1966. Effect of noncarcinogenic and carcinogenic hydrocarbons on antibody-forming cells measured at the cellular level in vitro. *J. Natl. Cancer Inst.* 36:1189–1195.

Street, J. C., and Sharma, R. P. 1975. Alteration of induced cellular and humoral immune responses by pesticides and chemicals of environmental concern: Quantitative studies of immunosuppression by DDT, aroclor 1254, carbaryl, carbofuran and methylparathion. *Toxicol. Appl. Pharmacol.* 32:587–602.

Subba Rao, D. S. V., and Glick, B. 1977. Pesticide effects on the immune response and metabolic activity of chicken lymphocytes. *Proc. Soc. Exp. Biol. Med.* 154:27–29.

Thaxton, P., and Parkhurst, C. R. 1973. Toxicity of mercury to young chickens. 3. Changes in immunological responsiveness. *Poultry Sci.* 52:761–764.

Thigpen, J. E., Faith, R. E., McConnell, E. E., et al. 1975. Increased susceptibility to bacterial infection as a sequela of exposure to 2,3,7,8-tetrachlorodibenzo-*p*-dioxin. *Infect. Immunity* 12:1319–1324.

Thomas, P. T., and Hinsdill, R. D. 1978. Effect of polychlorinated biphenyls on the immune response of rhesus monkeys and mice. *Toxicol. Appl. Pharmacol.* 44:41–51.

Thomas, P. T., and Hinsdill, R. D. 1979. The effect of perinatal exposure to tetrachlorodibenzo-*p*-dioxin on the immune responses of young mice. *Drug Chem. Tox.* 2:77–98.

Thomas, P., Fugmann, R., Aranyi, C., et al. 1985a. Effect of dimethylnitrosamine on host resistance and immunity. *Toxicol. Appl. Pharmacol.* 77:219–229.

Thomas, P., Ratajczak, H., Aranyi, C., et al. 1985b. Evaluation of host resistance and immune functions in cadmium-exposed mice. *Toxicol. Appl. Pharmacol.* 80:446–456.

Thomas, P. T., Fenters, J., Aranyi, C., et al. 1985c. Chemical-induced immunotoxicity, pp. 53–90. NIEHS Semiannual Report, L06133-8. Research Triangle Park, NC: National Institute of Environmental Health Sciences.

Thomas, P., Ratajczak, H., Eisenberg, W., et al. 1987. Evaluation of host resistance and immunity in mice exposed to the carbamate pesticide aldicarb. *Fund. Appl. Toxicol.* 9:82–89.

Thurman, G. B., Simms, B. G., Goldstein, A. L., et al. 1978. The effects of organic compounds used in the manufacture of plastics on the responsivity of murine and human lymphocytes. *Toxicol. Appl. Pharmacol.* 44:617–641.

Thurmond, L., Lauer, L., House, R., et al. 1986. Effect of short-term inhalation exposure to 1,3-butadiene on murine immune functions. *Toxicol. Appl. Pharmacol.* 86:170–179.

Turk, J. L., and Parker, D. 1982. Effect of cyclophosphamide on immunologic control mechanisms. *Immunol. Rev.* 65:99–113.

U.S. EPA (United States Environmental Protection Agency). 1982. Pesticide Assessment Guidelines. Subdivision M: Biorational Pesticides. EPA 540/9-82-028.

Vecchi, A. 1987. Some aspects of immune alterations induced by chlorodibenzo-*p*-dioxins and chlorodibenzofurans. In *Immunotoxicology,* eds. A. Berlin, J. Dean, M. Draper, E. Smith, and F. Sperafico, pp. 308–316. Boston: Martinus Nishoff.

Vecchi, A., Mantovani, A., Sironi, M., et al. 1980. The effect of acute administration of 2,3,7,8-tetrachlorodibenzo-*p*-dioxin (TCDD) on humoral antibody production and cell-mediated activities in mice. *Arch. Toxicol. Suppl.* 4:163–165.

Vireligier, J. 1975. Host defenses against influenza virus: The role of anti-hemagglutinin antibody. *J. Immunol.* 115(2):434–446.

Vos, J. G. 1977. Immune suppression as related to toxicology. *CRC Crit. Rev. Toxicol.* 5:67–101.

Vos, J. G., Kreeftenberg, J. G., Engel, H. W. B., et al. 1978. Studies on 2,3,7,8-tetrachlorodibenzo-*p*-dioxin-induced immune suppression and decreased resistance to infection: Endotoxin hypersensitivity, serum zinc concentrations and effects of thymosin treatment. *Toxicology* 9:75–86.

Vos, J. G., Van Logten, M. J., Kreeftenberg, J. G., et al. 1979. Effect of hexochlorobenzene on the immune system of rats following combined pre- and post-natal exposure. *Drug. Chem. Toxicol.* 2:61–76.

Walker, E. B., Van Epps, D. E., and Warner, N. 1981. Macrophage Chemiluminescence. In *Manual of Macrophage Methodology,* eds. H. Herscowitz, H. Holden, J. Bellanti, and A. Gjaffer, p. 389. New York: Marcel Dekker.

Ward, E. C., Murray, M., and Dean, J. H. 1985. Immunotoxicity of nonhalogenated polycyclic aromatic hydrocarbons. In *Immunotoxicology and Immunopharmacology,* eds. J. H. Dean, M. Luster, A. Munson, and H. Amos, pp. 291–303. New York: Raven Press.

Wasserman, M., Wasserman, D., Gershon, Z., et al. 1969. Effect of organochlorine pesticides on body defense systems. *Ann. N.Y. Acad. Sci.* 160:393–401.

Wasserman, M., Wasserman, D., Kedar, E., et al. 1972. Effects of dieldrin and gamma BHC on serum proteins and PBI. *Bull. Environ. Contam. Toxicol.* 8:177–185.

Winkelstein, J. A. 1981. The role of complement in the host defense against *Streptococcus pneumoniae. Rev. Infect. Dis.* 3:289–298.

Detection of Mutagens and Carcinogens by Physiochemical Techniques

George Bakale

Case Western Reserve University
Radiology Department
Biochemical Oncology Division
Cleveland, Ohio

I. INTRODUCTION

A. The Need to Identify Chemical Carcinogens

Recent legislation that demonstrates the public's concern with their being unknowingly exposed to carcinogens is California's Proposition 65, which went into effect March 1, 1987. The intent of Proposition 65 is to eradicate chemical biohazards from environmental sources through a series of steps that begins with the governor's issuing a list of the chemicals to be regulated. The controversy that this first step has already generated portends a myriad of difficulties that lay ahead in the enforcement of the law (Marshall, 1987). However, this storm of controversy may have the silver lining of raising public awareness to its exposure to carcinogens as well as focusing attention on the need for an efficacious means to identify such chemicals.

Recognition of a causal association between exposure to a chemical in the environment and the onset of cancer in an exposed population is generally attributed to Sir Percival Pott who, in 1775, observed that chimney sweeps, on reaching puberty, were "peculiarly liable to a most noisome, painful and fatal

The author thanks R. D. McCreary, who has made all of the pulse-conductivity measurements throughout the development of the k_e test. Thanks are also due Drs. K. P. Kundu, V. F. Rodriguez, and R. D. Daniels for their HPLC analyses of mixtures of carcinogens. Colleagues in the CWRU Biochemical Oncology and Radiation Physics Divisions of the Radiology Department, the CWRU Environmental Health Sciences Department, the Bereich Strahlenchemie of the Hahn-Meitner Institut, West Berlin, and the Interuniversitair Reactor Instituut, Delft, are also thanked. Special thanks are given to Margaret Satink Bakale for her patience, perseverance, and expertise in preparing this manuscript.

This work was supported by U.S. Department of Energy Contract DE-AC02-78EV04746, the National Institute for Occupational Safety and Health grants R01- and R02-OH-01331, and the North Atlantic Treaty Organization Collaborative Research Grant RG. 86/0135.

disease," which was cancer of the scrotum (Pott, 1775; cited by Higginson, 1976). Pott also attached significance to this cancer having a low incidence in the general population and inferred that the casual agent for the chimney sweeps' cancer was soot lodged in the scrotum. Three years after Pott's observations were published, regulations were enacted in Denmark that required chimney sweeps to bathe daily (Searle, 1976). This example illustrates that identifying a carcinogenic agent and taking measures to reduce exposure to it can produce a positive health effect without knowledge of the mechanism through which the agent induces cancer.

Approximately two centuries after Pott's study, epidemiological surveys of worldwide cancer incidences indicated that the majority of cancers were caused by environmental sources rather than genetic factors and were therefore, in theory, preventable (Higginson, 1969, 1976, 1983; Doll, 1980). Studies of this nature prompted nonepidemiologists to overstate these findings and conclude that 90 percent of cancers were caused by chemicals in the environment, a conclusion that was vigorously disclaimed by epidemiologists (Maugh, 1979) since their intended use of "environmental" was in the broadest sense and included a significant lifestyle component that was associated with cancers related to tobacco or alcohol use, dietary factors, reproductive and sexual behavior, etc. (Doll and Peto, 1981). However, whether cancer is induced by benzo[*a*]pyrene in the lung of a smoker or whether a teenage mother is protected from breast cancer by hormonal changes associated with the birth of her child, *chemicals* are involved. Identifying these chemicals and learning how they interact with genetic material should contribute to unraveling the multistep process of carcinogenesis.

Among the 44 chemicals or groups of chemicals for which the International Agency for Research on Cancer (IARC) has found limited or sufficient evidence of carcinogenicity to humans, approximately one-third are naturally occurring substances (Wilbourn, 1986). Not included in this list is the most thoroughly characterized human carcinogenic agent, ionizing radiation (Boice, 1981), which along with the naturally occurring substances shall be excluded from this discussion. Of the chemicals that are known to induce cancer in humans, the most ubiquitously distributed are the polycyclic aromatic hydrocarbons or PAHs which occur both naturally and as the incomplete combustion product of organic material (Shabad, 1980). Among these PAHs, the best characterized carcinogen is benzo[*a*]pyrene which is emitted into the atmosphere at a rate of about 1300 tons per year in the United States (Perera, 1981). Turning from these "natural" to man-made chemicals, Hammer estimated that "a new and potentially toxic chemical is introduced into industry every 20 minutes" (Hammer, 1985). If this estimate is combined with the findings of a study by Davis and Magee that the production of synthetic organic chemicals has been growing exponentially since about 1940 with a doubling time of 8 years (Davis and Magee, 1979), one surmises that the burden of potential carcinogens in our total environment is surely increasing. The critical question that arises is whether the technology that is needed to identify, contain, transport, control, and monitor these potentially biohazardous natural and synthetic chemicals has kept pace with the emission or production rates of the same chemicals so that human exposure is minimized. An optimistic answer to one aspect of this question is presented in the discussion that follows.

B. Classical Biological Means of Identifying Chemical Carcinogens

Three major methods are generally used in attempting to identify chemical carcinogens: epidemiological surveys, long-term animal tests, and short-term bioassays. The U.S. Interagency Staff Group on Carcinogens recently reviewed these three carcinogen-identification methods (Hart et al., 1986). Synopses of these reviews are as follows:

1. Epidemiological methods are the "only means of assessing directly the carcinogenic risk of environmental agents in humans," but the observational nature of the method, the paucity of data, the long latency period required to recognize causality, and frequent confounding influences of risk factors limits widespread application of the method.

2. Long-term animal tests "remain the best way of predicting (carcinogenic) effects in humans when epidemiological data are unavailable." However, questions remain regarding the optimal species, strain, dose, and route of exposure as well as the interpretation of tumor data and their relevance to humans. Also, the statistical analysis of the results is complex and requires the interaction of a toxicologist, pathologist, and statistician. These factors combine to place a 3-year, $500,000 cost per chemical tested according to a 1983 estimate (Weinstein, 1983).

3. Short-term bioassays are useful in screening chemicals to identify potential carcinogens and in obtaining mechanism-related information but cannot supplant long-term animal tests or epidemiological studies. Validation studies in which coded chemicals that include a greater number of noncarcinogens screened are recommended to complement those validation studies already conducted (see below). The test protocols used require a few days to several weeks and range in cost from $1000 to $20,000 per chemical.

In addition to these methods of identifying chemical carcinogens, which rely on the biological responses of living systems to test chemicals, nonbiological physicochemical methods have also evolved that are based on various structure-activity relationships (SARs) as well as chemical structure itself (Ashby, 1978).

A recent monograph (McKinney, 1985) and a review (Frierson et al., 1986) combine to provide a description of the historical development and the state of the art of these methods which range from lipophilicity-activity relationships (Dearden, 1985) to the application of artificial intelligence in developing SARs (Klopman, 1984). Various aspects of these SARs are cited in the ensuing discussion of another SAR, namely, electrophilicity-carcinogenicity, which is a common thread in many of the carcinogen-identification methods that were heretofore noted.

II. THEORETICAL BASIS
OF CARCINOGEN–SCREENING TESTS

A. Somatic Mutation Theory

The multistep nature of carcinogenesis which involves initiation, transformation, promotion, and tumorigenesis implies that focusing on only one step in the process can contribute but insignificantly toward elucidation of the overall mechanism. During the last two decades, however, the work of Elizabeth Miller and James Miller of the McArdle Laboratory and Bruce Ames at Berkeley have demonstrated that much insight concerning carcinogenesis can be obtained from elucidating one step, which in this case is initiation. The thread that is common to the studies of these two groups, which have different short-term objectives, is the somatic mutation theory of carcinogenesis; Miller and Miller (1971) have reviewed the theory dating to its founding in 1914 by Boveri.

Stated most simply, the somatic mutation theory of chemical carcinogenesis involves the interaction of an electrophilic chemical with a nucleophilic target that is generally assumed to be DNA, RNA, or another cellular component that processes genetic information (J. A. Miller, 1970; E. C. Miller, 1978). An updated version of the theory that illustrates its connection with oncogene activation was recently presented (Ramel, 1986) as well as evidence of somatic mutations in human lymphocytes in vivo (Turner et al., 1985). However, evidence that complex carcinogen-induced cellular changes other than simple point mutations also occur is abundant (e.g., Weinstein et al., 1985) as are alternative theories of carcinogenesis (e.g., Barrett, 1979; Holliday, 1979; Rubin, 1980; Fahmy and Fahmy, 1980; Cairns, 1981; Land et al., 1983; Bishop, 1987) including indications of an autoimmune component of cancer (Prehn and Prehn, 1987). This discussion, however, is confined to basic somatic mutation theory which has contributed greatly toward the development of short-term tests for screening carcinogens as is evident from the worldwide use of the Ames *Salmonella* bioassay in more than 3000 laboratories (Ames, 1984).

One of the most significant differences between the bioassay that Ames and co-workers developed (Ames, 1972; Ames et al., 1973, 1975) and earlier attempts to establish that bacterial mutagenicity was correlated with animal carcinogenicity was the incorporation of a mammalian microsomal system in the assay. This served to metabolically activate unreactive procarcinogens to a more electrophilic state. Ames' recognition of the need to incorporate such a step into the carcinogen screening bioassay that he developed was based predominantly on the work of Miller and Miller (see references above) who had conclusively demonstrated for many classes of chemicals that metabolic activation was needed to convert a biologically inert chemical, the procarcinogen, into a highly reactive state, the ultimate carcinogen. Thus, the common thread between the work of Ames and that of Miller and Miller again involves the electrophilic properties of carcinogens.

B. Electrophilicity-Carcinogenicity Correlation

The term electrophilicity is commonly (and conveniently) used among biologists to indicate a qualitative property of a chemical that is best defined by its literal translation, i.e., "electron loving." The author is unaware of attempts by biologists, chemists, or physicists to define electrophilicity more precisely, but electrophilicity is frequently and mistakenly equated with the well-defined physical properties of electron affinity and reduction potential, which also are often incorrectly interchanged. Comparisons of electron affinities and reduction potentials are available (Briegleb, 1964; Sjoberg and Eriksen, 1980) as are extensive discussions of gas-phase electron affinities (Christophorou, 1971), liquid-phase reduction potentials (Henglein, 1974), and liquid- and solid-phase Fermi levels (Reiss, 1985).

Since the time that the reduction potential of the hydrated electron was calculated (Baxendale, 1964), radiation chemists were aware that reduction potential could be utilized to clarify electron transfer reactions in aqueous systems. Consequently, pulse radiolysis techniques were developed to study reduction reactions using polarographic (Lilie et al., 1971), optical (Meisel and Czapski, 1975a), and ESR (Meisel and Neta, 1975b) methods of detection. These chemical studies complemented radiation biological studies in which the reduction potential was used to guide the development of hypoxic-cell radiosensitizers (Adams, 1980; Wardman, 1977). From such studies it was recognized that reduction potentials were correlated with cellular toxicity (Olive, 1979a; Adams et al., 1980) and metabolism (Olive, 1979b). These studies prompted a search for a correlation between the mutagenicity and the reduction potential of nitroaromatic compounds. Such a correlation was observed using the Ames test for nitroarenes (Klopman et al., 1983) but not for nitroheterocyclic compounds

(Hartman and Hartman, 1983). Bartsch et al. (1983) also conducted studies of the interrelationship among carcinogenicity, mutagenicity, and electrophilicity.

In addition to these preceding studies that indicate the involvement of electrons in the biological processes stated and the established involvement of electron transfer processes in respiration and metabolism (von Jagow and Engel, 1980), a significant effort has been directed toward obtaining evidence of the direct involvement of electrons in carcinogenesis. The central figure in much of this work was Nobel laureate Albert Szent-Györgyi (1968), who, from the time of his initial statements regarding an electronic theory of cancer in 1968 until his death in 1986, evoked considerable controversy (Holden, 1979). Two symposia dedicated to Szent-Györgyi attest to his having a loyal following that was receptive to his novel ideas (Kasha and Pullman, 1962; Wolstenholme et al., 1979).

Szent-Györgyi's (1976, 1977) cursory theory of electron-mediated carcinogenesis and subsequent refinements to this theory emerged from a theme that he had frequently expressed in the preceding decades (Szent-Györgyi, 1941a,b). This was that the unique activity and "subtlety" of many biological processes could not be effected by sluggish molecules chemically reacting but instead required the nimble transfer of energy and charge via an electronic mechanism. The mechanism of electron transport in biomolecules proposed by Szent-Györgyi was semiconduction; Shockley's "parking-garage" model of semiconduction (Shockley, 1950) was used by Szent-Györgyi to illustrate charge transport to audiences. Szent-Györgyi used a metal can with marbles as the insulator or full parking garage; removal of a marble, electron, or car permitted the marbles, electrons, and cars to be mobile. Also, the vacancy created was mobile in the three models, which will again be discussed in the section on excess electrons.

The preceding example and much of the work related to experimental and theoretical studies of charge conduction and of valence-band, conduction-band, and band-gap energies that was done from 1940 to 1970 have been reviewed from the viewpoint of an electrical engineer in the book by R. Pethig (1979a). Most frequently cited are the studies from the groups of D. D. Eley, H. Frohlich, J. J. Ladik, H. A. Pohl, S. Suhai, and Pethig's own work.

Although only briefly mentioned by Pethig, the quantum chemical studies of A. Pullman and B. Pullman are particularly noteworthy since their molecular orbital calculations of the K, L, and M regions of carcinogens and noncarcinogens significantly influenced cancer research for nearly three decades (A. Pullman and B. Pullman, 1955; B. Pullman and A. Pullman, 1963). This is well exemplified by Boyland's stating, in 1963, that "the general conclusion that only compounds with more than a minimum electron density at the K-Region were active has been universally accepted" (Boyland, 1964).

The K, L, and M regions of benzo[a]pyrene are illustrated and compared with the bay region of the same molecule in Fig. 1. The significance of the bay

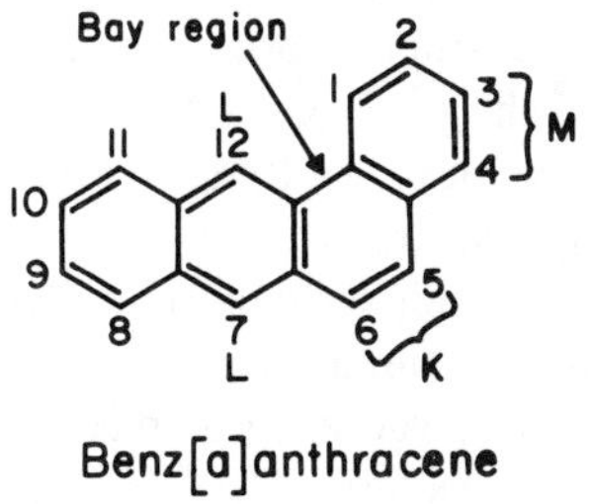

Benz[a]anthracene

FIG. 1. Comparison of K, L, M, and bay regions of benz[a]anthracene (see Pullman and Pullman, 1955, and Jerina et al., 1977, respectively).

region of PAHs to their mutagenic and carcinogenic activities merged from the groups of Jerina and Conney who identified bay-region diol-epoxides as having the greatest potency among the PAH metabolites formed (Jerina et al., 1977). Several comparisons of the mutagenic and carcinogenic potencies predicted from K, L, M, and bay region theories have been reported (see, for example Jerina et al., 1977; Smith et al., 1978; and Loew et al., 1985). Further discussion of the metabolic activation of procarcinogens is deferred to Section IV, carcinogen-target interaction.

Another study pertinent to the electrophilicity-carcinogenicity correlation was that of R. Mason who proposed that the initiation of carcinogenesis by PAHs was effected by the formation of a complex between the PAH and a target biomolecule that was assumed to be a protein (Mason, 1958, 1960). Mason further proposed that if an unfilled energy level in the PAH matched the highest filled level of the protein, electron transfer to the PAH occurred which induced positive-hole conduction in the protein and triggered the initiation of carcinogenesis. This theory differed significantly from other theories of PAH-induced carcinogenesis (see above) in that the PAH served as an electron acceptor rather than as an electron donor. With this theory, Mason correctly identified six carcinogens from 34 PAHs studied with the only apparent false positive being anthanthrene, which the IARC evaluated in 1983 and concluded that "there is limited evidence that anthanthrene is carcinogenic to experimental animals" (IARC, 1983). Classifications of the remaining 27 PAHs studied by Mason were too ambiguous for firm conclusions to be drawn, but this work provided a rationale for subsequent studies based on identifying chemical carcinogens by their electron attachment properties. One of these is the liquid-phase k_e test which is discussed in detail in the final three sections, and the other is the gas-phase analog of the k_e test which J. E. Lovelock developed two decades earlier.

Lovelock's initial studies of the electron attachment properties of molecules in the gas phase in the late 1950s stemmed from his interest in gas chromatography, or GC (Lovelock, 1958a,b, 1961a), which was a fast-growing analytical technique at that time. In the course of attempting to develop a detector for GC that was based on the electron-capture properties of the analyte, marked differences in the sensitivity of the electron-capture detector to various classes of chemicals were noted (Lovelock and Lipsky, 1960; Lovelock, 1961b). Low molecular weight aliphatic and aromatic hydrocarbons were found to capture electrons inefficiently, whereas higher molecular weight PAHs and halogen- or nitro-substituted compounds captured electrons with a much greater efficiency (Lovelock, 1961b). These observations prompted Lovelock and co-workers to suggest that the high electron-capture efficiencies might be biologically significant and that "any substance with a high electron absorption coefficient is potentially toxic . . ." with carcinogenic activity being only one manifestation by which the toxicity could be expressed (Lovelock et al., 1962).

Shortly thereafter, Lovelock adapted the electron-capture detector to study thermal electrons reacting with chemicals involved in oxidative metabolism (Lovelock, 1962) which was followed by a study of the electron-capture properties of steroids (Lovelock et al., 1963). Lovelock's electron-capture detectors were subsequently patented (Lovelock, 1966; Lovelock et al., 1972) and are widely used in analytical and environmental chemistry; however, it appears that no attempt was made to utilize these detectors in screening chemical carcinogens.

This discussion of the electrophilicity-carcinogenicity correlation is concluded by citing several pertinent observations by L. G. Christophorou, who has intensively studied electron-molecule interactions for nearly three decades. In the chapter on bioelectronics in his 1971 book, Christophorou reviewed much of the gas-phase work noted in this section and concluded that extending such measurements of the electron-accepting capacity of molecules to "the liquid or dissolved state would be most valuable." In a 1980 review of polyatomic anions, Christophorou proposed that future work could "lead to what we call an electron capture-based toxicity index." In the 1984 book that Christophorou edited, he and co-workers discussed both the extension of gas-phase reaction properties of electrons to the liquid phase (Christophorou and Siomos, 1984) and applications of electron-molecule interactions to the elucidation of biological processes (Christophorou and Hunter, 1984). In the former, much of the work on electron transport and attachment in model liquids was reviewed; in the latter, work on electron attachment to radiosensitizers and chemical carcinogens was described. Included in the discussion of radiosensitization was a description of a model in which the structure of water vicinal to the target biomolecule was proposed to play a key role in mediating charge transport (Bakale and Gregg, 1978). This effect of the microenvironment on electrophile-biomolecule interaction is discussed following a description of the influence of the medium on the transport and reaction properties of excess electrons.

III. EXCESS ELECTRONS AS PROBES OF ELECTROPHILICITY–CARCINOGENICITY

A. Discovery of Excess Electrons in Nonpolar Liquids

An excess electron in a liquid is simply a unit of excess negative charge that is immersed in but not associated with any single molecule of the surrounding solvent; as such it is the fundamental reducing agent. In a polar solvent, the electron solvation time τ_s depends on the solvation mechanism. For example, in alcohols at 21 °C, τ_s is correlated with the rotational time of the molecules τ_r, and τ_s ranges from 10 to 50 picoseconds (ps) as τ_r increases from 10 to 50 ps in alcohols increasing in size from methanol to decanol (Kenney-Wallace, 1977, 1980; Huppert et al., 1981). In contrast, τ_s in water at 21 °C is less than 0.3 ps from which it was inferred that prexisting deep traps dominate the electron hydration process (Wiesenfeld and Ippen, 1980). Alternative hydration mechanisms have also been proposed (e.g., Huppert et al., 1981) but will not be delineated since the interactions of excess electrons and carcinogens to be described are entirely confined to nonpolar media.

Two decades ago, excess electrons had been observed only in liquid rare gases under cryogenic conditions (Davis et al., 1962; Schnyders et al., 1966: Miller et al., 1968), but evidence obtained from several routes by radiation chemists pointed toward the existence of high mobility charge-carriers in liquid hydrocarbons (reviewed by Warman, 1982). Direct observation of these species was not forthcoming, however, until faster detection techniques and improved methods of purifying liquids were developed.

Tewari and Freeman (1968) were the first to combine a sufficiently fast detection system (1 ms time constant) and sufficiently pure liquids to observe excess electrons in their 1968 pulse-conductivity study. They observed a puzzling "overshoot" only in the radiation-induced conductances of the two most spherically symmetrical molecules studied, which were neohexane and neopentane. This observation was soon followed by two independent reports of accurate u_e measurement of excess electrons in several liquid hydrocarbons (Minday et al., 1969; Schmidt and Allen, 1969) from which it was evident that the overshoot observed by Tewari and Freeman was current induced by excess electrons having mobilities orders of magnitude greater than those of the ions that had been the object of their study. These three studies triggered numerous others that have been extensively reviewed, for example, Jortner and Kestner, 1973; Hummel and Schmidt, 1974; Kevan and Webster, 1976; Allen, 1976a; Schmidt, 1977; Schindewolf, 1978; Yakovlev, 1979; Warman, 1982; Freeman, 1983; Christophorou and Siomos, 1984; Holroyd, 1987.

B. Characterization of Physicochemical Properties of Excess Electrons

Only two physical properties of an excess electron are needed to convey considerable information regarding the medium in which the electron is immersed; these are its drift mobility μ_e and conduction-band energy V_0. As is implied by the term "mobility," μ_e provides a measure of how mobile an electron is in a liquid and, therefore, how strongly it interacts with the solvent molecules. V_0 also is a probe of electron–solvent interaction and can be viewed as a measure of the balance between short-range repulsion and long-range attraction energies (Jortner, 1972). Table 1 shows the good correlation between μ_e and V_0 that is expected. This correlation has been discussed in most of the reviews that conclude the preceding section. Further discussion of the relation of V_0 and μ_e and the structure of the medium is deferred to Section IV,A.

To those unfamiliar with electron or ion mobilities, the units conventionally used, cm²/Vs, may seem peculiar; however, if the charge-carrier mobility is thought of as the drift velocity in cm/s that the charged species attains in an electric field E measured in V/cm, then

$$\mu \ (\text{cm}^2/\text{Vs}) = \frac{v_d \ (\text{cm/s})}{E \ (\text{V/cm})} \tag{1}$$

where v_d is the drift velocity. Thus, the slope of a plot of v_d vs. E yields μ, and typical plots of this type are illustrated in Fig. 2 for electrons in the liquids tetramethylsilane [TMS; chemical formula, $Si(CH_3)_4$], isooctane (i-C_8), cyclo-

TABLE 1

Physical and Chemical Properties of Excess Electrons in Liquids at 21 ± 1 °C

Liquid	μ_e $(\text{cm}^2/\text{Vs})^a$	V_0 $(\text{eV})^a$	$k_e \times 10^{-12} \ (\text{M}^{-1}\text{s}^{-1})$ CCl$_4$	C$_2$H$_5$Br
n-Hexane	0.093	+0.1	1.2[b]	1.5[c]
c-Hexane	0.24	+0.01	3.0[b]	2.0[d]
i-Octane	5.6	−0.24	6.6[e]	5.5[f]
TMS	100	−0.55	52[g]	0.058[g]

[a]Holroyd, 1987.
[b]Bakale et al., 1977.
[c]Allen and Holroyd, 1974.
[d]Allen et al., 1975.
[e]Bakale et al., 1982.
[f]Beck, 1983.
[g]Bakale and Beck, 1986.

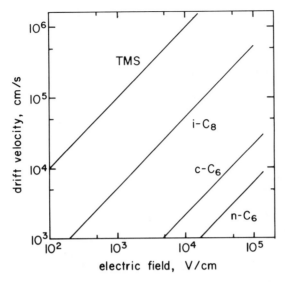

FIG. 2. Plot of electron drift velocity vs. electric field in the liquids tetramethylsilane (TMS), isooctane (i-C_8), cyclohexane (c-C_6), and normal hexane (n-C_6) at 21 °C. Slope of each plot is electron mobility.

hexane (c-C_6), and n-hexane (n-C_6). From Fig. 2 it is evident that a field of ~50 kV/cm is required in c-C_6 for electrons to reach the same v_d as that of electrons in TMS at only 100 V/cm. Nonproportional field dependences of v_d have been observed in many liquids (e.g., Bakale and Schmidt, 1973; Sowada et al., 1976) but are not pertinent to the carcinogen-screening results that will be discussed.

The v_d's plotted in Fig. 2 can be measured by several methods (Hummel and Schmidt, 1974; Schmidt, 1977; Warman, 1982), the most straightforward of which is a time-of-flight technique in which the time required for electrons to drift a known interelectrode distance in a known electric field is measured. A block diagram of a generic DC pulse-conductivity arrangement for measurements of either v_d or k_e is shown in Fig. 3a. A single pulse of ionizing radiation produces a homogeneous distribution of electrons and ions between two parallel electrodes separated a distance d in an ion chamber that contains the liquid or solution of interest. The radiolytic electrons are thermalized in any molecular liquid in ≪1 ns (Warman, 1982) and drift to the anode in the field across the electrodes provided by the voltage supply. As the electrons drift to the anode, their sibling cations drift to the cathode and each charge carrier induces a current, i_e or i_+, respectively, that is observed with the oscilloscope. The contribution of each charge carrier to the measured ion current i_m is proportional to

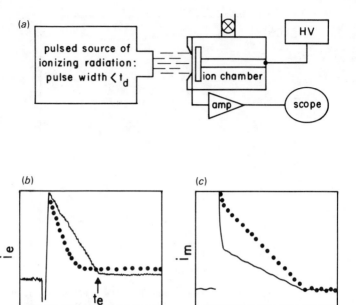

FIG. 3. (*a*) Block diagram of typical pulse-conductivity system with which μ_e's and k_e's are measured. (*b*) Oscillogram of trace of electron-current decay, i_e, in a pure liquid (solid line) and with an electron-attaching impurity present (dotted line); (*c*) Oscillogram of ion-current decay in the pure liquid of Fig. 3*b* (solid line) and with the same electron-attaching impurity present (dotted line).

the product of their concentration, n_e or n_+, and mobility, μ_e or μ_+ (Bakale et al., 1972). Since equal numbers of electrons and cations are produced and $\mu_e \gg \mu_+$ for all of the liquids to be discussed (Allen et al., 1976), $i_e \gg i_+$ until n_e is depleted to $n_e \ll n_+$ by neutralization at the anode.

The oscillogram in Fig. 3*b* illustrates that in a pure liquid the observed current is comprised of two linear segments; the initial faster decay is i_e which intersects the slower decaying i_+ at t_e. The equations governing these linear decays of each component are (Bakale and Schmidt, 1973):

$$i_e(t) = i_e(0) \left(1 - \frac{t}{t_e} \right) \tag{2a}$$

$$i_+(t) = i_+(0) \left(1 - \frac{t}{t_+} \right) \tag{2b}$$

At time $t = t_e$, all electrons have drifted to the anode, and v_d of the electrons $v_d(e)$ is given by

$$v_d = \frac{d}{t_e} \qquad (3)$$

An analogous equation applies to the linear decay of i_+ from which μ_+ can be extracted. The enhancement of i_+ in Fig. 3c compared to 3b is effected simply by increasing the terminating resistance R from 50Ω which is commonly used for t_e measurements at submicrosecond times to 5–500 kΩ for millisecond t_+ measurements. The slower response time of the detection system with the higher R can be tolerated for the slower decay of i_+.

The ideal linear decays of i_e and i_+ illustrated in Fig. 3 are readily obtainable provided that several conditions that govern the modes of electron decay other than drift to the anode are judiciously controlled. The first of these alternative modes is volume recombination, which is the neutralization of an electron and ion that have each escaped the coulombic field of their sibling charge. (Geminate recombination, the neutralization of a sibling electron-ion pair, is complete in $\ll 1$ ns in all liquids to be discussed and therefore does not contribute to the current decay that is observed [Warman, 1982; Holroyd, 1987]). The electron–ion recombination rate constant k_r increases linearly with μ_e as is expected from the Debye equation:

$$k_r = \frac{4\pi T}{\epsilon} \mu_e \qquad (4)$$

where ϵ is the dielectric constant of the liquid (Tezuka et al., 1983). The value of k_r measured in TMS is 5×10^{16} M^{-1}s^{-1} (Allen and Holroyd, 1974) which approaches the value expected from Eq. (4) and implies that the concentration of electrons and ions must be sub-nM in TMS if their recombination is to be negligible in the nanosecond time regime (Warman, 1982).

The second factor that can influence the decay of i_e and i_+ is comprised of three components related to E: space-charge, electrode-edge, and stray-current effects. The latter two can be controlled by carefully designing and shielding the ion chamber, respectively, but the first is inherent to all conductivity measurements and involves the perturbation of E by the fields of the charges as the electrons and ions separate in their drift toward the oppositely charged electrode. As is the case for electron-ion recombination, space-charge effects can be minimized by reducing n_e and n_+ (Boag, 1963; Gregg and Bakale, 1970).

The final factor that affects the decay of i_e and i_+ is the attachment of excess electrons to electron-accepting impurities. This is the same process that is involved in the measurement of k_e, which is the *chemical* property that character-izes excess electrons. The values of k_e for CCl_4 that are listed in Table 1 can be used to estimate the levels of impurities that can be tolerated in μ_e and μ_+ measurements. These k_e's were measured with known concentrations of the CCl_4 solute [S], which combined with the measured half-life of i_e, $t_{1/2}$, yields k_e (Bakale et al., 1972, 1975):

$$k_e = \frac{\ln 2}{[S]t_{1/2}} \tag{5}$$

The nonlinear i_e (dotted line) in Fig. 3b was calculated for the [S] needed to attach electrons to yield $t_{1/2}$ with respect to attachment of $3\mu s$. For the four solvents listed in Table 1, these values of [S] in units of nanomoles/liter and parts per billion (ppb) by weight are: *n*-hexane, 190/44; *c*-hexane, 77/15; *i*-octane, 35/7.8; and TMS, 4.4/1.1. The decay of i_e in Fig. 3b illustrates that t_e could not be accurately estimated with these levels of impurities present in the liquids and if an estimate of t_e were attempted from the nonlinear i_e, t_e would be underestimated by ~50% and the μ_e calculated would be overestimated by the same factor.

The presence of nanomolar quantities of electron-attaching solutes in these liquids also influences i_m, which is manifested by an approximate doubling that is more evident in Fig 3c than 3b. This increase is due to the conversion of excess electrons that would have drifted to the anode in a pure liquid to anions that contribute to the slower decaying ionic component in the impure liquid (Bakale et al., 1972). This doubling of the ion current is the basis of an electronic carcinogen monitor that is currently under development and is briefly described in Section VI.

A final note on the chemical reactivity of excess electrons concerns the relationship between liquid-phase k_e's and gas-phase electron affinities *EA*s. This subject has been discussed by Christophorou and Hunter (1984) who cautioned that the effects of the medium strongly influence k_e and that gas-phase properties should not be indiscriminantly extrapolated to the liquid phase. Similar conclusions had been reached by others, which is evident in several reviews (Warman, 1982; Christophorou and Siomos, 1984; Holroyd, 1987). To illustrate that k_e's are not correlated with *EA*s, the k_e's of 17 nitrobenzene derivatives measured in *c*-hexane (Bakale et al., 1977) are plotted in Fig. 4 vs. the relative gas phase *EA*s recently measured using a pulsed ion cyclotron resonance technique (Fakuda and McIver, 1985). It is evident that the k_e's and *EA*s are not correlated.

This introduction to the physicochemical properties of excess electrons can be summarized by the following: (1) μ_e is a readily obtainable measure of the

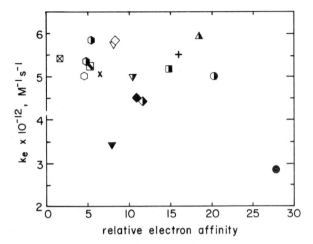

FIG. 4. Plot of k_e's measured in cyclohexane at 21 °C vs. relative gas-phase electron affinities of nitrobenzene and 16 derivatives; k_e's from Bakale et al. (1977) and electron affinities from Fukuda and McKiver (1985) (see either reference to identify chemicals).

electron's interaction with molecules of the surrounding medium; (2) k_e is also readily obtainable and provides a kinetic measure of the electron-accepting capacity of solutes dissolved in that medium; (3) conversion of high μ_e electrons to low μ_- anions via electron attachment to solutes is manifested by marked effects on i_e and i_m; and (4) liquid-phase k_e's are not correlated with gas-phase EAs.

C. Pulse-Conductivity Measurements of k_e

Although electron attachment to adventitious impurities was alluded to in the preceding section, the actual measurement of k_e's was not discussed. The pulse-conductivity system used in the bulk of our studies of electron attachment to carcinogens and noncarcinogens as well as sample preparation protocols are described in detail in our most recent work (Bakale and McCreary, 1987). Extracting k_e's from $t_{1/2}$'s measured from oscillograms of actual i_e decays, how-ever, was not discussed in detail, and these aspects of evaluating k_e's are de-scribed here for "pure" c-hexane and c-hexane containing phenanthrene at concentrations of 0.95 and 2.9 μM.

The decay of i_e's are shown in Fig. 5 for three samples that were irradiated under similar conditions; i.e., a 15-ns pulse of 1 MeV electrons deposited a dose of 1.5 rads in the liquids at 21 °C in a parallel-plate ion chamber having a 0.6-mm interelectrode distance across which 1800 V was applied. The high-energy electrons produced a concentration of secondary electrons of ~0.5 nM

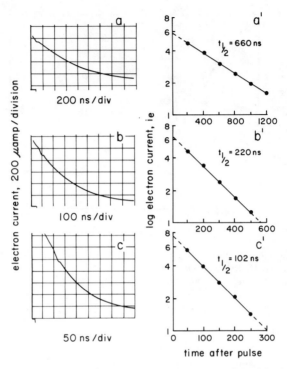

FIG. 5. Oscillograms of decay of electron current and first-order plots of the decay in: (*a, a'*) "pure" cyclohexane; (*b, b'*) 0.95 μM phenanthrene in cyclohexane; and (*c, c'*) 2.9 μM phenanthrene in cyclohexane.

which drifted in the field to produce the observed i_e's. In the "pure" *c*-hexane of Fig. 5*a*, the electrons attached to impurities with a $t_{1/2}$ of 660 ns; impurities having a k_e of 3×10^{12} $M^{-1}s^{-1}$ would have produced the observed $t_{1/2}$ at a concentration of 0.35 μM. Since this concentration of impurities was 700-fold greater than that of the electrons, pseudo-first-order kinetics should have governed the decay of i_e; the plot of log i_e vs. t in Fig. 5*a'* demonstrates that the decay was exponential, and this plot was used to obtain the $t_{1/2}$ of 660 ns. Electron decay via drift to the anode was $< 15\%$ based on $t_e = 4.5$ μs in the field of 30 kV/cm. Also, electron decay via electron-ion recombination was $< 4\%$.

The decays of i_e's in the 0.95 and 2.9 μM phenanthrene solutions are shown in Figs. 5*b* and 5*c*, respectively, and plots of log i_e vs. t in Figs. 5*b'* and 5*c'* are linear and yield respective $t_{1/2}$'s of 220 and 102 ns. Since electron attachment to impurities in the *c*-hexane contributes significantly to the decays of i_e observed in the phenanthrene solutions, Eq. (5) must be modified to account for this additional electron-decay mode in order to obtain the k_e of phenanthrene. This can be done simply by subtracting the impurity contribution, i.e.,

$$k_e = \frac{k_{obs} - k_0}{[\text{phenanthrene}]} \tag{6}$$

where k_{obs} is ln 2 divided by the observed $t_{1/2}$ in the phenanthrene solution, and k_0 is ln 2 divided by the $t_{1/2}$ observed in "pure" cyclohexane. Substituting numerical values into Eq. (6) for the example given yields, for the lower concentration of phenanthrene,

$$k_e = \frac{\ln 2[1/(220 \times 10^{-9}\text{s}) - 1/(660 \times 10^{-9}\text{s})]}{0.95 \times 10^{-6} \text{ M}} = 2.2 \times 10^{12} \text{ M}^{-1}\text{s}^{-1}$$

and for the higher concentration,

$$k_e = \frac{\ln 2[1/(102 \times 10^{-9}\text{s}) - 1/(660 \times 10^{-9}\text{s})]}{2.9 \times 10^{-6} \text{ M}} = 2.0 \times 10^{12} \text{ M}^{-1}\text{s}^{-1}$$

Thus, the k_e of phenanthrene in c-hexane at 21 °C is $2.1 \pm 0.1 \times 10^{12} \text{ M}^{-1}\text{s}^{-1}$, which was confirmed by analogous measurements conducted on a different day using a different phenanthrene solution that yielded a k_e of $2.0 \pm 0.1 \times 10^{12} \text{ M}^{-1}\text{s}^{-1}$.

The criterion that we use to determine whether a measured k_e is a positive or negative indication of carcinogenicity is the diffusion-controlled rate constant k_d; $k_e > k_d$ indicates that electron attachment occurs at every electron–solute encounter, which indicates that the solute is an electrophile and consequently is presumed to be a carcinogen (Bakale and McCreary, 1987). A value of $k_e < k_d$ indicates that electron attachment to the solutes does not occur at every encounter, indicating that an energy barrier to attachment exists and that the solute is not an electrophile. From this it is concluded that the solute is a noncarcinogen. Factors affecting both k_e and k_d and the basis for setting k_d at $3.0 \times 10^{12} \text{ M}^{-1}\text{s}^{-1}$ in c-hexane at 21 °C are discussed in Section III,D. Comparison of this value of k_d with the measured k_e of phenanthrene indicates that phenanthrene yields a negative response in the k_e test and is therefore predicted to be a noncarcinogen. Comparison of this test result with those of other short-term bioassays is deferred to Section V.

A concluding note on k_e measurements concerns other methods that are sometimes used to measure k_e's. Techniques other than pulse radiolysis such as laser photoionization (Beck and Thomas, 1972) may be used to generate thermal electrons whose attachment rate may be monitored by optical (Baxendale and Rasburn, 1974) or microwave (Warman et al., 1977) detection systems. Electron attachment has been studied in time regimes ranging from microseconds (Allen and Holroyd, 1974) to picoseconds (Beck, 1983a) in solutes that include multimolecular aggregates (Bakale et al., 1981). The same general principles as those outlined in the preceding discussion, however, still apply.

D. Influence of Physical Factors on k_e

From the values of μ_e and k_e listed in Table 1, it is evident that the transport and reaction properties of excess electrons depend strongly on the liquid in which electrons are immersed. Also, the k_e of CCl_4 appears to be correlated with the μ_e of the solvent in which attachment occurs, whereas the k_e of ethyl bromide (EtBr) is clearly not similarly correlated. Numerous studies of relationship between k_e and μ_e as well as the dependence of both properties on V_0 have contributed to a better understanding of the k_e-μ_e-V_0 interrelationship which is discussed in several reviews (Allen, 1976a, 1976b; Schmidt, 1977; Schindewolf, 1978; Yakovlev, 1979; Warman, 1982; Christophorou and Siomos, 1984; Holroyd, 1987). The effects of the solvent on μ_e, V_0, and k_e will now be summarized and this is followed by a discussion of k_d and the secondary physical factors that influence the transport, thermodynamic and reactivity properties of excess electrons in liquids.

Prior to Tewari and Freeman's observation of an overshoot in the conductivity currents measured in neohexane and neopentane (Tewari and Freeman, 1968) which was noted previously, studies of the free-ion yields or G_{fi}'s of liquids indicated that the number of electrons that escape their parent cation was markedly dependent on the structure of the liquid (reviewed by Allen, 1976b). This effect of structure or the shape of the solvent molecules was later found to markedly influence μ_e with the most spherical molecules such as methane, neopentane, neohexane, and TMS exhibiting the highest μ_e's (Schmidt and Allen, 1970; Dodelet and Freeman, 1972).

The most striking effect of the molecular shape of the solvent influencing μ_e and k_e is for the alkanes CH_4 and C_2H_6 which differ only by one methyl group but have extremely different shapes and μ_e's as well as k_e's. At $-160\,°C$, the μ_e of CH_4 is 400 cm^2/Vs (Bakale and Schmidt, 1973) whereas the μ_e of C_2H_6 is 1.3×10^{-3}cm^2/Vs (Schmidt et al., 1974), and the k_e's of the classical electron scavenger SF_6 in CH_4 and C_2H_6 are 8×10^{14} and 9×10^{10} $M^{-1}s^{-1}$, respectively (Bakale et al., 1975). These shape-related differences also extend to V_0 but a direct comparison of the two liquids at the same temperature is unavailable.

The shape of the solvent molecule that influences μ_e, V_0, and k_e also governs the electron transport mechanism in that liquid. In liquids having spherical molecules and μ_e's of 70–400 cm^2/Vs, the electron is delocalized or quasifree and the scattering of this electron cloud by the solvent molecules controls μ_e. In contrast, in liquids having asymmetric molecules and μ_e's < 1 cm^2/Vs, electron transport is assumed to occur via either a hopping or a trap-to-trap mechanism (see reviews cited previously). Which of these two transport mechanisms applies has not been clearly resolved nor has the mechanism of electron transport in liquids having μ_e's in the intermediate range between 1 and 70 cm^2/Vs been defined, but Jortner and Gaathon (1977) have attempted to categorize electron transport mechanisms in liquids over the range of μ_e's from 10^{-3} to > 100 cm^2/Vs.

From the strong dependence of μ_e on molecular shape, it is obvious that electron diffusion and, therefore, k_d will similarly depend on the shape of the solvent molecules. The relationship between μ_e and the electron diffusion coefficient D_e is given by the Nernst-Einstein equation, which is

$$D_e = \frac{\mu_e kT}{e} \qquad (7)$$

where k is the Boltzman constant, T is the absolute temperature, and e the charge of an electron. At 21 °C, D_e (cm^2/s) $= 0.025$ μ_e (cm^2/Vs).

The D_e's thus obtained can be used to calculate k_d's via the Smoluchowski equation, which in its most elementary form is

$$k_d = 4\pi R D_e \qquad (8)$$

where the remaining unknown variable R is the effective encounter radius of the reactants. In Eq. (8) the assumption was made that D_e is much greater than the diffusion coefficient of the electron-accepting solute which should be valid for all solvents and solutes that are discussed.

Evaluation of R requires that the electron–solute interaction potential be known; consequently, the most straightforward evaluation of R is for solutes for which the interaction energy is least (Bakale et al., 1977; Baird, 1977). Such a solute is CCl$_4$, which has a dipole moment of zero and a polarizability of 10.5 A^3 that yields a lower limit of R of 6.2 A in c-hexane (Bakale et al., 1977). This value of R and $D_e = 5.5 \times 10^{-3}$ cm^2/s substituted into Eq. (8) yields $k_d \geq 4.3 \times 10^{-9}$ cm^3/s or 2.6×10^{12} M^{-1}s^{-1} which is consistent with the measured k_e of CCl$_4$ of 3.0×10^{12} M^{-1}s^{-1}. The latter value is used as the boundary between a diffusion controlled and a nondiffusion controlled k_e, but it is emphasized that this k_e is only a lower limit of k_d since any interaction potentials between the electron and solute will be attractive and therefore increase R (Baird, 1977). The dynamics of electron–solute interactions are not sufficiently known to the point where R can be evaluated accurately and empirically determined R's can be interpreted differently (Yakovlev, 1979; Bakale and

Schmidt, 1981; Warman, 1982; Holroyd, 1987). Consequently, equating k_d with the k_e of CCl_4 is an arbitrary but pragmatic means of setting the boundary between the k_e's of electrophiles and nonelectrophiles.

An additional factor that influences k_e involves the energetics of the attachment process itself in which a thermal electron with potential energy V_0 attaches to a solute to form an anion with a polarization energy P^-, which in the nonpolar liquids being discussed is approximately -1 eV. The energetics of the attachment process has been treated by different approaches (Allen and Holroyd, 1974; Allen et al., 1975; Henglein, 1975, 1977; Funabashi and Magee, 1975; Christophorou, 1976), all of which have been thoroughly reviewed (Schindewolf, 1978; Yakovlev, 1979; Warman, 1982; Christophorou and Siomos, 1984; Holroyd, 1977, 1987). From such studies it appears that inferences from the gas-phase energy dependence of electron attachment can be cautiously made to liquid-phase k_e's which is illustrated by energy-dependent k_e's measured in liquid argon and xenon (Bakale et al., 1976) and recently in TMS (Bakale and Beck, 1986). The relevance of the influence of electron attachment energetics to screening chemical carcinogens with the k_e test is discussed in the next section.

IV. THEORETICAL BASIS OF THE k_e TEST

A. Electron Transfer at the Carcinogen-Target Interface

From the preceding discussions, it is clear that the structure of the medium that surrounds an excess electron intimately influences the electron's reactivity. In order to consider the physicochemical basis for the k_e-carcinogenicity correlation, which is the basis of the k_e test, it is necessary to consider first the nature of the medium in the microenvironment of the biomolecular target and an approaching electrophilic carcinogen and then to consider how this medium influences the target-electrophile interaction. Since the first point has been the subject of debate for more than half a century (Ling, 1972), no definitive conclusions to the second point can be expected. However, some new insight concerning both of these points has come from observations recently made in several disciplines. These new results will be outlined to demonstrate that k_e may serve as a measure of the capacity of a chemical to accept an electron from an electron-rich biomolecule which is proposed to be involved in the formation of a carcinogen-target adduct.

The literature on water in biological systems is replete with studies that have attempted to define to what degree cellular water is associated with biomolecules (Ling, 1972; Franks, 1979; Drost-Hansen and Clegg, 1979; Clegg, 1984), but perhaps the statement by Ma et al. (1981) most succinctly summarizes these studies, "Living cells are about three quarters water, and yet a person walking around does not slosh audibly, so the water must be integrated into the membrane in a network that has solidlike properties."

A more explicit summary of biologically associated water as well as bulk water was given by Pethig (1979b) who noted significantly different results obtained in nuclear magnetic resonance (NMR) studies. From subsequent NMR studies, the consensus appears to be that one or two layers of water associated with biomolecules have rotation and diffusion times about 5-10 times greater than that of bulk water (Lenk et al., 1980; Halle et al., 1981; Mank and Lebovka, 1983; Polnaszek and Bryant, 1984). In more recent microwave absorption studies by Van Zandt (1986), the first layer of biologically associated water was found to have a coupling time of 40 ps which also appears to be consistent with the NMR studies. A similar conclusion regarding water bound to DNA was also reached by a third independent route, viz. via the expert *ab initio* computations of Clementi (1985), who also noted that "the modeling of the interaction of a molecule Y intercalating DNA must consider DNA and counterions and the solvent."

This last point is particularly pertinent to the interaction of an electrophile and a target biomolecule encased in a sheath of bound water. Such water should appear to be "frozen" on the time scale at which an electron would probe its structure, which concerns the second point of discussion in this section. As was stressed in the preceding section, the μ_e of an excess electron is markedly dependent on the medium in which it diffuses. Consequently, if the sheath of water surrounding a biomolecule is highly structured or "ice-like," the μ_e of an electron in this sheath should be of the same order as the μ_e in ice. Although numerous attempts have been made to obtain an accurate estimate of μ_e in biomolecules or hydrated biomolecules (reviewed by Pethig, 1979c), only recently has a reliable value of the μ_e in hydrated DNA and collagen been measured (van Lith et al., 1986). The technique used in these measurements, pulse radiolysis with microwave-conductivity detection, does not permit the charge carrier to be identified unambiguously. However, from their results, van Lith et al. inferred that "a highly mobile 'dry' electron" migrates in the "ice-like water layer around the biopolymer" with a μ_e approximately that of an electron in ice, which is 25 cm^2/Vs (de Haas et al., 1983).

Although the μ_e's measured by van Lith et al. imply that excess electrons are highly mobile in the structured water vicinal to DNA and collagen, it must be stressed that this structure at the microscopic level is superimposed on macrostructured biomolecules that comprise the highly compartmentalized eukaryotic cells. Exterior to the well-defined cell membrane is the extracellular matrix which is known to have a definite influence on metabolism (Hay, 1983). Inside the cell is the cytoplasm, which is no longer considered to be only a proteinaceous soup but instead houses a highly structured cytoskeleton of microtubules, microfilaments, and intermediate filaments throughout which a web of the microtrabecular lattice is interwoven (Porter et al., 1983). The surface area of the cytoplasmic matrix is 80,000 μm^2 for a cell having a 16-μm diameter and a surface area of only 800 μm^2 (Gershon et al., 1985), which indicates that a significant quantity of water should be bound to the matrix elements in the cytoplasm.

All of these extranuclear structures, however, are primitive when compared to the multileveled structure of chromatin of the nucleus in which elegantly structured DNA helices coil into 10-nm "beads on a string" that wind into 30-nm solenoids (Pienta and Coffey, 1985). Pienta and Coffey propose that, in the human chromosome number 4, strings of these solenoids form loops having a length of 250 nm, and 18 of these loops assemble around the nuclear matrix to form a miniband with a diameter of 840 nm; there minibands then stack to form the chromosome. Nicolini (1986) recently described another model of chromatin organization in which organization culminates at the quintinary level to "drapery-like" structures.

What special electron-transport properties any of these structured cellular components have is unknown as is what role, if any, electrons play in processing and controlling genetic information. It appears, however, that these cellular components should have unique electron transport properties and that Nature did not invest in this architecture for no reason. It is known that gene transcription can be controlled by distant regulatory proteins, although the mechanism by which control occurs is not known (Ptashne, 1986). Also, the nonmutagenic carcinogen diethylstilbestrol, which yields a positive k_e response (Bakale et al., 1981b), interacts with cytoplasmic microtubules and thereby disrupts gene expression and leads to neoplastic transformation (Tucker and Barrett, 1986). Finally, Nicolini (1986) has discussed several examples of the influence of chromatin structure on gene expression in which a role for electron transfer could be inferred but the state of the art of the several disciplines involved precludes its observation.

B. Rational of Positive k_e Responses to Procarcinogens

As was described in Section II,A, many chemical carcinogens require metabolic activation to be sufficiently electrophilic to interact with a target nucleophile and cause a mutation that is proposed to be the potential initiating step in carcinogenesis (Miller and Miller, 1971; E. C. Miller, 1978; J. A. Miller, 1970). Such chemicals are classified as procarcinogens and their electrophilic metabolites as ultimate carcinogens. In the Ames bioassay, this metabolic conversion of procarcinogens to ultimate carcinogens is generally effected by rat liver microsomes (Ames, 1972; Ames, et al., 1973) although numerous other microsomal activation systems have been used (see, for example Cheh et al., 1980, and Raineri et al., 1981). The response of the tester strains to a test chemical is strongly dependent on the activating system used as well as the several variables associated with activation which has been viewed as one of the shortcomings of the *Salmonella* assay (Ashby and Styles, 1978).

In the carcinogen screening studies that we have conducted using the k_e test, no attempt was made to activate metabolically or chemically procarcinogens to their more electrophilic derivatives, the ultimate carcinogens (Bakale et al. 1981b, 1982; Bakale and McCreary, 1987); however, positive k_e responses were observed for most procarcinogens. A brief rationale of this apparent paradox was offered in our 1987 publication which will now be amplified since it is pertinent to the theoretical basis of the k_e test.

The effect of solvent structure on the k_e of a solute was delineated in Section III,D, and reference was made to the dependence of the k_e of a solute on the V_0 of a solvent. This k_e-V_0 dependence is itself dependent on the distribution of unoccupied electronic orbitals, D, that are available in the solute as a function of V_0. This nomenclature was used by Henglein in describing the k_e-V_0 dependence, or $k_e(V_0)$ (Henglein, 1975, 1977). The same concept has also been described in gas-phase electron physics terms by Christophorou (1976) using the dependence of the electron attachment cross section σ_e on the electron energy ϵ_e. The nomenclature of Henglein, which is more appropriate for liquid-phase electron transfer processes, was used in Fig. 6.

The two $D(V_0)$ curves sketched in Fig. 6 are for a hypothetical procarcinogen PC and its activated ultimate carcinogenic form, UC. With all other physical factors that influence k_e being equal, the maximum of the k_e-V_0 dependence would be at $V_0 = -0.15$ eV for PC and at -0.5 eV for UC. Consequently, a positive k_e response would be expected for PC in isooctane as well as in cyclohexane since a significant number of unoccupied orbitals are available in PC that favorably match the energy of excess electrons in either solution. In *n*-hexane and TMS, however, the k_e of PC may be significantly less than k_d. In addition, only in TMS is UC expected to yield a k_e that approaches k_d. The same model implies that if the $D(V_0)$ curve for the "hot spots" of an Ames tester strain is a maximum at -0.5 eV, a negative Ames test response would be observed for PC whereas a positive response would be found for UC. This may be restated more simply as "the excess electron in cyclohexane (or isooctane) is a better nucleophile than is the DNA of the Ames histidine-deficient auxotrophs" (Bakale and McCreary, 1987). Thus, the positive response to procarcinogens observed with the k_e test should not be regarded as being inconsistent with the somatic mutation theory of carcinogenesis.

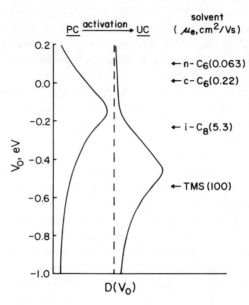

FIG. 6. Sketch of proposed change in the energy distribution of unoccupied electron orbitals in a procarcinogen before and after activation to an ultimate carcinogen. Conduction-band energies, V_0's, of n-hexane, c-hexane, i-octane, and TMS are indicated to provide frame of reference to ordinate.

V. VALIDATION STUDIES OF THE k_e TEST

In their assessment of the developmental status of the more than 100 short-term tests used to screen carcinogens, Brusick and Auletta (1985) used four criteria to classify the tests as experimental, transitional, or routine. Only 10 bioassays were classified as routine, and of these, the Ames test was by far the most widely used with more than 2500 tests being conducted in 46 laboratories in 1982. Applying the same criteria to the k_e test places its developmental stage at the transitional level, which was the approximate stage of development of the Ames test 15 years ago. During that period, the Ames test has been validated in several major studies which include the initial study of 300 chemicals in Ames' laboratory (McCann et al. 1975) and those of Sugimura et al. (1976), Purchase et al. (1978), Rosenkranz and Poirier (1979), Simmon (1979), Bartsch et al. (1980a,b), Kawachi et al. (1980), and DeFlora et al. (1984). For comparison, the k_e test has been limited to the screening of chemicals conducted only in our laboratory, the results of which were reported in Bakale et al. (1981b, 1982,

1983), Bakale and McCreary (1984), Bakale (1986), and Bakale and McCreary (1987).

In all of the k_e studies, results of the k_e responses R to test chemicals were compared with Ames-test mutagenicity responses, M. A similar comparison is again made between R and M in Table 2 for chemicals that we had reported in the earlier studies, and for which the classifications of carcinogenicity were recently reported in the U.S. Environmental Protection Agency Gene-Tox carcinogen data base (Nesnow et al., 1986), which are also included. In addition, the Ames test responses are not single-laboratory results such as those we had used in earlier comparisons of k_e with M but are the Ames test results reported in the survey of Palajda and Rosenkranz (1985). Thus, the results in Table 2 should be the least subjective comparison of R and M with test chemical carcinogenicity C that is currently possible.

Some explanation of the nomenclature used in Table 2 in addition to that provided in the legend seems warranted. The k_e responses of " + " and " − " indicate $k_e \geq k_d$ and $k_e < k_d$, respectively, in the studies listed in the column of references. Since k_e's that we measured in cyclohexane are more reliable than those measured in isooctane for reasons that were recently discussed (Bakale and McCreary, 1987), the k_e's measured in cyclohexane supercede those in isooctane where different R's were observed. The Ames test M's reported by Palajda and Rosenkranz (1985) are described in the legend of Table 2. Details of the criteria used by these authors in classifying M in these categories are presented in their survey. The classifications of C in the Gene-Tox data base are divided into eight categories using a modification of nomenclature used by the International Agency for Research on Cancer (see, for example, IARC, 1986); these categories are based on evidence that is a sufficient (S) or a limited (L) indication of a positive (P) or negative (N) carcinogenic response in test animals, which yield SP, SN, LP, or LN classifications of C. In addition to these four unequivocal classifications, evidence may be equivocal (E) or inadequate (I) and the latter are sometimes subdivided into evidence that suggests a positive, I(P), or a negative, I(N), indication of carcinogenicity. Additional details of the classification scheme are given in the Gene-Tox report (Nesnow et al., 1986).

Four criteria that are generally used in evaluating the predictivity of a short-term screening test for carcinogens are sensitivity, specificity, accuracy, and predictive value. These criteria have been discussed extensively by others (Cooper et al., 1979; Bartsch et al., 1980a; Purchase, 1982; Palajda and Rosenkranz, 1985) and are summarized in Table 3 in which evaluations of the four criteria for the k_e and Ames test results of Table 2 are also included.

Before any significance to the relative performances of the two tests is implied by the results in Table 3, it is stressed that the 62 chemicals having sufficient, limited, or suggested evidence of carcinogenicity or noncarcino-

TABLE 2

Comparison of k_e Test Response (R) and Ames Test Mutagenicity (M)
with Animal Test Evidence of Carcinogenicity (C) of 70 Chemicals
for which k_e and C Data Have Been Reported

Chemical[a]	CAS no.[b]	R[c]	Ref.[d]	M[e]	C[f]
Acetamide	60-35-5	−	1	−	SP
2-Acetylaminofluorene	53-96-3	+	4	+	SP
4-Aminoazobenzene	60-09-3	+	1, 2, 4	+	LP
o-Aminoazotoluene	97-56-3	+	1, 2, 4	+	SP
2-Aminobiphenyl	90-41-5	+	1, 4	+	SP
Aniline	62-53-3	−	1	−	I
Anthracene	120-12-7	−	4	−	I(N)
Benz[a]anthracene	56-55-3	+	1	+	SP
Benzo[a]pyrene	50-32-8	+	1, 2, 4	+	SP
Benzo[e]pyrene	192-97-2	+	1, 2, 4	+	I
Benzyl chloride	100-44-7	+	1, 2	?	LP
Butylated hydroxyanisole	25013-16-5	−	4	−	LP
Captan	133-06-2	+	4	NDA	LP
Carbon tetrachloride	56-23-5	+	1, 2	−	SP
Chlordane	57-74-9	+	3	NDA	LP
Chloroform	67-66-3	+	3	NDA	SP
Chrysene	218-01-9	+	1, 2	+	LP
p,p'-DDE	72-55-9	+	1, 2	−	LP
Diazinon	333-41-5	+	4	−	SN
Dibenz[a,h]anthracene	53-70-3	+	1	+	SP
1,2-Dibromo-3-chloropropane	96-12-8	+	3	+	SP
Dichlorvos	62-73-7	+	4	+	E
Diethylstilbestrol	56-53-1	+	1, 4	−	SP
Dimethoate	60-51-5	+	4	NDA	SN
7,12-Dimethylbenz[a]anthracene	57-97-6	−	4	+	SP
Dimethyl sulfoxide	67-68-5	−	1	−	I(N)
1,2-Epoxybutane	106-88-7	−	1	+	I
Ethanol	64-17-5	−	1, 2	−	LN
Ethylene dibromide	106-93-4	+	1, 2	+	SP
Ethynylestradiol	57-63-6	−	4	−	SP
Griseofulvin	126-07-8	+	3	−	LP
Hexachloroethane	67-72-1	+	3	NDA	LP
γ-Lindane	58-89-9	+	4	NDA	LP
Malathion	121-75-5	−	4	−	SN
Methyl carbamate	598-55-0	−	1	−	I
2-Methyl-4-dimethylaminazobenzene	54-88-6	−	4	NDA	LP
3'-Methyl-4-dimethylaminoazobenzene	55-80-1	−	4	−	SP
Methyl iodide	74-88-4	+	1, 2	NDA	SP
N-Methyl-N'-nitro-N-nitrosoguanidine	70-25-7	+	1	+	SP
Mirex	2385-85-5	+	3	NDA	SP
1-Naphthylamine	134-32-7	+	1, 2, 4	+	LN
2-Naphthylamine	91-59-8	+	1, 2, 4	+	SP

TABLE 2 (Continued)

Chemical[a]	CAS no.[b]	R[c]	Ref.[d]	M[e]	C[f]
4-Nitro-1,1'-biphenyl	92-93-3	+	1, 2	+	LP
Nitrofurantoin	67-20-9	−	4	+	I
2-Nitropropane	79-46-9	+	3	NI	SP
4-Nitroquinoline-N-oxide	56-57-5	+	1, 4	+	SP
N-Nitrosodibutylamine	924-16-3	+	1, 2, 4	NDA	SP
N-Nitrosodiethylamine	55-18-5	+	1, 2	+	SP
N-Nitrosodimethylamine	62-75-9	+	1, 2, 4	+	SP
N-Nitrosodiphenylamine	86-30-6	+	1, 2, 4	−	LP
N-Nitrosodipropylamine	621-64-7	+	1, 2	+	SP
N-Nitrosopiperidine	100-75-4	+	1, 2	+	SP
N-Nitrosopyrrolidine	930-55-2	+	1, 2	+	SP
Phenacetin	62-4-2	−	4	−	SP
Phenanthrene	85-01-8	−	4	−	I
Piperonyl butoxide	51-03-6	−	4	NDA	LN
Progesterone	57-83-0	+	4	−	SP
Pyrene	129-00-0	−	4	−	I
Reserpine	50-55-5	−	4	−	SP
Rhodamine	81-88-9	+	4	NDA	I(P)
Styrene	100-42-5	−	4	−	LP
Succinic anhydride	108-30-5	−	4	−	LP
Testosterone propionate	57-85-2	+	4	NI	SP
Thioacetamide	62-55-5	+	1, 2, 4	−	SP
4,4'-Thiodianiline	139-65-1	+	1, 2, 4	NDA	SP
Toxaphene	8001-35-2	+	3	NDA	SP
Tris(2,3-dibromopropyl)phosphate	126-72-7	+	3, 4	+	SP
Urethane	51-79-6	−	1, 2, 4	−	SP
Vinyl acetate	108-05-4	−	1, 2, 3	−	LP
Vinylidene chloride	75-35-4	+	3	+	LP

[a]Chemical nomenclature in Gene-Tox Data Base (Nesnow et al., 1986).

[b]Chemical Abstracts Service registry number.

[c]$R = $ " + " for $k_e \geq k_d$; $R = $ " − " for $k_e < k_d$.

[d]Reference from which R obtained: 1, Bakale et al. (1981); 2, Bakale et al. (1982); 3, Bakale et al. (1983); 4, Bakale and McCreary (1987).

[e]Ames test mutagenicity reported in survey by Palajda and Rosenkranz (1985): " + ", positive; " − ", negative; "NDA", no data available; "?", questionable. "NI" denotes chemical not included in survey.

[f]Classification of evidence of animal-test carcinogenicity by Nesnow et al. (1986): S (sufficient) or L (limited) evidence of positive (P) or negative (N) carcinogenic response; I (inadequate) evidence that for some chemicals may suggest either a positive, I(P), or negative, I(N), carcinogenic response; E indicates equivocal carcinogenic responses. Additional details given in text and in Nesnow et al. (1986).

TABLE 3

Sensitivity, Specificity, Accuracy, and Predictive Value of k_e and Ames Tests
for Chemicals Listed in Table 2[a]

Predictive criteria: definition	k_e test	Ames test
Sensitivity:		
$\dfrac{\text{Correct "}+\text{"}}{\text{Total carcinogens}} \times 100$	$\dfrac{42}{54} \times 100 = 78\%$	$\dfrac{22}{39} \times 100 = 56\%$
Specificity:		
$\dfrac{\text{Correct "}-\text{"}}{\text{Total noncarcinogens}} \times 100$	$\dfrac{5}{8} \times 100 = 63\%$	$\dfrac{5}{6} \times 100 = 83\%$
Accuracy:		
$\dfrac{\text{Total correct "}+\text{" and "}-\text{"}}{\text{Total chemicals}} \times 100$	$\dfrac{47}{62} \times 100 = 76\%$	$\dfrac{27}{45} \times 100 = 60\%$
Predictive value:		
$\dfrac{\text{Correct "}+\text{"}}{\text{Total "}+\text{"}} \times 100$	$\dfrac{42}{45} \times 100 = 93\%$	$\dfrac{22}{23} \times 100 = 96\%$

[a]Chemicals classified as SP, LP, or I(P) (see Table 2) are considered as carcinogens and those classified as SN, LN, or I(N) are considered as noncarcinogens in the evaluations of the predictive criteria.

genicity is too small of a data base upon which to draw any firm conclusions. This caution is particularly applicable to the eight putative noncarcinogens for which the evidence to be so classified is sufficient for only three, limited for three, and inadequate but suggested for the remaining two. The information summarized in Table 3 is provided as a convenience to those who would otherwise perform a similar evaluation and perhaps draw erroneous conclusions.

The results in Tables 2 and 3 indicate that the k_e's of a larger fraction of the 40 chemicals classified as SN in the 1986 Gene-Tox data base should be measured, and this is currently being done in our laboratory. The primary problem involved in identifying noncarcinogens is, of course, the impossibility of proving a negative (Shubik, 1984; Clayson and Krewski, 1986), which has resulted in a much large proportion of chemicals that have been identified as carcinogens. This greater proportion of carcinogens to noncarcinogens created a problem in designing batteries of short-term tests to predict the carcinogenicity of a chemical (Rosenkranz et al., 1984). However, steps have been taken by the National Toxicology Program to rectify this problem (Huff et al., 1985) which is evident from the animal test results reported in the last several years (Chu et al., 1981; Haseman et al., 1984; see also Shelby and Stasiewicz, 1984).

Another significant conclusion to be drawn from Table 2 concerns the overlap of false negative (FN) responses in the two tests. In our recent study of chemicals tested by Kawachi et al., only 4 of 17 FNs observed in the Ames test were also observed in the k_e test which implied that the k_e and Ames tests would complement each other efficaciously in a short-term test battery (Bakale and McCreary, 1987). The use of such batteries appears to show promise as a means of improving the accuracy of screening carcinogens (Heinz and Poulsen, 1983; Rosenkranz et al., 1984; Chankong et al., 1985; Ennever and Rosenkranz, 1986), and the use of cluster analysis may bring further improvement in predicting chemical carcinogenicity (Benigni and Guiliani, 1985; Pet-Edwards et al., 1985). From the results in Table 2, however, it is seen that 9 of the 12 FNs of the k_e test are also FNs in the Ames test. This leads to an opposite conclusion concerning the use of the two tests in a battery from that which we had drawn earlier. The overlap of FNs, however, adds support to the proposal made in the preceding section that excess electrons and the "hot spots" of the Ames tester strains interact with electrophiles analogously.

The subject of FNs raises the question of whether chemicals classified as having "limited positive evidence" of carcinogenicity in the 1986 Gene-Tox data base be included in analyzing the sensitivity of a short-term test in a validation study. Two chemicals that are particularly in question are styrene and vinyl acetate; the former was classified by Zeiger (1987) as having "equivocal" evidence of carcinogenicity in a recent study, and for the latter, there was concluded to be "inadequate evidence" for carcinogenicity in the IARC's (1986) most recent evaluation. If these two chemicals are deleted from consideration in calculating the sensitivity, the revised value is $42/52 \times 100 = 81\%$ for the k_e test and $22/37 \times 100 = 59\%$ for the Ames test. A third chemical that is negative in both tests and is also classified in the "limited positive" category is butylated hydroxyanisole which will surely be subjected to further evaluation since it is a commonly used food additive. The implication that the sensitivity of the k_e test is 81% is encouraging, particularly if one considers that using the carcinogenicity of chemicals in rats as a "short-term" test for the chemicals' carcinogenicity in mice has a sensitivity of only 83.8% (Purchase, 1980).

The only false positive response observed in the k_e and Ames tests is 1-naphthylamine which is classified in the "limited negative" category in the Gene-Tox data base and was classified in the 1985 Palajda-Rosenkranz survey as a carcinogen. This chemical has received much scrutiny because its analog, 2-naphthylamine, is a well-characterized human carcinogen (Vainio et al., 1985; Wilbourne et al., 1986). Deleting this false positive response from the calculation of the specificity for the k_e test increases its value to 71%, but as noted earlier, more putative noncarcinogens must be tested before any conclusions regarding the specificity of the k_e test are made.

VI. COST EFFECTIVENESS OF THE k_e
AND SHORT-TERM TESTS

One aspect of screening chemical carcinogens that has received increasing attention during the last few years has been the costs and time involved in various testing procedures, which may be at least partially related to the large expenses and the 2–3 years associated with animal testing. For example, Weinstein's 1983 estimate of conducting an animal test was $500,000. In 1986, Lave and Omenn estimated the animal test cost to exceed $1 million.

The expense and time involved with many routine "short-term" bioassays is also significant; some examples of these commercially available tests, their average cost, and the time involved until a report is issued on a per chemical basis are: Ames *Salmonella*/microsome, $1000, 3–8 weeks; sister-chromatid exchange (SCE) in vitro, $3600, 5–8 weeks; micronucleus (mice), $5800, 12–16 weeks; and the dominant lethal (mice), $17,000, 12–15 weeks (Brusick and Auletta, 1985). In our laboratory, the k_e's of two or three chemicals are routinely measured on a daily basis and the bulk of the time is involved with sample preparation, solvent purification, and filling and cleaning the ion chamber with the test solution; the actual measurement time is literally $\ll 1$ s. None of the sample-manipulation procedures has been optimized for maximum efficiency, but the number of samples analyzed could easily be increased by an order of magnitude if accuracy in the value of k_e were sacrificed for a simple plus-or-minus evaluation of R. Placing a cost per chemical on the k_e test is difficult since the down-time of the Van de Graaff generator has varied from 0 to 100% per year over the last several years. Using a pulsed source of ionizing radiation costing $100,000 is also not the most cost-efficient means of measuring k_e's. Portable x-ray sources having pulse widths of ~ 30 ns that deliver sufficient doses are available for less than $15,000, and generating excess electrons by photoionization using a pulsed laser (Beck and Thomas, 1972) is another cost-effective means of measuring k_e's.

The costs and time involved in using bioassays to screen carcinogens have prompted modifications to standard protocols in order to increase sample output. An example of this applied to the Ames test is the study by Zeiger et al. (1985) who recommended a sequential testing scheme in which the first step involves using tester strain TA100. A much more drastic modification to the Ames-test protocol involves the use of a COBAS Bact apparatus to automate the entire testing procedure (Arni et al., 1985).

Although the initial results obtained by Arni et al. appear to be encouraging, electrons are inherently more amenable to automation than are bacteria; consequently, work is in progress to develop a pulse-conductivity detector (PCD) that is compatible with high-performance liquid chromatography (HPLC). Rather than measuring the k_e of a test chemical, the PCD is designed to sample the ion current 1 μs to 1 ms after a pulse of ionizing radiation at a pulse rate of about 20 pulses per second. An increase in the ion current effected by an electron-attaching chemical in the solution flowing through the PCD and analogous to the i_m's illustrated in Fig. 3c can be measured and recorded. The use of the PCD in conjunction with HPLC permits the analysis of complex mixtures of test chemicals which are separated by the HPLC into component eluites that are sampled in the PCD. The constraint of the eluent being a nonpolar solvent restricts analyses to normal-phase HPLC separations, which does not appear to be a serious limitation since normal-phase separations of coal-tar oil into more than 15 resolved PAH's have been conducted in our laboratory.

VII. EPILOGUE

In all of our reports concerning the k_e's of chemical carcinogens, reference was made to a statement by Bridges (1976) in his review of short-term tests for screening chemical carcinogens; "even an empirical 'litmus paper' test, with no known theoretical basis, which gave an 80 to 90% predictiveness for carcinogenicity would be a powerful tool in the screening of chemicals for human toxicity." The k_e test may be the "litmus paper," and the PCD may be the "pH meter" for screening carcinogens, but far more important are the theoretical implications of the k_e-carcinogenicity correlation to the mechanism of the pre-chemical initiating step in the multistep process of carcinogenesis. Elucidation of these implications would require an interdisciplinary effort that could provide an affirmative answer to the question that Bailar and Smith (1986) raised in their study of cancer-mortality in the United States which is entitled "Progress Against Cancer?"

REFERENCES

Adams, G. 1981. Hypoxia-mediated drugs for radiation and chemotherapy. *Cancer* 48:696–707.

Adams, G. E., Stratford, I. J., Wallace, R. G., Wardman, P., and Watts, M. E. 1980. Toxicity of nitro compounds toward hypoxic mammalian cells in vitro: Dependence on reduction potential. *J. Natl. Cancer Inst.* 64:555–560.

Allen, A. O. 1976a. Drift mobilities and conduction band energies of excess electrons in dielectric liquids. *Natl. Stand. Ref. Data. Ser.*, Natl. Bur. Stand., U.S. Dept. Commerce.

Allen, A. O. 1976b. Yields of free ions formed in liquids by radiation. *Natl. Stand. Ref. Data. Ser,* Nat. Bur. Stand., U.S. Dept. Commerce.

Allen, A. O., and Holroyd, R. A. 1974. Chemical reaction rates of quasi free electrons in nonpolar liquids. *J. Phys. Chem.* 78:796–803.

Allen, A. O., Gangwer, T. E., and Holroyd, R. A. 1975. Chemical reaction rates of quasifree electrons in nonpolar liquids. II. *J. Phys. Chem.* 79:25–31.

Allen, A. O., deHaas, M. P., and Hummel, A. 1976. Measurement of ionic mobilities in dielectric liquids by means of concentric cyclindrical electrodes. *J. Chem. Phys.* 64:2587–2592.

Ames, B. N. 1984. Letter. *Science* 224:668–670, 757–760.

Ames, B. N. 1972. A bacterial system for detecting mutagens and carcinogens. In *Mutagenic Effects of Environmental Contaminants*, eds. E. Sutton, M. Harris, pp. 57–66. New York: Academic Press.

Ames, B. N., Durston, W. E., Yamasaki, E., and Lee, F. D. 1973. Carcinogens are mutagens: A simple test system combining liver homogenates for activation and bacteria for detection. *Proc. Natl. Acad. Sci. U.S.A.* 70:2281–2285.

Ames, B. N., McCann, J., and Yamasaki, E. 1975. Methods of detecting carcinogens and mutagens with the *Salmonella*/mammalian-microsome mutagenicity assay. *Mutat. Res.* 31:347–364.

Arni, P., Dollenmeier, P., and Müller, D. 1985. Automated modification of the Ames test with COBAS Bact. *Mutat. Res.* 144:137–140.

Ashby, J. 1978. Structural analysis as a means of predicting carcinogenic potential. *Br. J. Cancer* 37:904–923.

Ashby, J., and Styles, J. A. Factors influencing mutagenic potency in vitro. *Nature* 274:20–22.

Bailar, J. C. III, and Smith, E. M. 1986. Progress against cancer? *New Engl. J. Med.* 314:1226–1232.

Baird, J. K. 1977. Kinetics of electron capture by SF_6 in solution. *Can. J. Chem.* 55:2133–2143.

Bakale, G. 1986. A physico-chemical technique for identifying chemical carcinogens. In *Meeting Report on Eighth Interdisciplinary Cancer Research Workshop*, eds. P. Politzer, C. Parkanyi. *Cancer Res.* 46:1566–1568.

Bakale, G., and Beck, G. 1986. Field-dependent electron attachment in liquid tetramethylsilane. *J. Chem. Phys.* 84:5344–5350.

Bakale, G., and Gregg, E. C. 1978. Conjecture on the role of dry charges in radiosensitization. *Br. J. Cancer Suppl. III* 37:24–28.

Bakale, G., and McCreary, R. D. 1984. A pulse-radiolysis technique for screening carcinogens. In *Book of Abstracts*, 188th ACS National Meeting, abstract Phys-98. Washington, D.C.: American Chemical Society.

Bakale, G., and McCreary, R. D. 1987. A physico-chemical screening test for chemical carcinogens: The k_e test. *Carcinogenesis* 8:253–264.

Bakale, G., and Schmidt, W. F. 1973. Excess electrons and positive charge carriers in liquid methane. I. *Z. Naturforsch.* 28a:511–518.

Bakale, G., and Schmidt, W. F. 1981. Effect of an electric field on electron attachment to SF_6 in liquid ethane and propane. *Z. Naturforsch.* 36a:802–806.

Bakale, G., Gregg, E. C., and McCreary, R. D. 1972. Decay of quasifree electrons in pulse-irradiated liquid hydrocarbons. *J. Chem. Phys.* 57:4246–4254.

Bakale, G., Sowada, U., and Schmidt, W. F. 1975. Electron attachment to sulfur hexafluoride in nonpolar liquids. *J. Phys. Chem.* 79:3041–3044.

Bakale, G., Sowada, U., and Schmidt, W. F. 1976. Effect of an electric field on electron attachment to SF_6, N_2O and O_2 in liquid argon and xenon. *J. Phys. Chem.* 80:2556–2559.

Bakale, G., Gregg, E. C., and McCreary, R. D. 1977. Electron attachment to nitro compounds in liquid cyclohexane. *J. Chem. Phys.* 67:5788–5794.

Bakale, G., Beck, G., and Thomas, J. K. 1981a. Electron capture in water pools of reversed micelles. *J. Phys. Chem.* 85:1062–1064.

Bakale, G., McCreary, R. D., and Gregg, E. C. 1981b. Quasifree electron attachment to carcinogens in liquid cyclohexane. *Cancer Biochem. Biophys.* 5:103–109.

Bakale, G., McCreary, R. D., and Gregg, E. C. 1982. Quasifree electron attachment to carcinogens. *Int. J. Quantum Chem. Quantum Biol. Symp.* 9:15-25.

Bakale, G., McCreary, R. D., and Gregg, E. C. 1983. A pulse-conductivity screening test for carcinogens; 14th Annual Meeting of the Environmental Mutagen Society, San Antonio, TX, March 4-6, 1983. *Environ. Mutagen.* 5:471-472 (abstract).

Barrett, T. W. 1979. A theory of cancer induction by parametric excitation. *Cancer Biochem. Biophys.* 3:189-192.

Bartsch, H., Maleveille, C., Camus, A.-M., Brun, G., and Hautefeuille, A. 1980a. Validity of bacterial short-term tests for the detection of chemical carcinogens. In *Short Term Test Systems for Detecting Carcinogens,* eds. K. H. Norprath, R. C. Garner, pp. 58-73. Berlin: Springer-Verlag.

Bartsch, H., Maleveille, C., Camus, A.-M., Martel-Planche, G., Brun, G., Hautefeuille, A., Sabadie, N., Barbin, A., Kuroki, T., Drevon, C., Piccoli, C., and Montesano, R. 1980b. Validation and comparative studies on 180 chemicals with *S. typhimurium* strains and V79 Chinese hamster cells in the presence of various metabolizing systems. *Mutat. Res.* 76:1-50.

Bartsch, H., Terracini, B., Maleveille, C., Tomatis, L., Wahrendorf, J., Brun, G., and Dodet, B. 1983. Quantitative comparison of carcinogenicity, mutagenicity and electrophilicity of 10 direct-acting alkylating agents with carcinogenic potency in rodents. *Mutat. Res.* 110:181-219.

Baxendale, J. H. 1964. Effects of oxygen and pH in the radiation chemistry of aqueous solutions. *Radiat. Res. Suppl.* 4:114-140.

Baxendale, J. H., and Rasburn, E. J. 1974. Pulse radiolysis study of the kinetics of electron reactions in liquid n-hexane at room temperature. *J. Chem. Soc. Faraday Trans. I* 70:705-717.

Beck, G. 1983. A picosecond pulse-conductivity technique for the study of excess electron reactions. *Radiat. Phys. Chem.* 21:7-11.

Beck, G., and Thomas, J. K. 1972. Dynamics of electrons in nonpolar liquids. *J. Chem. Phys.* 57:3649-3654.

Benigni, R., and Giuliani, A. 1985. Cluster analysis of short-term tests: A new methodological approach. *Mutat. Res.* 147:139-151.

Bishop, J. M. 1987. The molecular genetics of cancer. *Science* 235:305-311.

Boag, J. W. 1963. Space charge distortion of the electric filed in a plane-parallel ionization chamber. *Phys. Med. Biol.* 8:461-467.

Boice, J. D. 1981. Cancer following medical irradiation. *Cancer* 47:1081-1090.

Boyland, E. 1964. Some aspects of the mechanism of carcinogenesis. In *Electronic Aspects of Biochemistry,* ed. B. Pullman, pp. 155-165. New York: Academic Press.

Bridges, B. A. Short term screening tests for carcinogens. *Nature* 261:195-200.

Briegleb, G. 1964. Electron affinity of organic molecules. *Angew. Chem.* (Intl. Ed. English) 76:617-632.

Brusick, D., and Auletta, A. 1985. Developmental status of bioassays in genetic toxicology. A report of Phase II of the U.S. Environmental Protection Agency Gene-Tox Program. *Mutat. Res.* 153:1-10.

Cairns, J. 1981. The origin of human cancers. *Nature* 28:353-357.

Chankong, V., Haimes, Y. Y., Rosenkranz, H. S., and Pet-Edwards, J. 1985. The carcinogenicity prediction and battery selection (CPBS) method: A Bayesian approach. *Mutat. Res.* 153:135-166.

Cheh, A. M., Hooper, A. B., Skochdopole, J., Henke, C. A., and McKinnell, R. G. 1980. A comparison of the ability of frog and rat S-9 to activate promutagens in the Ames test. *Environ. Mutagen.* 2:487-508.

Christophorou, L. G. 1971. *Atomic and Molecular Radiation Physics.* New York: Wiley-Interscience.

Christophorou, L. G. 1976. Electron attachment to molecules in dense gases ("Quasi-liquids"). *Chem. Rev.* 76:409-423.

Christophorou, L. G. 1980. Negative ions of polyatomic molecules. *Environ. Health Perspect.* 36:3–32.

Christophorou, L. G., and Hunter, S. R. 1984. From basic research to application. In *Electron-Molecule Interactions and Their Applications,* vol. 2, ed. L. G. Christophorou, pp. 317–422. New York: Academic Press.

Christophorou, L. G., and Siomos, K. 1984. Interphase physics: Linking knowledge on electron-molecule interactions in gases to knowledge on such processes in condensed matter. In *Electron-Molecule Interactions and Their Applications,* vol. 2, ed. L. G. Christophorou, pp. 221–316. New York: Academic Press.

Chu, K. C., Cueto, C., Jr., and Ward, J. M. 1981. Factors in the evaluation of 200 National Cancer Institute carcinogen bioassays. *J. Toxicol. Environ. Health.* 8:251–280.

Clayson, D. B., and Krewski, D. 1986. The concept of negativity in experimental carcinogenesis. *Mutat. Res.* 167:233–240.

Clegg, J. S. 1984. Properties and metabolism of the aqueous cytoplasm and its boundaries. *Am. J. Physiol.* 246:R133–R151.

Clementi, E. 1985. Ab initio computational chemistry. *J. Phys. Chem.* 89:4426–4436.

Cooper, J. A., III, Saracci, R., and Cole, P. 1979. Describing the validity of carcinogen screening tests. *Br. J. Cancer* 39:87–89.

Davis, D. L., and Magee, B. H. 1979. Cancer and industrial chemical production. *Science* 206:1356–1358.

Davis, H. T., Rice, S. A., and Meyer, L. 1962. Kinetic theory of dense fluids. XII. Electronic and ionic motion in liquid He^4I and liquid He^3. *J. Chem. Phys.* 37:1521–1527.

Dearden, J. C. 1985. Partitioning and lipophilicity in quantitative structure-activity relationships. *Environ. Health Perspect.* 61:203–228.

DeFlora, S., Zanacchi, P., Camoirano, A., Bennicelli, C., and Badolati, G. S. 1984. Genotoxic activity and potency of 135 compounds in the Ames reversion test and in a bacterial DNA-repair test. *Mutat. Res.* 133:161–198.

de Haas, M. P., Kunst, M., Warman, J. M., and Verberne, J. B. 1983. Nanosecond time-resolved conductivity studies of pulse-ionized ice. 1. The mobility and trapping of conduction-band electrons in H_2O and D_2O ice. *J. Phys. Chem.* 87:4089–4092.

Dodelet, J.-P., Freeman, G. R. 1972. Mobilities and ranges of electrons in liquids: Effect of molecular structure in C_5-C_{12} alkanes. *Can. J. Chem.* 50:2667–2679.

Doll, R. 1980. The epidemiology of cancer. *Cancer* 45:2475–2485.

Doll, R., and Peto, R. 1981. The causes of cancer: Quantitative estimates of avoidable risks of cancer in the United States today. *J. Natl. Cancer Inst.* 66:1191–1308.

Drost-Hansen, W., and Clegg, J. S. 1979. *Cell-Associated Water.* New York: Academic Press.

Ennever, F. K., and Rosenkranz, H. S. 1986. Short-term test results for NTP noncarcinogens: An alternate, more predictive battery. *Environ. Mutagen.* 8:849–865.

Fahmy, M. J., and Fahmy, O. G. 1980. Intervening DNA insertions and the alteration of gene expression by carcinogens. *Cancer Res.* 40:3374–3382.

Franks, F. 1979. Water. *A Comprehensive Treatise.* Vol. 6. Recent Advances. New York: Plenum.

Freeman, G. R. 1983. I. Electrons in fluids. II. Nonhomogeneous kinetics. *Ann. Rev. Phys. Chem.* 34:463–492.

Frierson, M. R., Klopman, G., and Rosenkranz, H. S. 1986. Structure-activity relationships (SAR's) among mutagens and carcinogens: A review. *Environ. Mutagen.* 8:283–327.

Fukuda, E. K., and McIver, R. T., Jr. 1985. Relative electron affinities of substituted benzophenones, nitrobenzenes and quinones. *J. Am. Chem. Soc.* 107:2291–2296.

Funabashi, K., and Magee, J. L. 1975. On the specific rates for electron scavenging in nonpolar solvents. *J. Chem. Phys.* 62:4428–4435.

Gershon, N. D., Porter, K. R., and Trus, B. L. 1985. The cytoplasmic matrix: Its volume and surface area and the diffusion of molecules through it. *Proc. Natl. Acad. Sci. U.S.A.* 82:5030–5034.

Gregg, E. C., and Bakale, G. 1970. Ionization currents in liquid ionization chambers: Low-conductivity liquids. *Radiat. Res.* 42:13–33.

Halle, B., Andersson, T. Forsén, S., and Lindman, B. 1981. Protein hydration from water oxygen-17 magnetic relaxation. *J. Am. Chem. Soc.* 103:500–508.

Hammer, W. 1985. *Occupational Safety Management and Engineering,* 3rd ed., p. 392. Englewood Cliffs, N.J.: Prentice-Hall.

Hart, R. W., and 19 permanent members of the U.S. Interagency Staff Group on Carcinogens. 1986. Chemical carcinogens: A review of the science and its associated principles. *Environ. Health Perspect.* 67:201–282.

Hartman, G. D., and Hartman, R. C. 1983. Non-mutagenic nitroheterocycles. The lack of correlation between bacterial mutagenicity and one-electron reduction potential. *Mutat. Res.* 117:271–277.

Haseman, J. K., Crawford, D. D., Huff, J. E. Boorman, G. A. and McConnell, E. E. 1984. Results from 86 two-year carcinogenicity studies conducted by the National Toxicology Program. *J. Toxicol. Environ. Health* 14:621–639.

Hay, E. D. 1983. Cell and extracellular matrix: Their organization and mutual dependence. In *Modern Cell Biology,* vol. 2. Spatial Organization of Eukaryotic Cells, ed. J. R. McIntosh, pp. 509–548. New York: Liss.

Heinz, J. E., and Poulsen, N. K. 1983. The optimal design of batteries of short-term tests for detecting carcinogens. *Mutat. Res.* 117:259–269.

Henglein, A. 1974. Estimated distributions of electronic redox levels in aq/e_{aq}^-, H_{aq}^+/H_{aq} and some other systems. *Ber. Bunsenges. Phys. Chem.* 78:1078–1084.

Henglein, A. 1975. Estimated distribution functions of electronic redox levels and the rate of chemical reactions of excess electrons in dielectric liquids. *Ber. Bunsenges. Phys. Chem.* 79:129–135.

Henglein, A. 1977. Electron reactivity as a function of phase. *Can. J. Chem.* 55:2112–2123.

Higginson, J. 1969. Present trends in cancer epidemiology. In *Proceedings of the Eighth Canadian Cancer Conference,* Honey Harbor, Ontario, 1968, ed. J. F. Morgan, pp. 40–75. Toronto: Pergamon.

Higginson, J. 1976. A hazardous society? Individual versus community responsibility in cancer prevention. *Am. J. Pub. Health* 66:359–396.

Higginson, J. 1983. Developing concepts on environmental cancer: The role of geographical pathology. *Environ. Mutagen.* 5:929–940.

Holden, C. 1979. Albert Szent-Györgyi, electrons, and cancer. *Science* 203:522–524.

Holliday, R. 1979. A new theory of carcinogenesis. *Br. J. Cancer* 40:513–522.

Holroyd, R. A. 1977. Equilibrium reactions of excess electrons with aromatics in non-polar solvents. *Ber. Bunsenges. Phys. Chem.* 81:298–304.

Holroyd, R. A. 1987. The electron: Its properties and reactions. In *Radiation Chemistry: Principles and Applications,* eds. Farhataziz and M. A. Rodgers, pp. 201–235. New York: VCH Publishers.

Holroyd, R. A., Tames, S., and Kennedy, A. 1975. Effect of temperature on conduction band energies of electrons in nonpolar liquids. *J. Phys. Chem.* 79:2857–2861.

Huff, J. E., McConnell, E. E., and Haseman, J. K. 1985. On the proportion of positive results in carcinogenicity studies in animals. *Environ. Mutagen.* 7:427–428.

Hummel, A., and Schmidt, W. F. 1974. Ionization of dielectric liquids by high energy radiation studied by means of electrical conductivity methods. *Radiat. Res. Rev.* 5:199–300.

Huppert, D., Kenney-Wallace, G. A., and Rentzepis, P. M. 1981. Picosecond dynamics of electron trapping in polar liquids. *J. Chem. Phys.* 75:2265–2269.

IARC. 1983. Polynuclear aromatic compounds; Part 1, Chemical, environmental and experimental data. *IARC Monogr.* 32:95–104. Lyon: IARC.

IARC. 1986. Some chemicals used in plastics and elastomers. *IARC Monogr.* 39:113–131. Lyon: IARC.

Jerina, D. M., Lehr, R., Schaeffer-Ridder, M., Yagi, H., Karle, J. M., Thakker, D. R., Wood, A. H., Lu, A. Y., Ryan, D., West, S., Levin, W., and Conney, A. H. 1977. Bay-region epoxides of dihydrodiols: A concept explaining the mutagenic and carcinogenic activity of benzo[a]pyrene and benzo[a]anthracene. In *Origins of Human Cancer,* eds. H. Hiatt, J. D. Watson, and I. Winsten, pp. 639–658. Cold Spring Harbor: Cold Spring Harbor Laboratory.

Jortner, J. 1972. Theoretical studies of excess electron states in liquids. *Ber. Bunsenges.* 75:696–714.

Jortner, J., and Gaathon, A. 1977. Effect of phase density on ionization processes and electron localization in fluids. *Can. J. Chem.* 55:1801–1819.

Jortner, J., and Kestner, N. R. (eds.). 1973. *Electrons in Fluids.* West Berlin: Springer-Verlag.

Kasha, M., and Pullman, B. (eds.) 1962. *Horizons in Biochemistry.* Albert Szent-Györgyi Dedicatory Volume. New York: Academic Press.

Kawachi, T., Yahagi, T., Kada, T., Tazima, Y., Ishidate, M., Sasaki, M., and Sugiyama, T. 1980. Cooperative programme on short term assays for carcinogenicity in Japan. In *Molecular and Cellular Aspects of Carcinogen Testing,* eds. R. Montesano, H. Bartsch, L. Tomatis, and W. Davis, pp. 323–330. Lyon: IARC.

Kenney-Wallace, G. A. 1977. Picosecond molecular relaxations: The role of the fluid in electron solvation. *Can. J. Chem.* 55:2009–2016.

Kenney-Wallace, G. A. 1980. Picosecond relaxation processes in liquids. *Phil. Trans. Roy. Soc. London* A299:309–319.

Kevan, L., and Webster, B. C. (eds.). 1976. Electron-Solvent and Anion-Solvent Interactions. Amsterdam: Elsevier.

Klopman, G. 1984. Artificial intelligence approach to structure-activity studies. Computer automated structure evaluation of biological activity of organic molecules. *J. Am. Chem. Soc.* 106:7315–7321.

Klopman, G., Tonucci, D. A., Halloway, M., and Rosenkranz, H. S. 1983. Relationships between polarographic reduction potential and mutagenicity of nitroarenes. *Mutat. Res.* 117:271–277.

Land, H., Parada, L. F., and Weinberg, R. A. 1983. Cellular oncogenes and multistep carcinogenesis. *Science* 222:771–778.

Lave, L. B., and Omenn, G. S. 1986. Cost-effectiveness of short-term tests for carcinogenicity. *Nature* 324:29–34.

Lenk, R., Bonzon, M., and Greppin, H. 1980. Dynamically oriented biological water as studied by NMR. *Chem. Phys. Lett.* 76:175–177.

Lilie, J., Beck, G., and Henglein, A. 1971. Pulsradiolyse und polarographie: Halbstufenpotentiale für die oxidation und reduktion von kurzlebigen organischen radikalen an der Hg-electrode. *Ber. Bunsenges. Phys. Chem.* 75:458–465.

Ling, G. N. 1972. Hydration of macromolecules. In *Water and Aqueous Solutions,* ed. R. A. Horne, pp. 663–700. New York: Wiley-Interscience.

Loew, G. H., Poulsen, M., Kirkijan, E., Ferrell, J., Sudhindra, S., and Rebagliati, M. 1985. Computer-assisted mechanistic structure-activity studies: Application to diverse classes of chemical carcinogens. *Environ. Health Perspect.* 61:69–96.

Lovelock, J. E. 1958a. A sensitive detector for gas chromatography. *J. Chromatogr.* 1:35–46.

Lovelock, J. E. 1958b. Meaurement of low vapour concentrations by collision with excited rare gas atoms. *Nature* 158:1460–1462.

Lovelock, J. E. 1961a. Ionization methods for the analysis of gases and vapors. *Anal. Chem.* 33:162–178.

Lovelock, J. E. 1961b. Affinity of organic compounds for free electrons with thermal energy: Its possible significance in biology. *Nature* 189:729–732.

Lovelock, J. E. 1962. Free electrons with thermal energy: Their generation for chemical use. *Nature* 195:488–489.

Lovelock, J. E. 1966. Gas analysis method and device for the qualitative and quantitative analysis of classes of organic vapors. U.S. Patent 3,247,375.

Lovelock, J. E., and Lipsky, S. R. 1960. Electron affinity spectroscopy—A new method for the identification of functional groups in chemical compounds separated by gas chromatography. *Nature* 82:431–433.

Lovelock, J. E., Simmonds, P. G., and Vandenheuvel. 1963. Affinity of steroids for electrons with thermal energies. *Nature* 197:249–251.

Lovelock, J. E., Zlatkis, A., and Becker, R. S. 1962. Affinity of polycyclic aromatic hydrocarbons for electrons with thermal energies: Its possible significance in carcinogenesis. *Nature* 193:540–541.

Lovelock, J. E., Davies, A. J., and Ferris, F. R. 1972. Method and apparatus for linearly measuring electron capture with an electron capture detector. U.S. Patent 3,634,754.

Ma, S.-M., Eyring, H., Veda, I., and Kaneshima, S. 1981. Modeling of biological reactions. *Int. J. Chem. Kinet.* 13:913–923.

Mank, V. V., and Lebovka, N. I. 1983. On the structure of water NMR spectra in membranes. *Chem. Phys. Lett.* 96:626–630.

Marshall, E. 1987. California's debate on carcinogens. *Science* 235:1459.

Mason, R. 1958. Electron mobility in biological systems and its relation to carcinogenesis. *Nature* 181:820–822.

Mason, R. 1960. Role of electron and exciton transfer in carcinogenesis. *Radiat. Res. Suppl.* 2:452–461.

Maugh, T. H. II. 1979. Cancer and the environment: Higginson speaks out. *Science* 205:1363–1366.

McCann, J., Choi, E., Yamasaki, E., and Ames, B. N. 1975. Detection of carcinogens as mutagens in the *Salmonella*/microsome test: Assay of 300 chemicals. *Proc. Natl. Acad. Sci. U.S.A.* 72:5135–5139.

McKinney, J. D. 1985. Monograph on Structure-Activity Correlation in Mechanism Studies and Predictive Toxicology. *Environ. Health Perspect.* 61:3–350.

Meisel, D., and Czapski, G. 1975a. One-electron transfer equilibria and redox potentials of radicals studied by pulse radiolysis. *J. Phys. Chem.* 79:1503–1509.

Meisel, D., and Neta, P. 1975b. One-electron redox potentials of nitro compounds and radiosensitizers. Correlation with spin densities of their radical anions. *J. Am. Chem. Soc.* 97:5198–5203.

Miller, E. C. 1978. Some current perspectives on chemical carcinogenesis in humans and experimental animals: Presidential address. *Cancer Res.* 38:1479–1496.

Miller, E. C., and Miller, J. A. 1971. The mutagenicity of chemical carcinogens: Correlations, problems, and interpretations. In *Chemical Mutagens—Principles and Methods for Their Detection,* ed. A. Hollaender, pp. 83–119. New York: Plenum.

Miller, J. A. 1970. Carcinogenesis by chemicals. An overview. GHA Clowes memorial lecture. *Cancer Res.* 30:559–576.

Miller, L. S., Howe, S., and Spear, W. E. 1968. Charge transport in liquid Ar, Kr, and Xe. *Phys. Rev.* 166:871–878.

Minday, R. M., Schmidt, L. D., and Davis, H. T. 1969. Free electrons in liquid hexane. *J. Chem. Phys.* 50:1473–1474.

Nesnow, S., Argus, M., Bergman, H., Chu, K., Frith, C., Helmes, T., McGaughy, R., Ray, V., Slaga, T., Tennant, R., and Weisburger, E. 1986. Chemical carcinogens. A review and analysis of the literature of selected chemicals and the establishment of the Gene-Tox Carcinogen Data Base. *Mutat. Res.* 185:1–195.

Nicolini, C. 1986. *Biophysics and Cancer.* New York: Plenum.

Olive, P. L. 1979a. Inhibition of DNA synthesis by nitroheterocycles. I. Correlation with half-wave reduction potential. *Br. J. Cancer* 40:89–93.

Olive, P. L. 1979b. Correlation between metabolic reduction rates and electron affinity of nitro-heterocycles. *Cancer Res.* 39:4512–4515.

Palajda, M., and Rosenkranz, H. S. 1985. Assembly and preliminary analysis of a genotoxicity data base for predicting carcinogens. *Mutat. Res.* 153:79–134.

Perera, F. 1981. Carcinogenicity of airborne fine particulate benzo[*a*]pyrene: An appraisal of the evidence and the need for control. *Environ. Health Perspect.* 42:163–185.

Pet-Edwards, J., Rosenkranz, H. S., Chankong, V., and Haimes, Y. Y. 1985. Cluster analysis in predicting the carcinogenicity of chemicals using short-term assays. *Mutat. Res.* 153:167–185.

Pethig, R. 1979a. *Dielectric and Electronic Properties of Biological Material.* New York: Wiley.

Pethig, R. 1979b. Water in biological systems. In *Dielectric and Electronic Properties of Biological Material,* pp. 100–149. New York: Wiley.

Pethig, R. 1979c. Electronic properties of biomacromolecules. In *Dielectric and Electronic Properties of Biological Material,* pp. 290–356. New York: Wiley.

Pienta, K. J., and Coffey, D. S. 1985. The nuclear matrix. An organizing structure for the interphase nucleus and chromosome. In *Structure and Function of the Genetic Apparatus,* eds. C. Nicolini and P. O. Ts'o, pp. 83–98. New York: Plenum.

Polnaszek, C. F., and Bryant, R. G. 1984. Self-diffusion of water at the protein surface: A measurement. *J. Am. Chem. Soc.* 106:428–429.

Porter, K. R., Beckerle, M., and McNiven, M. 1983. The cytoplasmic matrix. In *Modern Cell Biology,* vol. 2. *Spatial Organization of Eukaryotic Cells,* ed. J. R. McIntosh, pp. 259–302. New York: Liss.

Pott, P. 1963. Chirurgical observations relative to the cancer of the scrotum. (London, 1775), reprinted in *Natl. Cancer Inst. Monogr.* 10:7–13.

Prehn, R. T., and Prehn, L. M. 1987. The autoimmune nature of cancer. *Cancer Res.* 47:927–932.

Ptashne, M. 1986. Gene regulation by proteins acting nearby and at a distance. *Nature* 322:697–701.

Pullman, A., and Pullman, B. 1955. Electronic structure and carcinogenic activity of aromatic molecules. New developments. *Adv. Cancer Res.* 3:117–169.

Pullman, B., and Pullman, A. 1963. *Quantum Biochemistry.* New York: Academic Press.

Purchase, I. F. H. 1980. Inter-species comparisons of carcinogenicity. *Br. J. Cancer* 41:454–468.

Purchase, I. F. H. 1982. An appraisal of predictive tests for carcinogenicity. *Mutat. Res.* 99:53–71.

Purchase, I. F. H., Longstaff, E., Ashby, J., Styles, J. A., Anderson, D., Lefevre, P. A., and Westwood, R. F. 1978. An evaluation of 6 short-term tests for detecting organic chemical carcinogens. *Br. J. Cancer* 37:873–903.

Raineri, R., Poiley, J. A., Pienta, R. J., and Andrews, A. W. 1981. Metabolic activation of carcinogens in the *Salmonella* mutagenicity assay by hamster and rat liver S-9 preparations. *Environ. Mutagen.* 3:31–84.

Ramel, C. 1986. Deployment of short-term assays for the detection of carcinogens; Genetic and molecular considerations. *Mutat. Res.* 168:327–342.

Reiss, H. 1985. The Fermi level and the redox potential. *J. Phys. Chem.* 89:3783–3791.

Rosenkranz, H. S., Klopman, G., Chankong, V., Pet-Edwards, J., and Haimes, Y. Y. 1984. Prediction of environmental carcinogens: A strategy for the mid-1980's. *Environ. Mutagen.* 6:231–258.

Rosenkranz, H. S., and Poirier, L. A. 1979. Evaluation of the mutagenicity and DNA-modifying activity of carcinogens and noncarcinogens in microbial systems. *J. Natl. Cancer Inst.* 62:873–891.

Rubin, H. 1980. Is somatic mutation the major mechanism of malignant transformation? *J. Natl. Cancer Inst.* 64:995-1000.

Schindewolf, U. 1978. Physical and chemical properties of dissolved electrons (excess electrons). *Angew. Chem. Int. Ed. Engl.* 17:887-901.

Schmidt, W. F. 1977. Electron mobility in nonpolar liquids; The effect of molecular structure, temperature, and electric field. *Can. J. Chem.* 55:2197-2210.

Schmidt, W. F., and Allen, A. O. 1969. Mobility of free electrons in dielectric liquids. *J. Chem. Phys.* 50:5037.

Schmidt, W. F., and Allen, A. O. 1970. Mobility of electrons in dielectric liquids. *J. Chem. Phys.* 52:4788-4794.

Schmidt, W. F., Bakale, G., and Sowada, U. 1974. Excess electrons and positive charge carriers in liquid ethane. *J. Chem. Phys.* 61:5275-5278.

Schnyders, H., Rice, S. A., and Meyer, L. 1966. Electron drift velocities in liquified argon and krypton at low electric field strengths. *Phys. Rev.* 150:127-145.

Searle, C. E. 1976. Preface to *Chemical Carcinogens,* ACS Monograph 173, Washington, DC: American Chemical Society.

Shabad, L. M. 1980. Circulation of carcinogenic polycyclic aromatic hydrocarbons in the human environment and cancer prevention. *J. Natl. Cancer Inst.* 64:405-410.

Shelby, M. D., and Stasiewicz, S. 1984. Chemicals showing no evidence of carcinogenicity in long-term, two-species rodent studies: The need for short-term test data. *Environ. Mutagen.* 6:871-878.

Shockley, W. 1950. *Electrons and Holes in Semiconductors,* pp. 9-10. Princeton, N.J.: Van Nostrand.

Shubik, P., Interdisciplinary Panel on Carcinogenicity. 1984. Criteria for evidence of chemical carcinogenicity. *Science* 225:682-687.

Simmon, V. F. 1979. In vitro mutagenicity assays of chemical carcinogens and related compounds with *Salmonella* typhimurium. *J. Natl. Cancer Inst.* 62:893-899.

Sjöberg, L., and Eriksen, T. E. 1980. Nitrobenzenes: A comparison of pulse radiolytically determined one-electron reduction potentials and calculated electron affinities. *J. Chem. Soc. Faraday Trans. I* 76:1402-1408.

Smith, I. A., Berger, G. D., Seybold, P. G., and Servé, M. P. 1978. Relationships between carcinogenicity and theoretical activity indices in polycyclic aromatic hydrocarbons. *Cancer Res.* 38:2968-2977.

Sowada, U., Bakale, G., and Schmidt, W. F. 1976. Dependence of electron mobililty on electric field strength in nonpolar liquids. *High Energy Chem.* (Engl. ed.) 290-293.

Sugimura, T., Sato, S., Nagao, M., Yahagi, T., Matsushima, T., Sieno, Y., Takeuchi, M., and Kawachi, T. 1976. Overlapping of carcinogens and mutagens. In *Fundamentals in Cancer Prevention,* eds. P. N. Magee, S. Takayama, T. Sugimura, and T. Matsushima, pp. 191-215. Baltimore: University Park Press.

Szent-Györgyi, A. 1941a. Towards a new biochemistry? *Science* 93:609-611.

Szent-Györgyi, A. 1941b. The study of energy-levels in biochemistry. *Nature* 148:157-159.

Szent-Györgyi, A. 1968. Bioelectronics. Intermolecular electron transfer may play a major role in biological regulation, defense, and cancer. *Science* 161:988-990.

Szent-Györgyi, A. 1976. *Electronic Biology and Cancer.* New York: Marcel Dekker.

Szent-Györgyi, A. 1977. An electronic theory of cancer. *Int. J. Quantum Chem.* XII (Suppl 1):407-414.

Tewari, P. H., and Freeman, G. R. 1968. Dependence of radiation-induced conductance of liquid hydrocarbons and molecular structure. *J. Chem. Phys.* 49:4394-4399.

Tezuka, T., Namba, H., Nakamura, Y., Chiba, M., Shinsaka, K., and Hatano, Y. 1983. Free-ion yields, electron mobilities and electron-ion recombination rate constants in liquid and solid isooctane and several other nonpolar compounds. *Radiat. Phys. Chem.* 21:197-208.

Tucker, R. W., and Barrett, J. C. 1986. Decreased numbers of spindle and cytoplasmic microtubules in hamster embryo cells treated with a carcinogen, diethylstilbestrol. *Cancer Res.*
46:2088–2095.

Turner, D. R., Morley, A. A., Haliandros, M., Kutlaca, R., and Sanderson, B. J. 1985. In vivo
somatic mutations in human lymphocytes frequently result from major gene alterations. *Nature*
315:343–345.

Vainio, H., Hemminki, K., and Wilbourn, J. 1985. Data on the carcinogenicity of chemicals in the
IARC Monographs programme. *Carcinogenesis* 6:1653–1665.

van Lith, D., Warman, J. M., de Haas, M. P., and Hummel, A. 1986. Electron migration in
hydrated DNA and collagen at low temperatures. Part 1. Effect of water concentration. *J.
Chem. Soc. Faraday Trans. I.* 82:2933–2943.

Van Zandt, L. L. 1986. Resonant microwave absorption by dissolved DNA. *Phys. Rev. Lett.*
57:2085–2087.

von Jagow, G., and Engle, W. D. 1980. Structure and function of the energy-converting system of
mitochondria. *Angew. Chem.* (Int. Engl. Ed.) 19:659–748.

Wardman, P. 1977. The use of nitroaromatic compounds as hypoxic cell radiosensitizers. *Curr. Top.
Radiat. Res. Q.* 11:347–398.

Warman, J. M. 1982. The dynamics of electrons and ions in non-polar liquids. In *The Study of Fast
Processes and Transient Species by Electron Pulse Radiolysis,* eds. J. H. Baxendale, F. Busi,
pp. 437–527. Dordrecht: Reidel.

Warman, J. M., Infelta, P. P., de Haas, M. P., and Hummel, A. 1977. The study of primary and
secondary charge carriers in nanosecond pulse irradiated liquid dielectrics using a resonant
microwave cavity. *Can. J. Chem.* 55:2249–2257.

Weinstein, I. B., Arcoleo, J., Lambert, M., Hsiao, W., Gattoni-Celli, S., Jeffrey, A. M., and
Kirschmeier, P. 1985. Molecular mechanisms of multistage chemical carcinogenesis. In *Prog-
ress in Cancer Research and Therapy,* vol. 32: *Molecular Biology of Tumor Cells,* eds.
B. Wahren, G. Holm, S. Hammarstrom, P. Perlmann, pp. 55–70. New York: Raven.

Weinstein, M. C. 1983. Cost-effective priorities for cancer prevention. *Science* 221:17–23.

Wiesenfeld, J. M., and Ippen, E. P. 1980. Dynamics of electron solvation in liquid water. *Chem.
Phys. Lett.* 73:47–50.

Wilbourn, J., Haroun, L., Heseltine, E., Kaldor, J., Partensky, C., and Vainio, H. 1986. Response
of experimental animals to human carcinogens: An analysis based upon the IARC Monographs
programme. *Carcinogenesis* 7:1853–1863.

Wolstenholme, G. E. W., Fitzsimons, D. W., and Whelan, J. (eds.): 1979. *Submolecular Biology
and Cancer.* Amsterdam: Excerpta Medica.

Yakovlev, B. S. 1979. Excess electrons in non-polar molecular liquids. *Russ. Chem. Rev.* (Engl.
ed.) 48:1153–1179.

Zeiger, E. 1987. Carcinogenicity of mutagens: Predictive capability of the *Salmonella* mutagenesis
assay for rodent carcinogenicity. *Cancer Res.* 47:1287–1296.

Zeiger, E., Risko, K. J., and Margolin, B. H. 1985. Strategies to reduce the cost of mutagenicity
screening with the *Salmonella* assay. *Environ. Mutagen.* 7:901–911.

Modeling of Combined Toxic Effects of Chemicals

Erik R. Christensen

Department of Civil Engineering
University of Wisconsin-Milwaukee

Chung-Yuan Chen

School of Civil and Structural Engineering
Nanyang Technological Institute, Singapore

I. HISTORICAL PERSPECTIVE

The combined effects of several toxic chemicals on living organisms has been a subject of study for more than 50 years. It is of interest to scientists and professionals involved in the setting of water quality criteria for the protection of aquatic life against pollutants acting singly and in combination (EIFAC, 1980). The subject is also of concern to manufacturers and users of insecticides and herbicides where the goal often is to devise compounds that alone or in combination have the maximum efficiency (Hewlett and Plackett, 1979). Finally, the joint action of several chemicals is important to pharmacologists studying the action of drugs and to epidemiologists concerned with the role of several simultaneously acting factors, such as asbestos and tobacco smoking on human carcinogenesis (Reif, 1984). We are here primarily concerned with the joint effects of toxic chemicals on aquatic life, although many of the concepts and models described have much broader application.

Early work on the toxicity of mixtures of poisons was carried out by Southgate (1932) who distinguished between chemicals having different physiological action such as potassium cyanide and *p*-cresol, and those that are interchangeable, for example *p*-cresol and phenol. In the former case, there was little or no change in the toxicity when the chemicals were combined, while in the latter, *p*-cresol and phenol of equivalent toxicity were completely inter-

We thank Steven J. Broderius and Robert L. Spehar of the U.S. Environmental Protection Agency, Environmental Research Laboratory—Duluth, for reviewing the manuscript and for offering helpful advice and constructive criticism.

changeable and the two compounds appeared to have similar physiological actions on trout. It was also found that xylenol and p-cresol were only partly interchangeable. Although the quantitative treatment of the toxicity data was primitive by today's standards, the nature of the observations was crucial and is still fundamental to the modes of action that are now called concentration addition, where chemicals are interchangeable, and independent action, where the toxic chemicals affect separate biological systems.

A more developed quantitative approach to joint toxicity was presented by Bliss (1939). Bliss used probit analysis (Finney, 1971) for the dosage mortality curves of the individual toxicants and recognized similar joint action with parallel dose-response curves and independent joint action. It was realized that the action tolerances of individual organisms were completely correlated in similar joint action and that the coefficient of correlation for the tolerances could vary between 0 and 1 for independent action. Criteria for synergistic action were also discussed.

Plackett and Hewlett (1948) extended Bliss' statistical treatment by allowing a negative correlation of tolerances and by using a bivariate normal distribution function as a means of evaluating responses and the correlation of toxicant tolerances. They also considered the question of whether a mixture of poisons is more toxic if they act similarly or independently and concluded that no general rule could be formulated.

A major achievement of Hewlett and Plackett (1959) was to unify the similar and independent action models into a comprehensive model where the degree of similarity of two biological systems can vary between zero and full similarity, and where the coefficient of correlation of the action tolerances can vary between -1 and $+1$. They also dropped the requirement of parallel dose-response curves for similar joint action. Hewlett and Plackett's model is noninteractive, meaning that the combination of one toxicant with receptors of the organisms has no influence on the combination of the other with receptors or its intrinsic activity (Plackett and Hewlett, 1967). Models based on competitive action, following laws of mass action, were developed by Hewlett and Plackett (1964) and Ashford (1981). Christensen and Chen (1985) expanded Hewlett and Plackett's model to include n toxicants and an arbitrary tolerance distribution, for example, a logit or Weibull expression for individual toxicants.

Parallel with the development of models for the joint action of toxicants came efforts to characterize modes of joint action, either by the use of isobolograms or appropriate multiple toxicity indices. Isobolograms are curves showing combinations of toxicant concentrations giving a constant biological response, e.g., 50% fish mortality (Gaddum, 1949; Loewe, 1953). The isobole for similar joint action or concentration addition, considering two toxicants with the toxicant concentrations indicated linearly along orthogonal z_1 and z_2 axes, is a straight line intercepting these axes at Z_1 and Z_2 which are the concentrations of the toxicants acting singly giving the desired response. Zero concentrations are at the origin. The area below this line is characterized by the term potentiation or more than additive, while the area above it where $z_1 \leq Z_1$ and $z_2 \leq Z_2$ is called less than additive. Finally, the area for which $z_1 > Z_1$ and $z_2 > Z_2$ is characterized by the term antagonism. Concentration addition with parallel dose-response curves is reflected in isobolograms through parallel isoboles, while nonparallel dose-response curves are characterized by nonparallel isoboles (Hewlett, 1969).

The following relationship holds for concentration addition, considering n toxicants:

$$\frac{z_1}{Z_1} + \frac{z_2}{Z_2} + \cdots + \frac{z_n}{Z_n} = 1 \tag{1}$$

where z_i is the actual concentration of toxicant i and Z_i is the concentration of toxicant i, when acting alone, that gives the desired biological response.

Sprague and Ramsay (1965) called the terms of Eq. (1) toxic units. The sum M of toxic units for a given mixture is, therefore, an important toxicity index. More than additive action is obtained for $M < 1$, less than additive action for $M > 1$, and concentration addition for $M = 1$. Hewlett (1969) introduced a joint action ratio R which is also one for concentration addition. However, for $R > 1$ there is potentiation or more than additive action, and for $R < 1$, less than additive action. A disadvantage of the joint action ratio is that it can only be determined from a detailed knowledge of the isobologram.

Several other multiple toxicity indices have been proposed. Two of the most recent ones are the additive index of Marking (1977) and the multiple toxicity index (MTI) of Konemann (1981b). In our view, Konemann's index is particularly useful in that, as will be shown below, it is identical to the similarity parameter λ in Hewlett and Plackett's (1959) model under the assumption of equal toxic units of the individual components of the mixture and full correlation of toxicant tolerances. The MTI = 1 for concentration addition, 0 for independent action with full correlation of tolerances, >1 for more than additive action, and between 0 and 1 for less than additive action.

Fitting of the various multiple toxicity models to actual data has lagged behind the model development. Hewlett and Plackett (1950) fitted their independent action model (Plackett and Hewlett, 1948) to experimental data for the action of insecticides on flour beetles (*Triboleum castaneum*). For example, for direct spray of pyrethrins and DDT, there was a good model fit with a negative correlation coefficient $\rho = -0.74$ after 6 days' exposure. When these chemicals were applied as a film, the correlation coefficient was positive ($\rho = 0.66$) after 9 days' exposure. The action of γ-benzene hexachloride (lindane) and pyrethrins could not be described in terms of the independent action model ($\lambda = 0$). However, as shown by Christensen and Chen (1985), the results from the application of these chemicals could be fitted by a general bivariate Weibull model with a similarity parameter $\lambda = 0.58$ and a correlation coefficient $\rho = 1.00$. The second best fit was provided by a bivariate logit model with $\lambda = 0.64$ and $\rho = 0.70$.

Plackett and Hewlett (1963) fitted several sets of data to their bivariate toxicity model (Hewlett and Plackett, 1959) under the assumption of full correlation of toxicant tolerances. Also, Sawicki et al. (1962) demonstrated that a mixture of four insecticides occurring in pyrethrum flowers could be described by similar joint action in their insecticidal effect on house flies. This result could be expected in view of the fact that the four insecticides pyrethrins I and II and cinerins I and II are closely related chemically and toxicologically.

Results of mixture toxicity experiments with male guppies (*Poecilia reticulata*) were fitted to models for concentration addition and independent action by Anderson and Weber (1975b). They found that Cu-Ni mixtures were concentration additive with parallel dose-response curves, while mixtures of dieldrin and pentachlorophenate were characterized by independent action with zero correlation of tolerances. Other toxicant mixtures could not be described by any of these two models. The response parameter was fish mortality. A more detailed analysis of their data was carried out by Christensen (1987) who considered bivariate Weibull, logit, and probit models, partial similarity of biological systems, nonparallel dose-response curves for individual toxicants, and incomplete or negative correlation of toxicant tolerances.

Broderius and Smith (1979) studied lethal and sublethal effects of binary mixtures of several toxicants on fathead minnows (*Pimephales promelas*) and rainbow trout (*Salmo gairdneri*). For the experiments with mortality as the response parameter, the sum of toxic units M was significantly different from 1 ($0.7 < M < 1.3$) for binary mixtures of HCN with Cr(VI), Zn, or NH_3. The fact that the dose-response curves for the individual toxicants had significantly different slopes led the authors to test the mixtures for independent action, however, with a negative result. The sublethal action of the four toxicants was also not predictable from models for similar or independent action, and in most cases there was little increase in the effect over that expected from the toxicants acting singly. The authors concluded that there was a need for a general multiple toxicity approach.

Application of models for concentration addition with parallel dose-response curves for individual toxicants and for independent action to sublethal effects in mice were examined by Shelton and Weber (1981). They found that the action of carbon tetrachloride (CCl_4) and monochlorobenzene followed concentration addition. They also predicted concentration addition for CCl_4 and acetaminophen based on parallel dose-response curves for these toxicants. However, the experimental evidence did not support such a conclusion, thus demonstrating the unreliability of parallelism of dose-response curves as a criterion for concentration addition.

Other experimental efforts have been aimed at determining whether or not concentration addition occurred or if there was independent action with full correlation of tolerances. The latter mode of action implies that the toxicity is determined by the most toxic compound in the mixture so that there is no addition of toxicity. If this is the case, water quality criteria for toxicants acting singly will also provide adequate protection of aquatic life when the toxicants act in combination. The bioassays of Spehar et al. (1978) on the chronic effects of cadmium and zinc mixtures on flagfish (*Jordanella floridae*) indicated that the two metals did not act additively when combined as mixtures. This is in contrast to most other experimental evidence showing that several toxicants often are additive or nearly so when acting in combination. Wong et al. (1978) found that ten metals at the Great Lakes water quality objective levels were not toxic to algae if present individually, but strongly inhibited primary productivity when present together. These results clearly indicated that water quality criteria should not be based on levels valid for toxicants acting singly.

Christensen et al. (1985a) found Ni and Zn ions to be nearly concentration additive in their action on *Synechococcus leopoliensis*. The response parameter was algal growth rate. Addition of toxicity was also demonstrated by Biesinger et al. (1986) who found that Cd-Hg, Cd-Zn, and Zn-Hg mixtures all showed significant reduction in *Daphnia magna* reproduction at concentrations where the metals alone caused no significant effect. Similar results were obtained by Spehar and Fiandt (1986) regarding the acute and chronic effects of mixtures of six metals on three aquatic species. The response of fathead minnows was more than additive on an acute basis and less than additive on a chronic basis. The results for the biologically simpler daphnids (*Ceriodaphnia dubia*) were closer to strict concentration addition.

Recent studies of the multiple toxicity of mixtures of organic chemicals have included up to 50 compounds, often applied in equitoxic proportions (Konemann, 1981b; Hermens and Leeuwangh, 1982; Hermens et al., 1984a, 1984b, 1985; Broderius and Kahl, 1985). The interpretation of the results is aided by the concept of quantitative structure-activity relationships (QSARs) which are useful in predicting the LC_{50} levels for groups of similar chemicals primarily from the *n*-octanol/water partition coefficient, P_{oct} (Konemann, 1981a; Kaiser, 1984; Veith et al., 1985).

QSARs may be used as an indication of similarity in action such that chemicals with the same QSARs may be expected to act by similar joint action, that is concentration addition, while compounds from different QSARs are more likely to act via independent or partially similar action. The toxicity of mixtures of up to 50 chemicals from the same QSAR could in fact be described by concentration addition as indicated by a multiple toxicity index close to one (Konemann, 1981b; Hermens et al., 1984a). Even for mixtures of 8 and 24 chemicals with diverse modes of action, Hermens and Leeuwangh (1982) found that the toxicity to guppies (*Poecilia reticulata*) could be described by near concentration addition. The reason for this is not known. It is perhaps a result of multiple partial similar actions plus, possibly, potentiation between selected chemicals. Concentration addition also appears to apply to no-observed-effect concentrations as shown by Hermens et al. (1985) in a study on the joint toxicity of 10 and 25 compounds from the same QSAR on the growth of *Daphnia magna*.

From the above studies it is clear that chemicals contribute to the joint toxicity of mixtures, even at concentrations as low as 0.02 relative to the effect concentrations. This is remarkable and in contrast to the recommendations of the National Academy of Sciences and National Academy of Engineering (1972) and EIFAC (1980) which include lower limits of 0.2 and 0.1 for the relative concentrations, respectively, under which the contributions to the joint toxicity of individual chemicals should be ignored. Thus it appears that there is no threshold concentration under which a toxicant does not contribute to the joint toxicity. However, another aspect of the recommendations is confirmed by the above experiments, namely, the advisability of a water quality criterion based on concentration addition. This criterion can be formulated as a requirement for the sum of toxic units to be ≤ 1 in order to ensure that a given effect criterion is fulfilled, e.g., $\leq 50\%$ fish mortality.

Despite these advances in the understanding of multiple toxicity, official U.S. water quality criteria promulgated by the U.S. Environmental Protection Agency (U.S. EPA, 1985) are still based on levels valid for pollutants acting singly although there are provisions for toxicity based approach where the whole effluent is considered.

The most recent developments in multiple toxicity studies have been characterized by the introduction of transformations other than the probit equation for the individual toxicants, e.g., Weibull or logit transformations (Christensen, 1984; Christensen and Chen, 1985); by the fitting of models with partially similar joint action where the correlation of toxicant tolerances can assume values between -1 and $+1$ and the introduction of general models for n toxicants (Christensen and Chen, 1985); by the accumulation of a large experimental data base for two or more toxicants acting in combination; and finally, by a realization that many of these data sets cannot be fitted by the previous simple models for similar or independent joint action (Lewis and Perry, 1981; Durkin, 1981; Broderius, 1987).

In addition to providing a historical overview of multiple toxicity models and related experimental results, the objectives of the present chapter are as follows:

1. To introduce a consistent terminology for joint toxic action based on ρ, λ diagrams

2. To demonstrate general forms of isobolograms for two toxicants as a function of the similarity parameter, the correlation of tolerances, and the slopes of the dose-response curves for the two toxicants

3. To show that the multiple toxicity index introduced by Konemann (1981b) is identical to the similarity parameter λ of Hewlett and Plackett (1959) and Christensen and Chen (1985) for mixtures of toxicants of equal toxic units and full correlation of tolerances

4. To demonstrate that bivariate models including a similarity parameter λ and a correlation coefficient ρ can be fitted based on experimental isobologram points, and in particular, that most of the results of Broderius (1987) in fact can be fitted to models with partial similar action and complete correlation of tolerances

5. To demonstrate the application of multiple toxicity models with partial similar joint action and full correlation of tolerances in the case where the number of toxicants exceeds two

6. To show how these models can be applied in the setting of water quality criteria

II. BASIC MODELS AND CONCEPTS

A. Classification of Bioassays

Evaluation of the toxicity of chemicals acting singly and in combination is carried out by means of bioassays. These are usually conducted in the laboratory under controlled conditions but can also be arranged in the field under appropriate circumstances. Observations in the field of aquatic populations as a function of proximity to points of wastewater discharges, time of exposure, type of effluent, and so forth, constitute the ultimate check on theoretical and experimental toxicity predictions.

The bioassays considered here were carried out in the laboratory. A flow-through testing apparatus is usually preferable since it ensures constant toxicant concentrations and often is more realistic with respect to the simulation of actual field conditions in a lake or stream. Flow-through apparatus for multiple toxicity studies have been described by Anderson and Weber (1975b) and Broderius and Kahl (1985). Batch assays are convenient for testing the effects of chemicals on algal growth (Christensen et al., 1979). However, they have the disadvantage that conditions such as toxicant concentrations, pH, and light may change significantly throughout the duration of the test. Also, when it comes to simulating field conditions, they are usually less realistic than continuous-flow assays.

Possible forms of bioassays are shown in Table 1. Organisms can have a tolerance distribution for individual organisms (groups I and II) or originate from a single clone (groups III and IV). The response is either quantal or continuous. The most frequently adopted type of bioassay is one where the response is quantal and the individual organisms have a tolerance distribution (group I). This is, for example, the case when randomly selected male guppies are exposed to given toxicants and the response is taken as the mortality after a given period of exposure, e.g., 96 h (Anderson and Weber, 1975a, 1975b).

The meaning of a tolerance distribution for individual organisms is that, because of the natural variability among the test population, some organisms with high tolerance will survive at high toxicant concentrations or for long

TABLE 1

Populations of Organisms Considered in Bioassays

Genetic characterization	Type of response	
	Quantal	Continuous
	Group I	*Group II*
Tolerance distribution for individual organisms	Macroorganisms: Response: death of an organism (Anderson and Weber, 1975a, 1975b) Response: reduction in no. of young per female (Spehar and Fiandt, 1986) Classic probit analysis Binomial statistics	Macroorganisms: Response: reduction in weight (Spehar and Fiandt, 1986) Response at the organ system level: fraction of animals responding with significant activity of enzymes such as plasma alanine aminotransferase (ALT), an indicator of liver damage in mice (Shelton and Weber, 1981) Mixed cultures of microorganisms: Response: rate of growth, photosynthesis, and respiration; cell yield
	Group III	*Group IV*
All organisms from a single clone (no tolerance distribution for individual organisms)	Pure culture of microorganisms; Special case: synchronous growth. Response: growth rate based on cell number	Pure culture of microorganisms; General case. Response: rate of growth (Christensen and Nyholm, 1984), photosynthesis, and respiration; cell yield (Christensen et al., 1979)

times of exposure while others with low tolerance will not. The tolerance distribution may be considered to be the fraction of surviving organisms which is equal to the number of organisms having higher tolerance than the actual toxicant concentration divided by the total number of organisms at the start of the bioassay. However, when the organisms originate from a single clone they all have the same genetic material and will, therefore, respond in the same or nearly the same way to given toxicants. This situation occurs when algae or bacteria multiply by cell division from a single original cell.

With regard to the type of response obtained, it is often clearly either quantal, e.g., death of an organism, or continuous, as when the weight of surviving organisms is considered. For microorganisms, growth rate based on cell number in synchronous growth is also clearly a quantal parameter (group III organisms) because cell divisions take place at the same time so that the number of subsequent cell divisions completed always is an integer. However, in other cases it is more ambiguous, as, for example, when the response is measured in terms of the number of young per female. Each female producing offspring will have an integer number of young; however since various offspring-producing females do not obtain the same number of young, the average number of young per female, of, for example, daphnids, will generally not be an integer.

There is a similar ambiguity when the response is taken as the number of organisms responding with a significant plasma alanine aminotransferase (ALT) activity. Once the level of ALT that should be considered significant has been agreed upon, the number of organisms responding is quantal; however, this number depends on the somewhat arbitrarily chosen reference level of ALT since the ALT activity is a continuous parameter.

Acknowledging these uncertainties, reduction in the number of young per female has been listed under quantal response, and the fraction of animals responding with significant ALT activity is listed under continuous response (Table 1).

B. Dose-Response Curves for One Toxic Substance

The units of the linear transforms of the log-normal and logistic transformations are called probits and logits after *prob*ability un*its* and *log*istic un*its,* respectively. Following this practice we propose here that the units of the linear Weibull transformation be named weibits as an abbreviation for *Weibull un*its.

Log-linear response transformations of the weibut, probit, and logit types are given in Table 2. For a given toxicant concentration $z,$ the probability of response P indicates the number of dead animals divided by the initial total number of animals (group I organisms) or the reduction in growth rate based on cell number (group III organisms). The transformations are, therefore, toler-

TABLE 2

Comparison of the Weibit Transformation with Probit
and Logit Transformations

Type	Transformation[a]	Probability of response or relative inhibition
Weibit	$u = \ln k + \eta \ln z$	$P = 1 - \exp(-e^u)$
Probit	$Y = \alpha + \beta \log z$	$P = \dfrac{1}{2}\left(1 + \operatorname{erf}\dfrac{Y - 5}{\sqrt{2}}\right)$
Logit	$l = \theta + \phi \ln z$	$P = \dfrac{1}{1 + e^{-l}}$

[a] z is a toxicant concentration; $k, \eta, \alpha, \beta, \theta, \phi$ are constants; $A = \ln k$.

ance distributions. However, we will here assume that they also apply to the types of continuous responses listed in Table 1 (group II and IV organisms).

While the probit and logit transformations must be considered mainly empirical, we believe that in some cases there may be claimed a mechanistic-probabilistic basis for the Weibull model. This model can, in the case of algal growth, be derived from the basic assumption that the probability for a cell division to be blocked because of a differential increase in toxicant concentrations is proportional to the subsequent increase in the concentration of blocked receptors (Christensen and Nyholm, 1984). For group I organisms, the basic assumption leading to the Weibull model is that the hazard function is proportional to the product of a function of time and the concentration of blocked receptors (Christensen and Chen, 1985).

Comparisons of fits based on the three transformations listed in Table 2 have been carried out previously for macroorganisms and algae (Christensen, 1984, 1987; Christensen and Nyholm, 1984; Christensen et al., 1985b). The fits based on the Weibull model have generally been at least as good or better than fits based on the probit or logit models, thus supporting the suggested mechanistic-probabilistic basis of this model. A typical, not previously published, comparison is shown in Fig. 1 and Table 3, which summarize dosage-mortality data for fathead minnows exposed to uncoupler chemicals. It should be noted that the data in Fig. 1 represent a composite of tests with nine separate chemicals normalized according to the LC_{50} value of 2,4-dinitrophenol. Thus, the normalization is made with respect to the LC_{50} values only and not the slopes, meaning that a comparison based on χ^2 is not entirely correct. However, since the slopes of the individual dose-response curves for the nine chemicals are fairly similar, and since the responses are fairly evenly distributed between

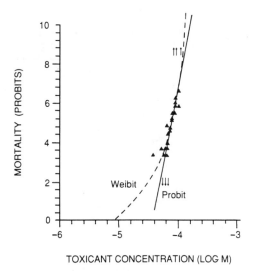

FIG. 1. Fitting of probit and Weibull functions to experimental dose-response data for juvenile fathead minnows exposed to uncoupler test chemicals. (Raw data from Broderius, 1987.)

0 and 100%, a comparison may still be valid. As may be seen from Fig. 1 and Table 3, the Weibull model appears indeed to provide the best fit.

The Weibull model was also compared to the probit and logit models with respect to its ability to predict LC_{50}'s with lower and upper 95% confidence limit for hypothetical test data promulgated by the American Society for Testing and Materials (ASTM) (Christensen, 1984). The results, which are reproduced in Table 4, indicate that the values obtained from the Weibull model are within the ranges recommended by ASTM (1980), except for the LC_{50} and lower 95%

TABLE 3

Fits of Weibit, Probit, and Logit Models to Dose-Response Data for Uncoupler Chemicals

Type	Intercept A (ln k, α, or θ)	Slope B	
		η, β, or ϕ	χ^2(28 df)[a]
Weibit	61.99	6.58	54.7
Probit	55.39	12.20	563.9
Logit	91.37	9.61	77.7

Source: Raw data from Broderius (1987).
[a]df: degrees of freedom.

TABLE 4

Fit of the Weibit, Probit, and Logit Models to Standard Test Data by the American Society for Testing and Materials (ASTM, 1980)

| Parameter | Case | Toxicant concentration (mg/L) | | | |
		ASTM	Weibit	Probit	Logit
LC_{50}	C	34.2–38.5	37.5	35.5	36.1
	D	29.1–30.0	31.1	29.5	29.6
	E	33.9–40.6	37.8	35.4	36.1
LL^a	C	26.1–30.7	30.1	28.8	28.9
	D	22.2–23.9	24.3	23.8	23.5
	E	26.1–30.8	29.8	28.2	28.5
UL^a	C	43.4–47.2	47.0	44.1	45.4
	D	36.3–38.9	38.3	36.6	37.0
	E	44.3–46.6	45.7	44.5	46.0
χ^{2b}	C		0.40	1.38	1.50
	D		0.39	0.30	0.57
	E		2.71	3.95	4.29

Source: Christensen (1984). Reprinted with permission from Pergamon Press, Ltd.
[a]LL, UL: lower and upper 95% confidence limits.
[b]Four degrees of freedom.

confidence limit in case D. The values are in almost all cases higher than those based on the probit and logit models.

Although the mortality data provided by ASTM are hypothetical, they are typical, for example, in fish bioassays, and it is therefore of some value to compare the goodness of fit as measured by χ^2. As may be seen from Table 4, none of the models shows a significant lack of fit. Nevertheless, in cases C and E where a difference in fit could be argued, the χ^2 values for the Weibull model are in fact lowest, thus suggesting the better fit of this model. Because of this, and the cumulative evidence of the merits of the Weibull model from other studies, we believe that a revision of the ranges of acceptable values given by ASTM should be contemplated. Specifically, we suggest that the ranges for LC_{50} and lower 95% confidence limit in case D should be changed to include the values found from application of the Weibull model, that is 31.1 and 24.3 mg/L, respectively.

C. Noninteractive Multiple Toxicity Model

A multiple toxicity model is noninteractive when the presence of one toxicant does not affect the combination of other toxicants with receptors or their intrinsic activity (Plackett and Hewlett, 1967). Consider the quantal response of organisms to two toxicants. The nonresponse probability Q (i.e., the survival fraction) can be expressed in the following form according to Hewlett and Plackett (1959):

$$Q = \Pr(\delta_1^{1/\lambda} + \delta_2^{1/\lambda} \leq 1) \tag{2}$$

where $\Pr$ = probability
    $\delta_i = z_i/\bar{z}_i$
    z_i = concentration of toxicant i
    $\bar{z}_i$ = concentration tolerance of an individual organism to toxicant i
    λ = similarity parameter for the action of two toxicants on two biological systems

We consider here cases where the response transform can be expressed in a log-linear form:

$$x_i = \alpha_i + \beta_i \log z_i$$
$$u_i = \alpha_i + \beta_i \log \bar{z}_i \tag{3}$$

where α_i, β_i = constants for toxicant i
    x_i = log transform of the concentration z_i
    u_i = log transform of the concentration tolerance $\bar{z}_i$

Hewlett and Plackett (1959) assumed that the distribution of log tolerances u_i was bivariate normal with a correlation coefficient ρ. Christensen and Chen (1985) expanded this model to include n toxicants and to allow an arbitrary monotone distribution, e.g., of the logit or the weibit type for the individual toxicants. For example, in the case of three toxicants, the nonresponse probability Q is calculated from

$$Q = \Pr(\delta_1^{1/\lambda_{12}} + \delta_2^{1/\lambda_{12}} \leq 1; \; \delta_1^{1/\lambda_{13}} + \delta_3^{1/\lambda_{13}} \leq 1;$$
$$\delta_2^{1/\lambda_{23}} + \delta_3^{1/\lambda_{23}} \leq 1; \; \delta_1^{1/\lambda_{123}} + \delta_2^{1/\lambda_{123}} + \delta_3^{1/\lambda_{123}} \leq 1) \tag{4}$$

where the symbols have the same meaning as above and where λ_{123} is the similarity parameter for the action of three toxicants on three biological systems.

The log tolerances are distributed according to the multivariate normal density function

$$f(\tilde{U}) = \frac{1}{(2\pi)^{n/2}|A|^{1/2}} \exp\left(-\frac{1}{2} U^* A^{-1} U\right) \tag{5}$$

where A is the correlation matrix

$$\mathbf{A} = \begin{bmatrix} 1 & \rho_{12} & \cdots & \rho_{1n} \\ & 1 & \cdots & \rho_{2n} \\ \text{Symmetric} & & \cdots & 1 \end{bmatrix} \tag{6}$$

and $\mathbf{U}^*$ is a vector of normal equivalent deviates (NEDs) of log tolerances

$$\mathbf{U}^* = (U_1, U_2, \ldots, U_n) \tag{7}$$

In order to evaluate Q by equations such as (4), it is necessary to integrate the density function, Eq. (5), over the appropriate region in the n-dimensional space of normal equivalent deviates. To facilitate this integration, the correlation matrix $\mathbf{A}$ is diagonalized by an orthogonal transformation such that the density function Eq. (5) can be written as a product of n univariate normal density functions.

A requirement for the diagonalization is that the correlation matrix (6) have positive eigenvalues, meaning that it should be positive definite. We believe that this is a general requirement for the multiple toxicity model which also for other transformations is dependent on the density function, Eq. (5). This does not mean, however, as was suggested by Konemann (1981b) that the concept of negative correlation has no application for mixtures with more than two toxicants. We shall illustrate this by an example with three toxicants. Consider the following correlation matrices:

$$\mathbf{A}_1 = \begin{bmatrix} 1 & 0.93 & 0.81 \\ & 1 & 0.79 \\ \text{Sym.} & & 1 \end{bmatrix} \tag{8}$$

$$\mathbf{A}_2 = \begin{bmatrix} 1 & -0.70 & 0.91 \\ & 1 & -0.65 \\ \text{Sym.} & & 1 \end{bmatrix} \tag{9}$$

$$\mathbf{A}_3 = \begin{bmatrix} 1 & -0.49 & -0.66 \\ & 1 & -0.38 \\ \text{Sym.} & & 1 \end{bmatrix} \tag{10}$$

The first matrix $\mathbf{A}_1$ indicates positive correlation of tolerances between the three toxicants. The eigenvalues are positive (2.6882, 0.2426, 0.0692) and the biological meaning is clear in that organisms that are very susceptible to toxicant 1 also are fairly susceptible to toxicants 2 and 3.

In the second case, Eq. (9), the tolerances to toxicants 1 and 2 are negatively correlated, and the third toxicant correlates positively with toxicant 1 but negatively with toxicant 2. The eigenvalues (2.5126, 0.4004, 0.0870) here are also positive, and the result is biologically acceptable. Thus, it is indeed possible to have negative correlations for more than two toxicants. However, considering the third case, Eq. (10), given the negative correlation between toxicants 1 and 2, it is conceivable to have highly negative correlations between a third toxicant and toxicant 1 or 2, but not both of them at the same time. In this case one of the eigenvalues (1.6754, 1.3536, -0.0291) is negative. In conclusion, some correlations can be negative as long as the eigenvalues are positive.

From Eqs. (2) and (3) the integration region

$$\delta_1^{1/\lambda} + \delta_2^{1/\lambda} \leq 1$$

can be written

$$\text{hexp} \frac{x_1 - u_1}{\lambda\beta_1} + \text{hexp} \frac{x_2 - u_2}{\lambda\beta_2} \leq 1 \tag{11}$$

where hexp a means 10 raised to the power a. In the bivariate normal case, the nonresponse probability Q is then determined by integrating the bivariate normal density function Eq. (5) with $n = 2$, over the region given by inequality (11).

Bivariate distributions of other types, e.g., weibit or logit, are constructed according to the procedure given by Christensen and Chen (1985). For example, in case of a bivariate weibit distribution, the nonresponse probability Q is calculated from Eq. (2) using the log-linear Weibull transforms:

$$x_i = \ln k_i + \eta_i \ln z_i$$
$$u_i = \ln k_i + \eta_i \ln \bar{z}_i \tag{12}$$

so that the integration region in a bivariate Weibull plane will be given by

$$\exp \frac{x_1 - u_1}{\lambda\eta_1} + \exp \frac{x_2 - u_2}{\lambda\eta_2} \leq 1 \tag{13}$$

This region R is then mapped on a bivariate NED plane into a region R' and Q is determined by integration of the bivariate normal function Eq. (5) with $n = 2$, over R'. It can be shown that bivariate distributions constructed according to this procedure satisfy six fundamental requirements of noninteractive tolerance distributions. Similar procedures can be followed in the n-dimensional case.

Results of calculations carried out by Christensen and Chen (1985) using a

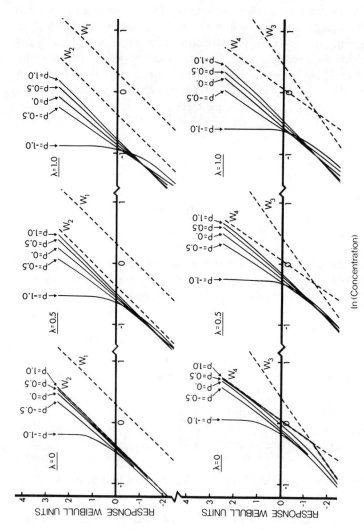

FIG. 2. Response weibits *vs.* ln (concentration) for a bivariate Weibull model with parallel (W_1, W_2) and nonparallel (W_3, W_4) dose-response lines for the individual toxicants. (Reproduced from E. R. Christensen and C.-Y. Chen, "A General Noninteractive Multiple Toxicity Model Including Probit, Logit, and Weibull Transformations." *Biometrics* 41:711–725, 1985. With permission from The Biometric Society.)

140

bivariate Weibull model are reproduced in Fig. 2. The dose-response curves are either parallel (W_1, W_2) or nonparallel (W_3, W_4):

$$W_1: u = -0.8372 + 2.651 \ln z \tag{14}$$

$$W_2: u = 0.9940 + 2.651 \ln z \tag{15}$$

$$W_3: u = -0.8372 + 1.856 \ln z \tag{16}$$

$$W_4: u = 0.2264 + 3.712 \ln z \tag{17}$$

The concentrations of each toxicant in the mixture are equal and given by the logarithmic abscissa. The response curves follow the same general trend as that noted by Hewlett and Plackett (1959) for the bivariate normal case. Because the slopes are >1, the response for concentration addition ($\lambda = 1$, $\rho = 1$) is higher than that for response multiplication ($\lambda = 0$, $\rho = 0$). Also, the response for response addition ($\lambda = 0$, $\rho = -1$) is higher than that for response multiplication which, in turn, exceeds that for no addition ($\lambda = 0$, $\rho = 1$). It may also be verified that for a given correlation coefficient ρ, the response increases with increasing overlap of the two biological systems affected, that is for increasing similarity coefficient λ.

We shall, in the following, demonstrate how analytical expressions for isobolograms can be derived for noninteractive multiple toxicity models in the case where the tolerances are fully correlated. Consider a general log-linear transformation:

$$x_i = a_i + b_i \ln z_i$$
$$u_i = a_i + b_i \ln \bar{z}_i \tag{18}$$

where a_i, b_i are constants and x_i, u_i are log-linear response transforms for the ith toxicant. This system is equivalent to those given by Eqs. (3) or (12), except that there is here no requirement for the response transform to follow any particular function such as Weibull, probit, logit, and so forth. The only condition is that the probability of response must be a monotone function of u.

In the case of two toxicants the integration region in Eq. (2) can be written as follows, using Eq. (18);

$$\exp \frac{x_1 - u_1}{\lambda b_1} + \exp \frac{x_2 - u_2}{\lambda b_2} \leq 1 \tag{19}$$

Full correlation of the tolerances to toxicant 1 and 2 means that the integration region degenerates into a straight line forming a $45°$ angle with the u_1 and u_2 axes. The line is limited to the left by the point of intersection between the line $u_1 = u_2$ and the curve defined by Eq. (19), omitting the inequality sign. The corresponding value of the response transform is called $u' = u_1 = u_2$. In order to determine isoboles, we must maintain a constant response u'. Thus the isoboles are determined by inserting $u' = u_1 = u_2$ into Eq. (19) and omitting the inequality sign:

$$\exp \frac{x_1 - u'}{\lambda b_1} + \exp \frac{x_2 - u'}{\lambda b_2} = 1 \tag{20}$$

To the response u' correspond the following individual tolerances $\bar{z}_1$ and $\bar{z}_2$ of the two toxicants according to Eq. (18):

$$\bar{Z}_i = \exp \frac{u' - a_i}{b_i} \qquad i = 1, 2 \tag{21}$$

However, these expressions also define the concentrations of the two toxicants acting singly giving the desired response because of the definition of the probability of response and because all organisms with higher tolerances will survive. Thus we may write $Z_1 = \bar{Z}_1$ and $Z_2 = \bar{Z}_2$, where Z_1 and Z_2 are the concentrations of the two toxicants acting singly giving the desired response transform u'. From Eq. (18), the toxicant concentrations z_i are

$$z_i = \exp \frac{x_i - a_i}{b_i} \qquad i = 1, 2 \tag{22}$$

The expression for the isoboles is now determined by inserting Eqs. (21) and (22) with $\bar{Z}_i = Z_i; i = 1, 2$ into Eq. (20):

$$\left(\frac{z_1}{Z_1}\right)^{1/\lambda} + \left(\frac{z_2}{Z_2}\right)^{1/\lambda} = 1 \tag{23}$$

where z_1, z_2 are the actual values of toxicant concentrations giving the desired response, e.g., 50% mortality, and Z_1, Z_2 are the concentrations of the toxicants acting singly giving this response.

Equation (23) is valid only for full correlation of the toxicant tolerances ($\rho = 1$). For noninteractive models, the similarity parameter λ is restricted to assume values between 0 and 1. However, the expression is very general in that it depends neither on the type of log-linear transform, e.g., weibit, probit, or logit, nor on the magnitude of the slopes for the individual dose-response curves. Thus there is no requirement of parallelism for these curves. The isoboles are also independent of the level of response selected as the reference level, e.g., 10 or 50% mortality. Note that the isobole for concentration addition is obtained by setting $\lambda = 1$:

$$\frac{z_1}{Z_1} + \frac{z_2}{Z_2} = 1 \tag{24}$$

For n toxicants, Eq. (23) becomes

$$\left(\frac{z_1}{Z_1}\right)^{1/\lambda} + \left(\frac{z_2}{Z_2}\right)^{1/\lambda} + \cdots + \left(\frac{z_n}{Z_n}\right)^{1/\lambda} = 1 \tag{25}$$

where λ is the overall similarity parameter, e.g., λ_{123} of Eq. (4); z_i is the concentration of toxicant i in the mixture giving the desired response; and Z_i is the concentration of toxicant i, acting alone, giving this response. It is assumed that the integration region in equations such as Eq. (4) is specified only in terms of the overall similarity parameter λ.

D. Interactive Multiple Toxicity Models

The starting point for developing interactive models has often been isobolograms (Hewlett, 1969). It is interesting to note that Hewlett proposed the following isobole for two toxicants exhibiting potentiation:

$$\left(\frac{z_i}{Z_1}\right)^{\gamma} + \left(\frac{z_2}{Z_2}\right)^{\gamma} = 1 \tag{26}$$

where $\gamma < 1$. Several sets of experimental data on the action of insecticides were successfully fitted to this model. However, no data were found to fit the equation for $\gamma > 1$.

Equation (26) is, of course, equivalent to Eq.(23) for $\gamma = 1/\lambda > 1$. It is apparent that Hewlett (1969) or Hewlett and Plackett (1979) did not realize that their parameter γ in fact is equal to the reciprocal of the similarity parameter λ when there is full correlation of tolerances. As we shall see later, Eq. (23) indeed fits several recent experiments with $0 < \lambda < 1$. These toxicants may therefore be characterized as noninteractive.

There is a single case for which $\lambda > 1$ or $\gamma < 1$, and this will be called interactive. In the latter case, the toxicants are more than additive and the isobole is situated below the concentration addition line. In view of Eq. (23), we may therefore say that the two biological systems affected not only overlap fully but, since λ is greater than one, that they are folded onto themselves and therefore are subjected to increased toxic action relative to concentration addition.

In order to account for deviations from concentration addition, Finney (1942) proposed the following isobole

$$V_1 + \kappa (V_1 V_2)^{1/2} + V_2 = 1 \tag{27}$$

where $V_1 = z_1/Z_1$, $V_2 = z_2/Z_2$, and κ is a constant. A positive value of κ means potentiation (or more than additive action) and a negative value less than additive action. Finney (1942) arrived at the isobole for concentration addition by assuming parallel probit lines for the individual toxicants. However, as shown by Hewlett and Plackett (1959) and, also, in the above derivation of Eq. (24), such an assumption is unnecessary. Equation (27) is symmetrical in V_1 and V_2. Asymmetrical versions have been proposed by Hewlett and Plackett (1979) and by Durkin (1981). While Eq. (27) and related asymmetrical versions may be useful to describe experimental data, they are purely empirical and therefore of limited value compared to Eq. (23) which has more of a theoretical foundation, even for $\lambda > 1$.

Models for competitive action based on the law of mass action for the combination of agonists and antagonists with receptors of the organism were developed by Hewlett and Plackett (1964). In their terminology, an agonist is a toxicant that, taken separately, can induce a given response, while an antagonist decreases the effect of the agonist without being able to produce the toxic response by itself. One version of their model for the response x in NEDs to the combined action of an agonist and an antagonist is

$$x = \alpha + \beta \log \frac{z_1}{1 + \psi z_2} \tag{28}$$

where α, β and ψ are constants, z_1 is the concentration of the agonist (toxicant), and z_2 that of the antagonist. It is clear from Eq. (28) that a high concentration z_2 of the antagonist will decrease the response, and that the response will reduce to the single toxicant probit equation (Table 2) when the concentration of the antagonist $z_2 = 0$. Hewlett and Plackett (1964) also showed that the competitive model for two agonists (toxicants), which both can produce the same response, is equivalent to that for similar action, Eq. (2) with $\lambda = 1$:

$$Q = \Pr(\delta_1 + \delta_2 \le 1) \tag{29}$$

Thus, Eq. (29) can be arrived at from the assumption of both noninteractive as well as interactive toxicants.

According to Hewlett and Plackett (1979), synergism occurs when a compound (synergist) lowers the concentration of a toxicant for which a certain biological response takes place. However, the synergist itself is unable to produce the toxic response. Hewlett and Plackett (1979) suggested the following empirical model for synergism:

$$Y = \alpha + \beta_1 \log z_1 + \frac{\beta_2 z_2}{\omega + z_2} \tag{30}$$

where α, β_1, β_2, and ω are constants; z_1 and z_2 are the concentrations of toxicant and synergist, respectively; and the response Y is expressed in terms of probits. When the synergist concentration z_2 is zero, the expression reduces to the probit equation. Also, for constant response Y', the necessary toxicant concentration decreases to the limiting value

$$z_1' = \text{hexp} \frac{Y' - \alpha - \beta_2}{\beta_1} \tag{31}$$

for high values of the concentration z_2 of the synergist. Equation (30) was fitted to data for the toxicity to flour beetles of the insecticide pyrethrin with the synergist piperonyl butoxide by Hewlett (1969) and Finney (1971).

E. Exposure Time as a Variable

The response variable in most toxicity tests is percent mortality within a fixed time of exposure, e.g., 96 h (EIFAC, 1980). For the purpose of comparing the toxicity of various chemicals and to evaluate the nature of the joint action of mixtures, this type of bioassay is often quite adequate.

Nevertheless, in some cases it is of interest to consider the response as a function not only of toxicant concentration but also of time of exposure. For example, in order to simulate the exposure of fish to toxicants in wastewater discharged into a river or a lake, the effective time of exposure becomes an important variable. One way of taking time into account in the expressions for the probability of response (Table 2) is to require that the following time-concentration relationship be fulfilled for constant response:

$$t^n C = K \tag{32}$$

where t is the exposure time, C is the toxicant concentration, and K and n are constants. This relationship has been found to be approximately valid under a variety of circumstances, such as disinfection (Watson, 1908) and induction of tumors in laboratory animals (Druckrey, 1967). The exponent n is often in the range between 0.5 and 3. From Eq. (32), a given response can be obtained either with a high concentration C of the toxicant and short exposure time t or with a long exposure time and a low level of the toxicant. By taking logarithms of Eq. (32) we obtain

$$n \ln t + \ln C = \ln K \tag{33}$$

Therefore, the obvious extensions of the weibit, probit, and logit transformations (Table 2) to include time are

$$\text{Weibit: } u = \ln k + \eta_1 \ln z + \eta_2 \ln t \tag{34}$$

$$\text{Probit: } Y = \alpha + \beta_1 \log z + \beta_2 \ln t \tag{35}$$

$$\text{Logit: } l = \theta + \phi_1 \ln z + \phi_2 \ln t \tag{36}$$

All of these transforms are linear in $\ln z$ and t and, therefore, represent planes in the $\ln z$, $\ln t$ response transform space.

The procedure to evaluate the response to two or more toxicants, considering time as as a variable, is the same as the one outlined above through Eqs. (2) through (13) except that the response transforms, for example for the weibit model, rather than being given by Eq. (12), now are expressed by

$$x_i = \ln k_i + \eta_{1i} \ln z_i + \eta_{2i} \ln t$$
$$u_i = \ln k_i + \eta_{1i} \ln \bar{z}_i + \eta_{2i} \ln t \tag{37}$$

where η_{1i} and η_{2i} are constants for toxicant i, and the other symbols have the same meaning as above.

III. JOINT ACTION OF TOXICANTS ILLUSTRATED BY ρ, λ DIAGRAMS

Once the slopes and intercepts for the log-linear transformations of two toxicants are known, the bivariate toxicity model represented by Eq. (2) is characterized by two parameters: the correlation coefficient ρ and the similarity parameter λ. Estimation of these parameters by maximum likelihood procedures is addressed in a subsequent section.

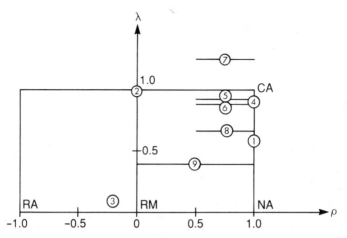

FIG. 3. Fitted values of the correlation coefficient ρ and the similarity parameter λ from nine bivariate toxicity experiments (Table 6). The position of the circled indices indicate the ρ, λ values with estimated uncertainties equal to the extent of the circles, or the horizontal bars (indices 5–9).

To illustrate the estimated parameter values, we find it convenient to plot them in a ρ, λ diagram (Fig. 3). In this figure the symbols CA, NA, RM, and RA stand for concentration addition, no addition, response multiplication, and response addition, respectively. The symbols are placed at the corresponding values of ρ and λ, which are also indicated in Table 5.

Pairs of estimated ρ, λ values from various experimental data (Table 6) are indicated by circled index numbers running from 1 to 9. Parameters from bioassays with indices 1 through 4 were determined from dose-response data including widely different response levels, while values from bioassays 5 through 9 were estimated from experimental isobologram points. The latter data had

TABLE 5

Definition of Basic Modes of Action of the Bivariate Noninteractive
Toxicity Model Represented by Eq. (2)

Type of action	Abbreviation	Parameter values	
		ρ	λ
Concentration addition	CA	1	1
No addition	NA	1	0
Response multiplication	RM	0	0
Response addition	RA	−1	0

TABLE 6

List of Bioassays the Results of Which Have Been Fitted to the Noninteractive
Bivariate Toxicity Model, Eq. (2)[a]

Index	Toxicants	Organisms	Reference in which fitting is documented
1	β-Benzene hexachloride, pyrethrins	Flour beetles (*Triboleum castaneum*)	Christen and Chen, 1985
2	Ni, Cu	Male guppies (*P. reticulata*)	Christensen, 1987
3	Pentachlorophenol, dieldrin	Male guppies (*P. reticulata*)	Christensen, 1987
4	Ni^{2+}, Zn^{2+}	Green algae (*Selenastrum capricornutum*)	Christensen, 1987
5	Octanol, 2-octanone	Fathead minnows (*P. promelas*)	Present work
6	Hydrocyanic acid, rotenone	Fathead minnows (*P. promelas*)	Present work
7	Octanol *n*-octyl cyanide	Fathead minnows (*P. promelas*)	Present work
8	2,4-Dinitrophenol, rotenone	Fathead minnows (*P. promelas*)	Present work
9	Octanol, 2,4-pentanedione	Fathead minnows (*P. promelas*)	Present work

[a]The results of the fitting are shown in Fig. 3.

markedly greater uncertainties in the estimated ρ values. This is shown in the figure by the inclusion of horizontal bars indicating the estimated uncertainty in ρ. In other cases, the extent of the circles surrounding the index numbers gives an approximate indication of the uncertainty in ρ and λ. Most of the nine data points are in the square for which $0 \leq \rho \leq 1$ and $0 \leq \lambda \leq 1$, with an apparent increase in frequency around concentration addition. Point 3 is close to response multiplication, and point 7 indicates more than additive action.

The probability of response P can be visualized as a response surface in a three-dimensional plot of P vs. ρ and λ (Fig. 4). We have here assumed that the two individual toxicants follow weibit transformations (Table 2) with the following values of the parameters: $k_1 = k_2 = 1$; $\eta_1 = 5$, $\eta_2 = 3$; $z_1 = 0.7794$, $z_2 = 0.7552$. Thus, the log tolerances corresponding to these values are

$$u_1 = 5 \ln 0.7794 = -1.2462$$

$$u_2 = 3 \ln 0.7552 = -0.8423 \tag{38}$$

The corresponding response probabilities for the individual toxicants are

$$P_1 = 1 - \exp\left[-\exp\left(-1.2462\right)\right] = 0.25$$

$$P_2 = 1 - \exp\left[-\exp\left(-0.8423\right)\right] = 0.35 \tag{39}$$

The response surface in Fig. 4 was calculated by means of the computer program MULTOX (Chen and Christensen, 1986) in which the nonresponse probability Q is calculated from Eq. (2).

We can verify that the responses of Fig. 4 for no addition (NA), response multiplication (RM), and response addition (RA) are in accordance with the above values of P_1 and P_2 and the rules for calculating the resulting probability of response P (Hewlett and Plackett, 1979):

$$\text{NA: } P = \max (P_1, P_2) = P_2 = 0.35 \tag{40}$$

$$\text{RM: } P = 1 - (1 - P_1)(1 - P_2) = 1 - 0.75 \times 0.65 = 0.51 \tag{41}$$

$$\text{RA: } P = \min (1, P_1 + P_2) = P_1 + P_2 = 0.60 \tag{42}$$

The response for RA is always greater than that for RM which, in turn, is greater than the response for NA. It may also be noted that the response for CA

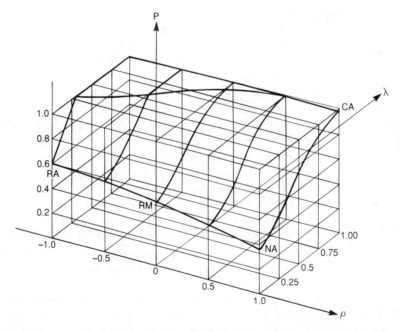

FIG. 4. Three-dimensional illustration of the response P as a function of the correlation coefficient ρ and the similarity parameter λ. The two individual toxicants follow weibit transformations. Toxicant concentrations and model parameters are given in the text.

is greater than that for RM as predicted by Christensen and Chen (1985) when the slopes η_1, η_2 of the weibit transformations both are > 1.

Finally, it is clear that the response always increases for fixed ρ and increasing λ. This is a result of the nature of the integration region as expressed, for example, by Eqs. (12) and (13). When λ increases, say from λ_1 to λ_2 ($\lambda_2 > \lambda_1$), the corresponding integration region R_2 will be fully contained within the first region R_1 such that the nonresponse probability Q_2 will be smaller than the original value Q_1. Therefore, the probability of response P_2 will be greater than P_1 where $P = 1 - Q$.

IV. JOINT ACTION OF TOXICANTS ILLUSTRATED BY ISOBOLOGRAMS

It is often convenient to express the results of bioassays with two toxicants in the form of points in an isobologram (Broderius and Kahl, 1985; Broderius, 1987). Muska and Weber (1977) constructed isobolograms for CA, RA, RM, and NA based on hypothetical parallel dose-response curves for individual toxicants. Their terminology differs from ours in that they call all three modes of action RA, RM, and NA response addition. Their isobologram looks approximately like the top middle one of Fig. 5 for which index = 1 and $Q = 0.5$. The indices relate to the slopes of the individual log-linear transformations of the two toxicants (Table 7). We are here using the weibit transformation, but the isobolograms for other transformations are not very different from the ones shown here provided that the slopes are scaled appropriately.

Muska and Weber's isobologram was reproduced by Durkin (1981) and Broderius (1987), and without further information one might be led to think that the isobologram always should look like theirs, i.e., similar to the top middle one of Fig. 5. That this is not the case becomes evident from an inspection of all 12 isobolograms of Fig. 5. The isoboles for RA and RM are strongly dependent on the slopes of the weibit transformations such that they may be situated above the CA line when the slopes are greater than one (index 1) and below this line when these slopes are less than one (index 3). When one slope is > 1 and the other < 1 (index 4), the isoboles for RA and RM are both above and below the CA line. As previously noted by Christensen and Chen (1985) and by Kodell (1987), the isoboles for RM and CA become equivalent when both of the slopes are 1 (index 2).

From Fig. 5 we further note that the NA and CA isoboles are independent of the slopes and the response levels. For a particular set of slopes (index value), the RA isobole is the only one that depends on the level of response (Q), and the RA isobole always indicates a more toxic mixture than the RM isobole. The NA isobole, given by the two sides of the square opposite to the coordinate axes,

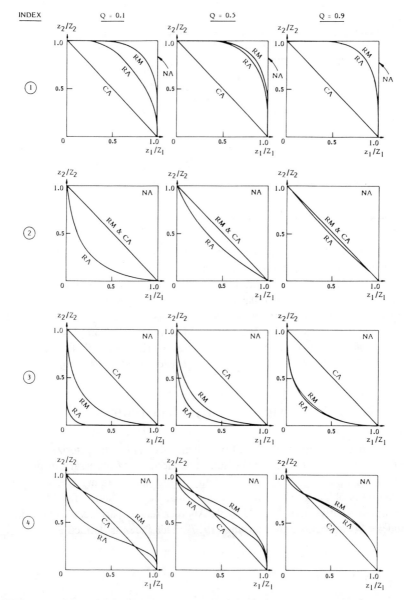

FIG. 5. Isobolograms for pairs of toxicants characterized by indices 1–4 at nonresponse levels $Q = 0.1$, 0.5, and 0.9. Weibit parameters for the individual toxicants of these pairs, along with effect concentrations of the toxicants acting singly, are given in Table 7.

TABLE 7

Weibit Parameters and Effect Concentrations for the Two Toxicants Acting Singly, Used in the Calculation of the Isobolograms Shown in Fig. 5

				Effect concentrations for toxicants acting singly at given response levels					
	Weibit parameters			$Q = 0.1$		$Q = 0.5$		$Q = 0.9$	
Index	k_1	k_2	η_1	η_2	Z_1	Z_2	Z_1	Z_2	Z_1	Z_2
1	1	1	5	3	1.182	1.321	0.929	0.885	0.638	0.472
2	1	1	1	1	2.303	2.303	0.693	0.693	0.105	0.105
3	1	1	0.5	0.5	5.302	5.302	0.481	0.481	0.0111	0.0111
4	1	1	0.5	3	5.302	1.321	0.481	0.885	0.0111	0.472

indicates in all cases the least toxic combination. The toxicity is here determined by only one of the two toxicants.

Most bioassays are characterized by slopes > 1 in the weibit transformations for individual toxicants so that the isoboles of the type labeled index 1 would apply. However, there are cases where the slope is < 1, e.g., around 0.5 such that the other types of isobolograms could be relevant. Examples are bioassays with rainbow trout fry exposed to cadmium (Christensen, 1984) and bioassays with algae inhibited by various metals (Christensen et al., 1985b).

V. MULTIPLE TOXICITY INDICES

The degree of joint action of toxic chemicals can be characterized by means of a suitable toxicity index. The sum of toxic units M necessary to produce a given response, e.g., 50% fish mortality, is one such parameter. Concentration addition is obtained when $M = 1$, more than additive action for $M < 1$, and less than additive action for $M > 1$. Another related parameter is the additive index (AI) introduced by Marking (1977). This index is defined as

$$AI = \frac{1}{M} - 1 \qquad \text{for } M \leq 1 \tag{43}$$

$$AI = 1 - M \qquad \text{for } M > 1 \tag{44}$$

The additive index is zero for concentration addition, positive for more than additive action, and negative for less than additive action. When AI is positive,

$1 + AI$ indicates the toxicity magnification factor relative to CA, and when AI is negative, $1 - AI$ is the toxicity reduction factor relative to CA.

Although M and AI are useful as toxicity indices in many cases they have the following disadvantages. Their values depend on the relative proportion of chemicals in the mixture, and although they have constant values for CA this is not the case for NA since their values for the latter mode of action depend on the number of chemicals in the mixture. For example, mixtures of 10 and 15 chemicals can have M values of 10 and 15, respectively, for no addition (NA). Furthermore, because of the logarithmic form of the concentration in the log-linear transformations for the individual toxicants (Table 2), it would be desirable to have a toxicity index which is logarithmic in toxicant concentration.

In order to overcome these difficulties, Konemann (1981b) introduced a multiple toxicity index (MTI) as follows:

$$MTI = 1 - \frac{\log M}{\log M_0} \qquad (45)$$

where $M_0 = M/f_{max}$
$\quad f_{max}$ = largest value of z_i/Z_i in the mixture
$\quad z_i$ = concentration of toxicant i in the mixture
$\quad Z_i$ = concentration of toxicant i, acting singly, giving the desired response
$\quad M = \Sigma_{i=1}^n z_i/Z_i$ = sum of toxic units giving the desired response
$\quad n$ = number of chemicals in the mixture

When the concentration z_i of each chemical relative to its effect concentration Z_i, when acting alone, is a constant f for all chemicals, $f = z_i/Z_i$, Eq. (45) reduces to

$$MTI = 1 - \frac{\log M}{\log n} \qquad (46)$$

The application of these toxicity indices is illustrated in Tables 8 and 9 which summarize selected toxicity data from Broderius and Kahl (1985). According to these authors, the chemicals of Table 8 can all be modeled by a structure-toxicity relationship characteristic of a narcosis type of action. One might therefore expect near concentration addition when these chemicals are combined into mixture numbers 9, 11, and 13 (Table 9), and this mode of action is in fact indicated by the values of the three toxicity indices: M and MTI are close to one, and AI is approximately zero.

We shall in the following show that Konemann's MTI as defined by Eq. (46) is identical to the overall similarity parameter λ, e.g., λ_{123} of Eq. (4), when the

TABLE 8

Organic Chemicals Tested in Multiple Toxicity Experiments[a]

No.	Class	Name
4	Alcohol	1-Octanol
8	Ketone	2-Octanone
12	Ether	Diisopropyl ether
16	Alkyl halide	Tetrachloroethylene
19	Benzene	1,3-Dichlorobenzene
22	Nitrile	n-Octyl cyanide
25	Tertiary aromatic amine	n,n-Dimethyl-p-toluidine

[a]Classification adopted from Broderius and Kahl, 1985.

toxicant tolerances are fully correlated. In this case we obtain the following relationship from Eq. (25), the definition of M, and the requirement of equal toxic units, i.e., $z_i/Z_i = M/n$:

$$n \left(\frac{M}{n} \right)^{1/\lambda} = 1 \tag{47}$$

By taking logarithms of this equation, we have

$$\lambda = 1 - \frac{\log M}{\log n} \tag{48}$$

TABLE 9

Examples of Mixtures of Organic Chemicals and Corresponding Toxicity Indices for Tests with Juvenile Fathead Minnows (*Pimephales promelas*)

	Chemicals in mixture (Table 8)			Toxicity indices		
Mixture no.	Nos.	Total (n)		Sum of toxic units (M)	Additive index (AI)	Mixture toxicity index (MTI)[a]
9	4, 8, 12	3		0.960	0.0417	1.037
11	4, 8, 12, 22, 25	5		1.04	−0.0440	0.976
13	4, 8, 12, 16, 19, 22, 25	7		0.891	0.122	1.059

Source: Data from Broderius and Kahl, 1985.
[a]Calculated from Eq. (46).

This expression for λ is identical to Eq. (46) for MTI and the proof is therefore completed. Thus MTI is not just a toxicity index, but is also a model parameter which can be used in conjunction with the noninteractive multiple toxicity model

$$Q = \Pr(\delta_1^{1/\lambda} + \delta_2^{1/\lambda} + \cdots + \delta_n^{1/\lambda} \leq 1) \qquad (49)$$

with full correlation of toxicant tolerances to predict the response for any combination of the toxicant concentrations. This application will be demonstrated below in the sections on mixtures of more than two toxicants and on water quality criteria.

VI. EXAMPLES OF MODEL FITTING WITH TWO TOXICANTS

The development of theoretical models has traditionally been further advanced than more practical but equally important topics such as parameter estimation and goodness of fit tests based on given experimental data. For example, the general bivariate toxicity model, Eq. (2), in which the tolerances follow a bivariate normal function was developed by Hewlett and Plackett in 1959 but was still not fitted to experimental data 20 years later (Hewlett and Plackett, 1979). However, recently we have demonstrated how such a fitting can be accomplished using our general noninteractive multiple toxicity model (Christensen and Chen, 1985).

We discuss here some of these results where the fitting is carried out based on dose-response data including several different response levels. Also included are new results regarding the fitting of the model to experimental isobologram data obtained by Broderius (1987), as well as information on the relationship between the similarity parameter λ and quantitative structure-activity relationships (QSARs) for various groups of chemicals.

A. Fitting Based on Dose-Response Curves

We consider here the experimental data of Hewlett and Plackett (1950) for the toxicity of γ-benzene hexachloride (γ-BHC) and pyrethrins, acting singly (Tables 10 and 11) and in combination (Table 12) on flour beetles (*Triboleum castaneum*). Hewlett and Plackett (1950) fitted univariate probit expressions to the mortality data for these toxicants acting singly, and a bivariate normal model for independent action ($\lambda = 0$) to the data for joint action. By contrast, we considered univariate weibit, probit, and logit expressions for the individual toxicants, as well as bivariate versions of these models where λ can assume any

TABLE 10

Toxicity to *Tribolium castaneum* of Films Formed by γ-BHC and Pyrethrins in Shell Oil P31

Insecticide	Deposit (mg per 10 cm^2)	Number of beetles	Observed percent mortality after 6 days
γ-BHC (0.1% w/v)	16.21	50	46
	17.30	50	56
	18.54	49	40.8
	20.49	48	60.4
	22.29	50	56
	24.44	50	62
Pyrethrins (1.6% w/v)	15.84	50	4
	17.71	50	8
	18.52	51	11.8
	20.41	50	40
	22.00	50	56
	23.82	50	94
Control		200	0

Source: Hewlett and Plackett (1950).

TABLE 11

Fits of the Models to the Experimental Data of Table 10

Insecticide	Model	Parameters[a] Intercept, A	Slope, B	χ^2(4 df)
γ-BHC	Weibit	−3.2408	0.99766	3.86
	Probit	2.4288	2.0563	3.88
	Logit	−4.1122	1.4282	3.88
Pyrethrins	Weibit	−34.764	11.249	2.75
	Probit	−20.068	18.968	9.71
	Logit	−44.373	14.568	6.85

Reproduced from E. R. Christensen and C.-Y. Chen, "A General Noninteractive Multiple Toxicity Model Including Probit, Logit, and Weibull Transformations." *Biometrics* 41:711–725, 1985. With permission from The Biometric Society.

[a]$A = \ln k$, α, or θ; $B = \eta$, β, or ϕ.

TABLE 12

Toxicity to *Tribolium castaneum* of Films Formed by a Two-Component Insecticide
(0.1% γ-BHC and 1.6% Pyrethrins) in Shell Oil P31

Deposit [mg/(10 cm^2)]	No. of beetles	Probit $\lambda = 0.63$ $\rho = 0.72$ $\chi^2 = 4.42$	Logit $\lambda = 0.64$ $\rho = 0.70$ $\chi^2 = 4.36$	Weibit $\lambda = 0.58$ $\rho = 1.00$ $\chi^2 = 5.59$	Observed % mortality[a]
		Calculated percent mortality for maximum likelihood			
16.18	50	68.3	68.6	65.0	66
17.09	50	72.5	72.8	69.3	78
18.29	50	77.9	78.0	75.3	82
20.34	49	86.3	86.1	85.6	79.6
22.13	50	92.2	91.9	93.6	90
23.32	50	95.2	94.7	97.4	98

Reproduced from: E. R. Christensen and C.-Y. Chen, "A General Noninteractive Multiple Toxicity Model Including Probit, Logit, and Weibull Transformations." *Biometrics* 41:711–725, 1985. With permission from The Biometric Society.
[a]Hewlett and Plackett, 1950.

value between 0 and 1 for the combined toxicants (Christensen and Chen, 1985).

In the fitting of a bivariate Weibull model such as Eqs. (2) and (12) there are, in principle, six parameters to be determined: k_1, k_2, η_1, η_2, λ, and ρ. From one point of view, it is desirable to determine all six parameters in one large experiment including at least six and preferably of the order of 20 experimental data points covering a wide range of toxicant concentrations and response levels. However, in reality such large experiments tend to be cumbersome and will often yield unreliable estimates of some of the parameters. Instead, we have chosen to determine k_1, k_2, η_1, and η_2 from experiments with toxicants acting singly (Table 11), and then assume that these values would be valid in experiments with two toxicants. Therefore, only two parameters, namely λ and ρ, would remain to be determined in later experiments.

Fitting of univariate data for the purpose of determining k_1, k_2, η_1, η_2 is based on the log-linear weibit transformation (Table 2):

$$u = \ln k + \eta \ln z \qquad (50)$$

and is carried out either using calculations based on maximum likelihood (Christensen and Chen, 1985) or minimum χ^2 (Christensen et al., 1986). The goodness of fit is calculated by χ^2:

$$\chi^2 = \sum_{i=1}^{N} n_i \frac{(Q_i - q_i)^2}{Q_i(1 - Q_i)} \tag{51}$$

where n_i = number of test organisms in trial i
  q_i = experimental survival fraction in trial i, e.g., 54% in the first case and 44% in the second (Table 10)
  Q_i = calculated survival fraction in trial i
  N = total number of trials (6)

The starting point in the bivariate fitting is the calculation of the tolerances in weibits corresponding to the concentrations of the two toxicants. Consider, for example, 20.34 mg/(10 cm^2) of 0.1% γ-BHC and 1.6% pyrethrins (Table 12). The associated weibits are

$$\gamma\text{-BHC: } x_1 = -3.2408 + 0.99766 \ln 20.34 \tag{52}$$

$$\text{Pyrethrins: } x_2 = -34.764 + 11.249 \ln 20.34 \tag{53}$$

Assume now that $\lambda = \lambda_1$. The integration region in the Weibull plane is then given by Eq. (13). Next, this region R is mapped on the NED plane into region R' and the nonresponse probability $Q = Q_1$ is calculated by integrating the bivariate normal function of correlation coefficient $\rho = \rho_1$ over the region R'. This calculation is then done also for the five remaining concentrations, e.g., 22.13 mg/(10 cm^2) of 0.1% γ-BHC and 1.6% pyrethrins. The resulting nonresponse probabilities are called $Q_2, Q_3, \ldots, Q_6$. We then calculate the log-likelihood (LL):

$$\text{LL} = \sum_{i=1}^{N} n_i \left[q_i \log Q_i + (1 - q_i) \log (1 - Q_i) \right] \tag{54}$$

where the symbols have been defined previously for Eq. (51).

The above procedure is now repeated with different λ, ρ pairs giving generally different LL values. In this way we systematically vary λ and ρ until the pair producing maximum likelihood (LL) has been identified. The step size was 0.01 for λ and 0.02 for ρ. The optimal values of λ and ρ are displayed in Table 12 along with χ^2 calculated from Eq. (51). From Table 12, the bivariate probit and logit models provide a somewhat better fit than the Weibull model to the bivariate data. However, as can be seen from Table 11, the probit and logit models do not fit the univariate pyrethrins data very well, and based on χ^2 from the totality of data we conclude that the bivariate Weibull model provides the only acceptable fit at the 23% level of probability.

B. Fitting Based on Isobolograms

In addition to using mixture data including several different response levels, model parameters can be estimated by fitting the model to experimental isobologram points. We demonstrate here how such a fitting may be carried out using experimental results obtained by Broderius (1987) for the mortality of juvenile fathead minnows (*Pimephales promelas*) affected by five different pairs of organic toxicants (Table 13). In each case, seven ratios of two test chemicals were used to determine experimental isobologram points with four concentrations following an 80% dilution factor at each mixture ratio. The isobologram points indicated the concentrations of the two toxicants producing 50% mortality within an exposure time of 96 h. These points, along with their 95% confidence limits, are reproduced in Figs. 6–10 for each pair of toxicants considered.

The bivariate probit model is used to fit the experimental data. Slopes and intercepts of the probit expressions for the individual toxicants are given in Table 13. The fitting is accomplished by a maximum likelihood procedure similar to the one outlined above. The log-likelihood given by Eq. (54) is maximized using the following values of the variables: n_i = 20; N = 5; q_i = 0.5. The survival fractions Q_i are calculated for the five mixture concentrations defined by the experimental isobologram points. For example, in the case of hydrocyanic acid and rotenone (Fig. 7) these concentrations are (4.32, 35); (3.53, 62); (2.76, 87); (1.57, 112); (0.77, 126) where the units are micrograms per liter (μg/L) and the first numbers in each pair are for rotenone, and

TABLE 13

Slopes and Intercepts of Individual Probit Equations for the Chemicals Considered
in Acute Joint Toxicity Tests with Juvenile Fathead Minnows

Figure	Chemicals	LC_{50}	$\log LC_{50}$	Slope[a] $B = \beta$	Intercept[b] $A = \alpha$
6	Octanol	14 mg/L	1.146	13.5	− 10.471
	2-Octanone	63 mg/L	1.799	13.5	− 19.287
7	Hydrocyanic acid	150 μg/L	2.176	12.0	− 21.112
	Rotenone	5.15 μg/L	0.7118	22.0	− 10.660
8	n-Octyl cyanide	0.104 mM	− 0.9830	13.5	18.271
	Octanol	0.104 mM	− 0.9830	13.5	18.271
9	2,4-Dinitrophenol	13.5 mg/L	1.130	8.5	− 4.605
	Rotenone	4.6 μg/L	0.6628	22.0	− 9.582
10	Octanol	13.5 mg/L	1.130	13.5	− 10.255
	2,4-Pentanedione	122 mg/L	2.086	13.6	− 23.370

Conducted by Broderius (1987), Figs. 6–10.
[a]S. J. Broderius, personal communication (1986).
[b]Intercept A calculated from $5 = A + B \log z$, e.g., $5 = A + 13.5 \times 1.146$; $A = -10.471$.

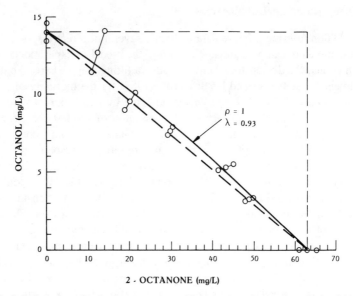

FIG. 6. Isobologram showing concentrations of 2-octanone and octanol producing 96-h, 50% mortality for juvenile fathead minnows. The experimental points with 95% confidence limits are from Broderius (1987). The curve indicates the fit of our bivariate noninteractive model (Tables 13 and 14).

the second are for hydrocyanic acid. Values of the log-likelihood (LL) are calculated for several combinations of λ and ρ using step sizes of $\Delta\lambda = 0.01$ and $\Delta\rho = 0.2$, and the pair yielding maximum LL is retained. This pair then represents the final estimate of these parameters. In the present example, we find $\lambda = 0.87$ and $\rho = 1.0$.

The estimated values of λ and ρ are next used to determine the calculated isoboles which are drawn in full line in Figs. 6–10. From these figures, the fits are good in three cases (Figs. 6, 7, and 10) and somewhat unsatisfactory in two (Figs. 8 and 9). These qualitative impressions are confirmed by χ^2 as calculated from Eq. (51) (Table 14).

It is remarkable that the estimated value of the correlation coefficient ρ in all five cases is equal to one. This means that the isoboles are all described by Eq. (23). For example, in the case of octanol and 2,4-pentanedione (Fig. 10), the equation for the isobole is

$$\left(\frac{z_1}{122}\right)^{1/0.39} + \left(\frac{z_2}{13.5}\right)^{1/0.39} = 1 \qquad (55)$$

where z_1 and z_2 are the concentrations (in mg/L) in the mixture of 2,4-pentanedione and octanol, respectively.

The use of expressions such as Eq. (55) for the isoboles would greatly simplify the estimation of parameter(s) in that there is only one, namely λ, to be determined and in that this parameter can be determined from the analytical expression, Eq. (23). If the toxic units z_1/Z_1 and z_2/Z_2 are different, λ must be determined by iteration. However, for equal toxic units $z_1/Z_1 = z_2/Z_2 = M/2$, we obtain

$$\lambda = 1 - \frac{\log M}{\log 2} \tag{56}$$

which, of course, is a special case of Eq. (48). As an example of the application of Eq. (56), consider Fig. 10. The point for which the concentrations of 2,4-pentanedione and octanol are 92 mg/L and 10.1 mg/L, respectively, approxi-

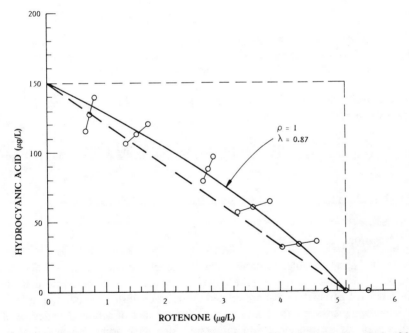

FIG. 7. Isobologram showing concentrations of rotenone and hydrocyanic acid producing 96-h, 50% mortality for juvenile fathead minnows. The experimental points with 95% confidence limits are from Broderius (1987). The curve indicates the fit of our bivariate noninteractive model (Tables 13 and 14).

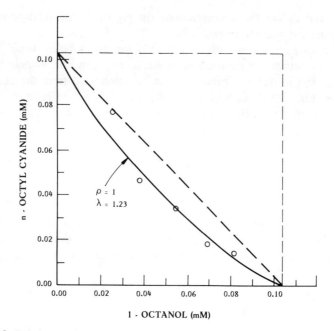

FIG. 8. Isobologram showing concentrations of 1-octanol and n-octyl cyanide producing 96-h, 50% mortality for juvenile fathead minnows. The experimental points are from Broderius (1987). The curve indicates the fit of our bivariate noninteractive model (Tables 13 and 14).

mately fulfills the condition of Eq. (56). Thus, we obtain the following approximate value of λ:

$$\lambda = 1 - \frac{\log (92/122 + 10.1/13.5)}{\log 2} = 1 - \frac{\log 1.502}{\log 2} = 0.41 \quad (57)$$

which is close to the value of 0.39 obtained by maximum likelihood.

C. Relationship with QSAR

To be most valuable, multiple toxicity models should be more than just statistical descriptions of the joint action of toxicants. It is desirable that the models have parameters that are related to the mechanisms of action or at least indicate whether or not the toxicants have similar sites of action. The degree of correlation of tolerances should preferably also have some physiological or biochemical basis. These features would give the models predictive capability. We describe in the following why the above noninteractive bivariate model has a significant measure of these characteristics, and how the similarity parameter

λ can be interpreted in terms of quantitative structure-activity relationships (QSARs) for various groups of chemicals. The cases shown in Figs. 6–10 will be used for illustration.

An important example of QSARs for organic chemicals acting singly is the almost linear correlation between the logarithm of the octanol-water partition coefficient P_{oct} and logarithm of the toxicant effect concentration, expressed for example as the 96-h LC_{50}. Similarly, there is often a linear correlation between the logarithm of the water solubility and the logarithm of the 96-h LC_{50} value. This applies to a specific class of chemicals with similar structural properties such as certain alcohols.

Several classes of chemicals can often be combined into a common group which can be modeled by a single QSAR characteristic of one type of toxic action such as narcosis (Broderius and Kahl, 1985). Broderius (1987) distinguished between four different groups of toxicants each with its specific QSAR. The designations for these groups are: narcosis I, narcosis II, uncoupler, and electron inhibitor.

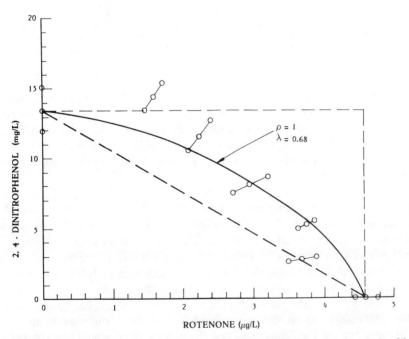

FIG. 9. Isobologram showing concentrations of rotenone and 2,4-dinitrophenol producing 96-h, 50% mortality for juvenile fathead minnows. The experimental points with 95% confidence limits are from Broderius (1987). The curve indicates the fit of our bivariate noninteractive model (Tables 13 and 14).

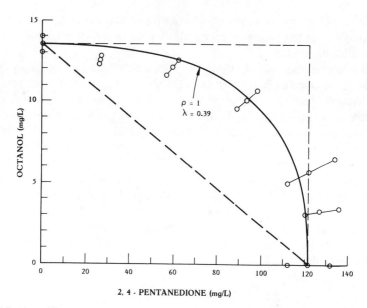

FIG. 10. Isobologram showing concentrations of 2,4-pentanedione and octanol producing 96-h, 50% mortality for juvenile fathead minnows. The experimental points with 95% confidence limits are from Broderius (1987). The curve indicates the fit of our bivariate noninteractive model (Tables 13 and 14).

The toxic action of narcosis I chemicals is general membrane perturbation resulting from reversible retardation of cytoplasmic activity caused by absorption of foreign molecules into biological membranes. Chemicals of the narcosis II group exemplified by certain phenols and anilines exert their toxic action through membrane depolarization which is a more sensitive mechanism than the one described above.

The mode of action of uncoupler chemicals is uncoupling of oxidative phosphorylation. For the latter two groups, narcosis II and uncoupler chemicals, other factors than log P_{oct} are often involved in the QSARs. These include molecular descriptors such as electronic and steric factors which reflect the polarity of the chemicals. A frequently used electronic descriptor is pK_a. The last group, electron inhibitors, includes chemicals such as HCN and rotenone which inhibit electron transport in the mitochondria of cells.

Chemicals from the same group, e.g., narcosis I, would be expected to act by similar joint action that is concentration addition, while chemicals from different groups, e.g., uncoupler and electron inhibitor, would be expected to be less than additive. While this was realized by Broderius (1987), we have, as may be seen from Figs. 6–10 and Table 14, extended this qualitative statement by demonstrating that the bivariate noninteractive model can be fitted to the experimental isoboles, that the correlation coefficient in all cases is equal to one, and that the cases where λ is close to one (Figs. 6–8) involve chemicals from the same group while they otherwise come from different groups (Figs. 9 and 10). The use of λ as a similarity parameter is, therefore, consistent with the general classification of chemicals in QSARs. Chemicals from the same QSAR have similar sites of action, and λ is therefore close to one, while chemicals from different QSARs have different or partially overlapping sites of action, meaning that λ is between zero and one.

With regard to the correlation coefficient ρ, it should be noted that the precision of the estimates obtained from experimental isobologram points (Figs. 6–10) is considerably less than in the cases where the fitting is based on the response curves involving several different response levels (Fig. 3).

The fact that the tolerances are positively correlated may be explained from the values of λ which are between 0.39 and 1.23 (Table 14). Thus there is in all cases a fair amount of overlap of the biological systems affected. The two chemicals would therefore have to reach the same general area in the organism to exert their toxic action. Because of this they are likely to be similarly affected by random variations in the makeup of the organisms. For example, if an organism has a particularly high tolerance to one toxicant because of a thick protective layer of epithelium, it is likely also to have a high tolerance to the other toxicant since the latter also must penetrate this layer to exert its toxic action.

TABLE 14

Fitting of the Bivariate Probit Model to Experimental Isobologram Points Obtained by Broderius (1987) for the Acute Toxicity of Pairs of Organic Chemicals to Juvenile Fathead Minnows (Figs. 6–10)

Figure	Chemicals	Toxic action	Similarity parameter (λ)	Correlation coefficient (ρ)	χ^2(3 df)
6	Octanol	Narcosis I	0.93	1.00	3.85
	2-Octanone	Narcosis I			
7	Hydrocyanic acid	Electron inhibitor	0.87	1.00	2.17
	Rotenone	Electron inhibitor			
8	n-Octyl cyanide	Narcosis I	1.23	1.00	9.42
	Octanol	Narcosis I			
9	2,4-Dinitrophenol	Uncoupler	0.68	1.00	28.63
	Rotenone	Electron inhibitor			
10	Octanol	Narcosis I	0.39	1.00	4.90
	2,4-Pentanedione	Electrophilic (?)			

VII. MIXTURES OF MORE THAN TWO TOXICANTS

The chemicals listed in Table 8 all belong to a group of 27 compounds characterized by Broderius (1987) as narcosis I test chemicals. Chemicals from this group are, therefore, expected to act jointly through a mode of action that is near concentration addition. Because of this, and based on the above findings regarding the correlation coefficient ρ for pairs of chemicals from this group (Figs. 6 and 8), we shall assume that the correlation matrix A, Eq. (6), has 1's in all positions when the number n of toxicants is >2. This assumption would also appear to be reasonable in view of the fact that the response to two toxicants generally is rather insensitive to variations in ρ, as long as ρ is not negative (Fig. 4).

In order to make the task of calculating Q from equations such as Eq. (4), a tractable one, it is also necessary to make assumptions regarding the similarity parameters, e.g., λ_{123} and λ_{12}. Our assumption here is that the similarity in toxic action can be adequately characterized by a single similarity parameter $\lambda = \lambda_{12...n}$, common for all n toxicants. Thus we shall calculate Q from Eq. (49) assuming full correlation of the toxicant correlations.

With these assumptions, the probit u' of the joint response can be calculated by iteration from an extension of Eq. (20)

$$\text{hexp}\,\frac{x_1 - u'}{\lambda\beta_1} + \text{hexp}\,\frac{x_2 - u'}{\lambda\beta_2} + \cdots + \text{hexp}\,\frac{x_n - u'}{\lambda\beta_n} = 1 \quad (58)$$

where $x_i = \alpha_i + \beta_i \log z_i; i = 1, n$

α_i, β_i = intercept and slope in the probit transformation for toxicant i (Table 2)

z_i = concentration of toxicant i

Solution of this equation emerges as a special case of MULTOX (Chen and Christensen, 1986). We shall apply Eq. (58) to mixtures no. 9, 11, and 13 including a total of 3, 5, and 7 narcosis I test chemicals, respectively (Table 9). The results are presented in Tables 15–17. In these tables, the ranges and average values of the probit slopes were obtained from Broderius (personal communication, 1986).

In all cases, application of λ determined from Eq. (48), using equal toxic units the sum of which equals M, gives as expected the standard response $P = 0.5$ (NED = 0). By changing λ to 1 and keeping the toxicant concentrations, the new response u^* (probits or NEDs) can be calculated as follows. Consider the sum M of toxic units:

TABLE 15

**Prediction of the Response of Fathead Minnows to Mixture No. 9 of Table 9
from the Computer Program MULTOX Assuming Various Slopes
and Concentrations of the Individual Toxicants**

Similarity parameter (λ)	Chemical no.						Response probability (P)	Response NEDs
	Probit slopes			Concentrations[a]				
	4	8	12	4	8	12		
1.037	13.5	13.5	13.5	0.32	0.32	0.32	0.499	−0.003
1	13.5	13.5	13.5	0.32	0.32	0.32	0.407	−0.236
1.037	13.5	13.5	13.5	0.40	0.25	0.50	0.852	1.044
1	13.5	13.5	13.5	0.40	0.25	0.50	0.793	0.817
1.037	9	9	9	0.40	0.25	0.50	0.757	0.697
1.037	18	18	18	0.40	0.25	0.50	0.919	1.398
1.037	10	13	15	0.40	0.25	0.50	0.833	0.964

[a]In units of concentrations giving $P = 0.5$.

$$M = \sum_{i=1}^{n} \frac{z_i}{Z_i} \tag{59}$$

where $z_i/Z_i = M/n$

z_i = concentration of toxicant i

Z_i = concentration of toxicant i, acting singly, giving the standard response X (probits or NEDs)

When λ has a value as determined from Eq. (48), we have

$$u' = X = X_i = \alpha_i + \beta_i \log Z_i \qquad i = 1, n \tag{60}$$

where u' (probits) is the joint response. Now let $\lambda = 1$ and we obtain

$$x_i = \alpha_i + \beta_i \log \left(\frac{M}{n} Z_i \right)$$

$$= \alpha_i + \beta_i \log \left(\frac{M}{n} \right) + \beta_i \log Z_i$$

$$= X + \beta_i \log \frac{M}{n} \tag{61}$$

The new response $u*$ can now be determined by iteration from

TABLE 16

Prediction of the Response of Fathead Minnows to Mixture No. 11 of Table 9
from the Computer Program MULTOX Assuming Various Concentrations
of the Individual Toxicants (All Probit Slopes = 13.5)

Similarity parameter (λ)	Concentrations[a] for chemical no.					Response probability (P)	Response NEDs
	4	8	12	22	25		
0.976	0.208	0.208	0.208	0.208	0.208	0.502	0.004
1	0.208	0.208	0.208	0.208	0.208	0.589	0.224
0.976	0.1	0.07	0.3	0.5	0.15	0.683	0.477
1	0.1	0.07	0.3	0.5	0.15	0.747	0.664

[a]In units of concentrations giving $P = 0.5$.

$$\text{hexp} \frac{X + \beta_1 \log (M/n) - u^*}{\beta_1}$$

$$+ \text{hexp} \frac{X + \beta_2 \log (M/n) - u^*}{\beta_2}$$

$$+ \cdots + \text{hexp} \frac{X + \beta_n \log (M/n) - u^*}{\beta_1} = 1 \qquad (62)$$

In case of identical slopes $\beta_1 = \beta_2 = \cdots = \beta_n = \beta$, Eq. (62) is reduced to

$$n \, \text{hexp} \frac{X + \beta \log (M/n) - u^*}{\beta} = 1$$

or

$$u^* = X + \beta \log M \qquad (63)$$

Thus the new response u^* is equal to the standard response plus the product of β and log M. This is the same result as obtained from MULTOX, as may be verified from the second lines of Tables 15–17. For example, from Table 15, $u^* = -0.236$ NEDs $\approx \beta \log M = 13.5 \log 0.960 = -0.239$. The small deviation is caused by numerical errors.

Other aspects of model predictions listed in Tables 15–17 are as follows. The third and fourth lines contain data pertaining to arbitrary concentration of the toxicants, e.g., 0.40, 0.25, and 0.50 with $\lambda = 1.037$, determined from Eq. (48), and $\lambda = 1$ (Table 15). The corresponding difference in NEDs of response, $0.817 - 1.044 = -0.227$ has the same sign as $\beta \log M$ but an absolute value which is less than this quantity. The deviation from $\beta \log M$ appears to be particularly significant when some toxicant concentrations in the mixtures are very small or large compared to the values listed in the first two lines. An example of this are the response NEDs of Table 16 where the difference is $0.664 - 0.477 = 0.187$ compared to $\beta \log M = 13.5 \log 1.04 = 0.230$.

Finally, the effect of variations in individual probit slopes may be seen from the last three lines of Table 15. In the present case, high values (18) give a high response, 1.398 NEDs, while low values (9) produce a much lower response, 0.697. Intermediate values (10, 13, 15) give a response, 0.964, which is close to the one (1.044) obtained for average slopes (13.5).

The single most important fact emerging from Tables 15–17 is that we indeed are able to make predictions for the toxicity of the three mixtures under conditions that differ from the standard conditions as represented by the first line in each table. These standard conditions are those under which the standard response $P = 0.5$; NED $= 0$ was obtained experimentally (Broderius and Kahl, 1985).

We demonstrated earlier that the multiple toxicity index MTI under these circumstances is identical to the similarity parameter λ for full correlation of the toxicant tolerances. Thus, the standard conditions may in fact be perceived as guidelines for conducting an experiment from which not only MTI but also the then identical model parameter λ may be determined. Although a standard response of NED $= 0$ was used here for convenience, any constant reference level may be used as discussed previously. From the knowledge of λ and assuming full correlation of toxicant tolerances, we are then able to use the multiple toxicity model to calculate the biological response for any given combination of toxicant concentrations in the mixture.

While the predictions of Tables 15–17 have not yet been confirmed experimentally, we are fairly optimistic that such confirmations eventually may be

TABLE 17

Prediction of the Response of Fathead Minnows to Mixture No. 13 of Table 9 from the Computer Program MULTOX Assuming Various Concentrations of the Individual Toxicants (All Probit Slopes = 13.5)

Similarity parameter (λ)	Concentrations[a] for chemical no.							Response probability, (P)	Response NEDs
	4	8	12	16	19	22	25		
1.059	0.1273	0.1273	0.1273	0.1273	0.1273	0.1273	0.1273	0.499	−0.003
1	0.1273	0.1273	0.1273	0.1273	0.1273	0.1273	0.1273	0.250	−0.676
1.059	0.1	0.02	0.12	0.04	0.2	0.05	0.09	0.0143	−2.19
1	0.1	0.02	0.12	0.04	0.2	0.05	0.09	0.00276	−2.77

[a]In units of concentrations giving $P = 0.5$.

made. This belief is based in part on the encouraging results demonstrated in a previous section regarding the fitting of the model to experimental isobologram points for pairs of organic chemicals from the narcosis I group (Figs. 6 and 8) to which all of the here-considered chemicals belong.

VIII. SUBLETHAL ENDPOINTS

The application of sublethal endpoints in toxicity assays has gained increased attention in recent years. There are several reasons for this. Certain categories of tests for carcinogenicity of chemicals are (or were until recently) conducted with a large number of higher animals which eventually would be sacrificed. Thus consideration of the welfare of the animals dictates their minimal use in such bioassays. Recent research has shown that equivalent knowledge regarding the carcinogenic potential of chemicals in many cases can be obtained from smaller sublethal experiments in which the measurements of molecular predictors provide the desired information.

In bioassays where the aquatic toxicity is evaluated using, for example, fathead minnows as test organisms, the principal reason for the application of sublethal endpoints is that such testing often is more relevant with regards to reflecting actual effects of pollutants in aquatic ecosystems (Sprague, 1986). Algal assays where the maximum yield or the growth rate is measured as a function of toxicant concentrations are inherently sublethal in that only the inhibition in cell division is measured.

While lethal tests are quantal by their very nature, sublethal tests have usually a continuous response (Table 1). An exception to this rule is growth rate of microorganisms based on cell number assuming synchronous growth (group III organisms).

The models described here for the response of organisms to one or several toxicants are based on quantal responses. However, as described above, it is possible to convert a continuous response to a quantal response by requiring that the continuous response exceed a certain critical value c in each organism in order for the quantal response to occur. Such a procedure was proposed by Hewlett and Plackett (1956) and implemented by Shelton and Weber (1981) who considered the activity of plasma alanine aminotransferase (ALT) as an indicator of liver damage in mice (Table 1). One difficulty with this procedure is that the response would depend on the often fairly arbitrary selection of the magnitude of the critical value c necessary to trigger the quantal response in each organism.

When the sublethal continuous response is in the form of growth rate of microorganisms, there is no need to convert a continuous response to a quantal one. The relative inhibition or reduction in growth rate is a direct measure of the reduction in the number of consecutive cell divisions caused by the toxicants. In this case, the relative inhibition P can be modeled by probit, logit, or weibit distributions (Table 2). In the bivariate case, the correlation ρ between cell division tolerances would be one when all organisms originate from the same clone while λ as before indicates the degree of overlap between two biological systems affected by the toxicants.

An example of fitting of the bivariate logit model to experimental results for the growth rate of the green alga *Selenastrum capricornutum* affected by nickel and zinc is shown in Fig. 11. The algal assays were conducted according to the protocol of the U.S. EPA (1978), and ionic concentrations of nickel and zinc were calculated by means of the computer program MINEQL (Westall et al., 1975). It is seen that the model fits the experimental points well for $\rho = 1$, $\lambda = 0.9$, thus confirming the hypothesis that Ni^{2+} and Zn^{2+} approximately act on similar systems (enzymes) via concentration addition.

The account given above concerning the use of sublethal endpoints in toxicity tests is necessarily brief and of introductory nature. The main reason for this is that the information available in the literature on this subject is scarce. However, considering the relevance of sublethal endpoints, we expect that this topic will become an important consideration in future investigations.

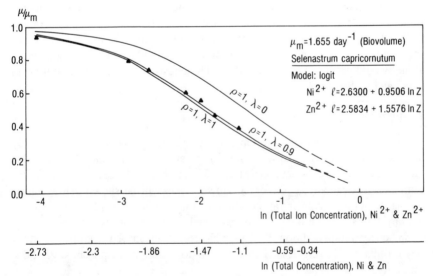

FIG. 11. Relative growth rate of *Selenastrum capricornutum* affected by mixtures consisting of Ni^{2+} and Zn^{2+} in equitoxic proportions. The best fit of the bivariate logit model is obtained for near concentration addition, that is, $\rho = 1$, $\lambda = 0.9$.

IX. WATER QUALITY CRITERIA

As mentioned in the introduction under historical perspective, current water quality criteria (U.S. EPA, 1985) are still based on limits for toxicants acting alone assuming no addition despite documents pointing to the desirability of using a criterion based on concentration addition (National Academy of Sciences and National Academy of Engineering, 1972; EIFAC, 1980). We shall, in the following, illustrate how different the effluent limits can be when either no addition, a set of actual multiple toxicity parameters, or concentration addition is used as a basis for the setting of water quality criteria under the assumption of a given required level of protection of aquatic organisms.

We consider as example the toxicity of the mixture of toxic compounds in the proposed effluent from the Exxon Crandon zinc-copper mine project (Department of Natural Resources, 1986). The investigation was conducted by one of the authors (ERC) for the Wisconsin Department of Justice, and with its permission we reproduce here some of the relevant findings.

A. Example: Crandon Mining Project

The basis for the Crandon study was a concern that even if the water quality criteria for each individual toxicant in the effluent were fulfilled, the combined toxicity of the mixture might not be acceptable. In fact, Spehar and Fiandt (1986) found significant adverse effects on fathead minnows and daphnids of metal mixtures in which the concentrations of individual metals, when acting alone, would produce little or no effect. These tests were conducted both on an acute and a chronic basis.

1. Approach

Projected effluent concentrations, along with daily maximum and monthly average limits for principal toxic substances are given in Table 18. From this table, the projected effluent concentrations c_E are below the respective limits for individual toxicants ($c_{d,\max}$ or $c_{m,\mathrm{av}}$). However, as will be shown below, under the assumption of concentration addition, the combined effluent exceeds the toxicity limits both on a daily maximum and a monthly average basis.

In concentration addition, the following condition derived from Eq. (1) must be fulfilled for having less than or equal to a certain response, e.g., 50% mortality:

$$\frac{z_1}{Z_1} + \frac{z_2}{Z_2} + \cdots + \frac{z_n}{Z_n} \leq 1 \tag{64}$$

TABLE 18

Proposed Effluent Limitations and Projected Effluent Quality for Surface Water Discharge to Swamp Creek with Indication of Toxic Units for Each Substance[a]

Toxicant	(1) Daily maximum $c_{d,max}$ (mg/L)	(2) Toxic units $c_E/c_{d,max}$	(3) Monthly average limits based on effluent flows of 2000–3000 gpm $c_{m,av}$ (mg/L)	(4) Toxic units $c_E/c_{m,av}$	(5) Projected effluent quality (daily average) c_E (mg/L)
Arsenic	1.48	0.034	0.436	0.115	0.05
Cadmium	0.073	0.008	0.0032	0.188	0.0006
Chromium[6+]	0.058	0.207	0.036	0.333	0.012
Copper	0.025	0.4	0.019	0.526	0.01
Cyanide	0.093	0.065	0.010	0.6	0.006
Lead	0.6	0.067	0.082	0.488	0.04
Mercury	0.002	0.085	0.0002	0.85	0.00017
Nickel	2.58	0.008	0.072	0.278	$<0.02^b$
Selenium	1.0	0.06	0.115	0.522	0.06
Zinc	0.44	0.136	0.10	0.6	0.06
Sum		1.070		4.500	

[a]Data in columns (1), (3), and (5) are from Department of Natural Resources, State of Wisconsin, Public Service Commission, 1986.

[b]Here assumed to be 0.02 mg/L.

where $z_1, z_2, \ldots, z_n$ are the actual concentrations of the toxicants, and $Z_1, Z_2, \ldots, Z_n$ are the concentrations of these toxicants producing the given response, when acting alone. In the present case, z is c_E and Z is $c_{d,max}$ or $c_{m,av}$. Based on concentration addition, the sum of toxic units must therefore be less than or equal to one in order to ensure that the effect criterion, e.g., less than 50% mortality, is fulfilled.

A note about the significance of the toxicant concentrations of Eq. (64) is in order here. These concentrations refer to levels of individual toxic compounds in a water of given pH, dissolved oxygen content, level of suspended solids, and so forth. As to the nature and levels of the toxic compounds, these can in many cases not be directly derived from the basic constituents of the water because of chemical reactions and the differential toxicity of various ions or complexes formed from the basic constituents. For example, cyanide is known to complex with metals with a following reduction in toxicity of both the metal and cyanide. Also, as was implied in Fig. 11, the important quantities are not total concentrations of metals such as nickel and zinc, but rather their ionic concentrations.

While the formation of toxic ions or complexes cannot be directly deduced from the original constituents of the water, the nature and concentrations of these compounds may often be estimated by using a chemical equilibrium computer program such as MINEQL (Westall et al., 1976) as was in fact done in Fig. 11.

Because all individual limits, except for Cr^{6+}, are given in terms of total concentrations (Table 18), we find it most appropriate also to consider total concentrations in the evaluation of multiple toxicity effects. Interactive chemical effects such as complexation with cyanide are ignored in the following analysis. This simplification is made here for the purpose of illustration, and from this point of view, the multiple toxicities calculated below represent a worst-case scenario. However, there is in principle no reason why these effects could not be incorporated in the procedure for calculation of multiple toxicity to be described in the following. Also, if individual limits in the future should be based on concentrations of actual toxic compounds such as ions, e.g., Pb^{2+}, or complexes, the basic procedure, outlined below, for the evaluation of the joint toxicity remains valid.

The toxic units have been calculated and summed in Table 18 for both daily maximum (column 2) and monthly average (column 4) limits. From this table, the sums of toxic units are > 1, indicating that the limits are exceeded for the mixture of toxicants, particularly on a chronic basis (4.5), but also for the acute response (1.07).

The parameters in the computer program MULTOX (Chen and Christensen, 1986) were determined as follows. LC_{50}'s (daily maximum limits) and EC_{50}'s (monthly average limits) for individual toxicants were taken from Table 18, while slopes for individual toxicants (As, Cd, Cr^{6+}, Cu, Hg, Pb) using fathead minnows and daphnids (*C. dubia*) were determined from data presented by Spehar and Fiandt (1986). The EC_{50} is the effective concentration giving 50% response. The resulting parameters of the univariate probit equations are given in Table 19. We used Eq. (49) with full correlation of toxicant tolerances as the multivariate version of MULTOX. With the possible exception of cyanide, this would appear to be a reasonable assumption since many metals or metalloids (selenium and arsenic) have fairly similar sites of toxic action.

The similarity parameter λ was determined by calibrating the model to fit the experimental results of Spehar and Fiandt (Table 20). As may be seen from this table, the toxic units within each experiment are approximately equal, except perhaps for *C. dubia* on a chronic basis, and the necessary condition for using Eq. (48) for calculation of λ is, therefore, fulfilled. The resulting values are listed in column 3 of Table 20. Depending on the value of λ, the action is more than additive ($\lambda > 1$), additive ($\lambda = 1$), or less than additive ($\lambda < 1$). The acute response is in terms of percent mortality, and the chronic response either in terms of percent reduction in weight for fathead minnows exposed to metals

TABLE 19

Intercepts α and Slopes β of Probit Transformations for Individual Toxicants Considered
in the Setting of Water Quality Criteria for Surface Water Discharge to Swamp Creek

	Fathead minnows			Daphnids (C. dubia)		
	α			α		
Toxicant	Daily maximum	Monthly average	β	Daily maximum	Monthly average	β
Arsenic	4.45	6.17[a]	3.25	4.15	6.80	5.0
Cadmium	11.0	18.1	5.25	7.27	9.99	2.00
Chromium[6+]	8.71	9.33	3.00	12.4	13.7	6.00
Copper	9.01	9.30	2.50	11.0	11.4	3.75
Cyanide	17.1	28.4	11.7[b]	17.1	28.4	11.7[b]
Lead	5.67	8.26	3.00	5.78	8.80	3.50
Mercury	14.5	17.9	3.50	23.2	30.0	6.75
Nickel	2.40	12.2	6.32[b]	2.40	12.2	6.32[b]
Selenium	5.00	9.23	4.50[c]	5.00	9.23	4.50[c]
Zinc	6.87	10.3	5.25[d]	5.71	7.00	2.00[d]

[a]Calculated from $5 = \alpha + 3.25 \log 0.436$.
[b]From Anderson and Weber (1975b) using male guppies (*Poecilia reticulata*).
[c]Based on data from Spehar (personal communication, 1986).
[d]Assumed equal to the cadmium values.

for 32 days, or percent reduction in number of young per female of *C. dubia* exposed to metals for 7 days.

In running the program we have assumed that the daily maximum limits represented LC_{50} values, and the monthly average limits were effective concentrations giving 50% response on a chronic basis. Thus, in evaluating the results, any response higher than 50% should be considered unacceptable. The program was run for the following concentration factors f of the effluent: 3, 1, 0.3, 0.1, 0.03. For example, a factor of 1 means undiluted effluent, and a factor 0.03 means that the effluent is diluted 33 times.

An effluent concentration factor $f = 1$, $\log f = 0$, means that the projected effluent quality is characterized by the actual concentrations c_E of column 5, Table 18. The concentration factor $f = 3$, $\log f = 0.48$, indicates a hypothetical effluent with concentrations $3c_E$. For example, the concentrations of copper would be 0.03 mg/L and that of mercury 0.00051 mg/L. If this situation ever occurs, it would be under extreme conditions such as failure of the wastewater treatment plant to perform properly, or unusually high loading conditions for the treatment plant.

The meaning of an effluent concentration factor < 1, for example $f = 0.3$,

TABLE 20

Toxic Units in Acute and Chronic Test with Fathead Minnows and *C. dubia*[a]

| Test | Organism | (1) Toxicant | | | | | | (2) Sum M | (3) Similarity parameter λ |
		As	Cd	Cr^{6+}	Cu	Hg	Pb		
Acute	Fathead minnow	0.10	0.09	0.12	0.08	0.08	0.06	0.53	1.35
	C. dubia	0.24	0.22	0.24	0.26	0.26	0.25	1.47	0.78
Chronic	Fathead minnow	0.37	0.61	0.58	0.54	0.50	0.71	3.31	0.33
	C. dubia	0.22	0.10	0.13	0.32	0.25	0.06	1.08	0.96

[a]Data in columns (1) and (2) are from Spehar and Fiandt, 1986.

$\log f = -0.52$, is that all calculations are carried out for a hypothetical projected effluent with concentrations $0.3c_E$. Thus, as an example, the concentrations of copper and mercury would be 0.003 and 0.000051 mg/L, respectively. The purpose of including such a case is to estimate the toxicity of the effluent in the case where its quality could be upgraded by improved treatment technology or decreased input load to the treatment plant.

2. Results

The results of the computer runs are shown in Fig. 12. The response is expressed on a probit and percent scale. Undiluted effluent is indicated by the abscissa $\log f = 0$. It is clear that the toxicity limits are substantially exceeded by the mixture, particularly for *C. dubia* on a chronic basis, and less so for fathead minnows. From Fig. 12 it may also be seen that a "safe" effluent can be obtained for $\log f = -0.61$, or by diluting the effluent by a factor 4.1 (*C. dubia*, $\lambda = 0.96$). This number is consistent with the sum of toxic units for monthly average limits, 4.5 (Table 18) which applies for strict concentration addition ($\lambda = 1$).

Generally, one would expect the chronic response to be above the acute response, as is indeed the case for *C. dubia*. The partial reversal for fathead minnows reflects, on one hand, the fact that the limits in Table 18 were not necessarily derived from toxicity data for the same organism, and on the other, that fish on an acute basis may be affected by more than additive action ($\lambda = 1.35$), while on a long-term basis they are able to detoxify such that they then are affected by less than additive action ($\lambda = 0.33$). The lower chronic toxicity would also arise from acclimation or different chronic mechanisms. This is in contrast to the biologically simpler daphnids where the action is closer to strict concentration addition ($\lambda = 0.78$ or 0.96).

Because daily maximum limits (Table 18) are based on LC_{50}'s for the most sensitive organisms, and the toxic units used for calibration (Table 20) refer to LC_{50}'s and EC_{50}'s for acute and chronic tests, respectively, it is reasonable to use a 50% response level as a reference, which indeed was done in Fig. 12. However, as can be seen from Fig. 13, the results change insignificantly even if a different reference level such as 6.7% (probit = 3.5) were to be chosen.

The results of uncalibrated model runs are shown in Fig. 14. It is here assumed that there is strict concentration addition ($\lambda = 1$). A "safe" effluent is obtained for $\log f = -0.65$, or by diluting the effluent a factor of 4.5 (chronic and acute basis). On a daily maximum basis alone we obtain $\log f = -0.03$ corresponding to a dilution of only 1.07 times. As expected from concentration addition, these dilution factors are identical to the sums of toxic units given in Table 18.

Also, in contrast to the results obtained with the calibrated model (Figs. 12

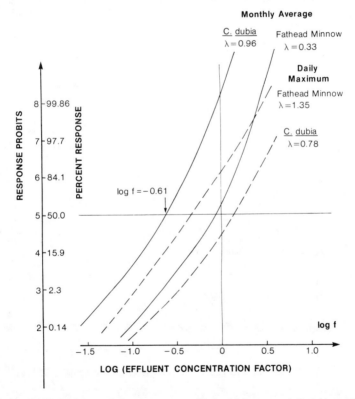

FIG. 12. Calculated acute and chronic response of fathead minnows and *C. dubia* vs. the log of the effluent concentration factor. Values of λ have been determined by calibrating the model to the experimental results of Spehar and Fiandt (Table 20). The reference response level is 50%. Undiluted effluent, i.e., actual projected effluent quality is characterized by $\log f = 0$, $f = 1$. A safe effluent is obtained at $\log f = -0.61$, that is, by diluting the effluent 4.1 times.

and 13), there is a clear separation in Fig. 14 between curves representing chronic and acute results, with the former lying above the latter. The main conclusion from Fig. 14, i.e., that a safe effluent is obtained by diluting 4.5 times, is not very different from the one obtained by the calibrated model (Fig. 12) where a dilution factor of 4.1 was required. Thus, in the case considered, the application of calibration data has little influence on the required overall dilution, even though individual acute and chronic responses are strongly affected by it.

B. Implications of the Crandon Study

The Crandon example shows clearly that limits valid for toxicants acting alone, assuming no addition, may be inadequate to protect aquatic life against adverse effects from several toxicants in a mixture. This is generally true, and

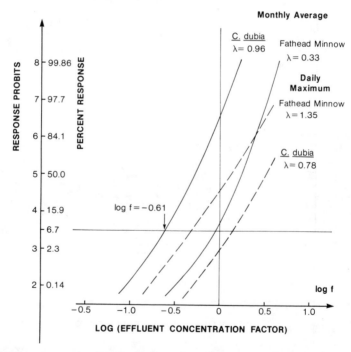

FIG. 13. Calculated acute and chronic response of fathead minnows and *C. dubia* vs. the log of the effluent concentration factor. Values of λ have been determined by calibrating the model to the experimental results of Spehar and Fiandt (Table 20). The reference response level is 6.7%. Undiluted effluent, i.e., actual projected effluent quality is characterized by $\log f = 0$, $f = 1$. A safe effluent is obtained at $\log f = -0.61$, that is, by diluting the effluent 4.1 times.

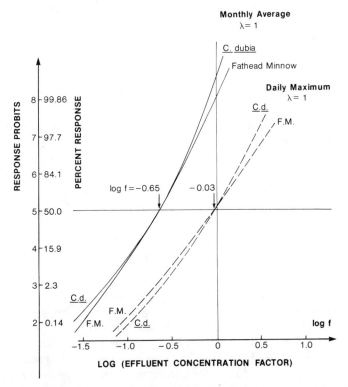

FIG. 14. Calculated acute and chronic response of fathead minnows and *C. dubia* vs. the log of the effluent concentration factor. Uncalibrated model run with $\lambda = 1$ (concentration addition). The reference response level is 50%. Undiluted effluent, i.e., actual projected effluent quality is characterized by $\log f = 0$, $f = 1$. A safe effluent is obtained at $\log f = -0.65$, $f = 4.5$ (chronic and acute basis) or $\log f = -0.03$, $f = 1.07$ (acute basis).

particularly if, as in this example, the toxicants in the mixture act via a mode of action that is near concentration addition. Since chemicals often contribute to the joint toxicity via near-concentration addition, even for a large number of chemicals with equal and proportionately small toxic units (Konemann, 1981b; Hermens and Leeuwangh, 1982), it would, therefore, appear that there is no threshold under which a chemical does not contribute to the mixture toxicity.

If calibration information is available, as in Figs. 12 and 13, it should be incorporated as shown in these figures in the setting of water quality criteria. The corresponding water quality criterion is derived from Eq. (25)

$$\left(\frac{z_1}{Z_1}\right)^{1/\lambda} + \left(\frac{z_2}{Z_2}\right)^{1/\lambda} + \cdots + \left(\frac{z_n}{Z_n}\right)^{1/\lambda} \leq 1 \qquad (65)$$

where the symbols have the same meaning as above. If n toxicants are known to follow response multiplication with weibit transformations for the individual toxicants, the water quality criterion becomes (Christensen and Chen, 1985):

$$\left(\frac{z_1}{Z_1}\right)^{\eta_1} + \left(\frac{z_2}{Z_2}\right)^{\eta_2} + \cdots + \left(\frac{z_n}{Z_n}\right)^{\eta_n} \leq 1 \qquad (66)$$

where $\eta_1, \eta_2, \ldots, \eta_n$ are the individual slopes in the weibit transformations.

In the absence of calibration data, we believe that the water quality criterion, Eq. (64), based on concentration addition, is appropriate. It is more protective for aquatic life against adverse effects from chemicals in a mixture than individual limits for toxicants based on no addition, and because near-concentration addition is so common, it is also more generally applicable.

X. FUTURE RESEARCH

The weibit transformation for individual toxicants has proven to be generally as good as or better than the probit and logit transformations. Its proposed mechanistic-probabilistic basis makes it further attractive and it should therefore be further studied, for example, by comparing it with other transformations with regard to its ability to fit large experimental data sets such as those collected by the Center for Lake Superior Environmental Studies (1986).

A related question is whether the LC_{50}, which is almost universally used in QSAR studies to characterize the toxicity of compounds, indeed is an optimal parameter in such studies. The intercept $\ln k$ in the weibit transformation (Table 2) may be a more appropriate parameter. The reason is that k may have chemical-toxicological significance as being proportional to the stability constant of a toxicant-receptor complex (Christensen and Chen, 1985). To test whether LC_{50} or $\ln k$ is the better parameter to use in QSAR studies, one should run correlations between these parameters and a typical structure parameter such as P_{oct} using data sets such as the one mentioned above (Center for Lake Superior Environmental Studies, 1986). The parameter providing the best correlation would be preferable.

The multiple toxicity model incorporating exposure time, e.g., Eqs. (4), (5), and (34), needs to be tested experimentally. Also, the idea of plotting the correlation coefficient ρ and the similarity parameter λ as points in ρ, λ diagrams deserves further study. It may be possible to classify regions in such diagrams in terms of QSARs of the chemicals.

The isobolograms of Fig. 5 for which the slope η in the weibit transformation is ≤ 1 should be explored experimentally. We suggest that such isoboles would be obtained when primitive organisms, such as algae or trout fry, are exposed to metals or combinations of metals and organics.

The successful fitting of the bivariate model to isobolograms for pairs of organic chemicals form the same or different QSARs (Figs. 6–10) is very encouraging. Such studies should be expanded to other groups of chemicals and include applications of the derived water quality criterion, Eq. (65) with $n = 2$. Predictions, such as those of Tables 15–17, for the response of organisms to n toxicants where $n > 2$ should be checked experimentally along with implementations of the associated water quality criterion, Eq. (65).

There is a need to improve the theoretical foundation for models with sublethal endpoints. For higher organisms it may be possible to develop models for the toxic continuous response of an individual organism, rather than a population, exposed to two or more toxicants.

In the case of microorganisms such as algae it would be desirable to model the growth rate not only as a function of toxicants but also limiting nutrients. Such a model, which is as yet untested experimentally, has recently been formulated by Chen (1987) through an extension of a probabilistic theory for multiple nutrient limitation (Chen and Christensen, 1985).

For all types of organisms, a variety of biological responses should be considered. These include death, enzymes, e.g., ALT indicative of liver damage in mice, reduction in growth rate, weight, respiration, and teratogenic and carcinogenic effects. To ensure that the multiple toxicity models are valid not only in the laboratory but also under more realistic conditions, it would be desirable to have some field testing. Thus it would, for example, be possible to make realistic estimates of the effects of aluminum and other metals made biologically available through acid rain reaching actual lake or river ecosystems.

We have shown here that a certain type of interactive model can be derived as a logical extension of the noninteractive model, that is when the similarity parameter λ is >1. Our proposed qualitative explanation that two or more biological systems are folded onto themselves needs to be reviewed and given a more quantitative basis. In cases where the noninteractive model gives a less than satisfactory fit, as in Fig. 9, one should examine if the noninteractive model can and should be extended to cover nonsymmetrical cases, involving two similarity parameters λ_1 and λ_2, or if it would be necessary to invoke an interactive model.

We expect that the fitting of the very general noninteractive model given by Eqs. (3)–(7) will be impractical, although it should be possible to tackle the three-dimensional problem. One way of doing this would be to experimentally determine at least seven points on a three-dimensional isobole surface and then calculate λ_{12}, λ_{13}, λ_{23}, λ_{123}, ρ_{12}, ρ_{13}, and ρ_{23} by maximizing the log-likelihood given by Eq. (54).

The ultimate value to aquatic ecology of modeling the biological effects of multiple toxicants lies in the acquired predictive capability enabling us to estimate the toxic effects from a given wastewater discharge, and in the following formulation of water quality criteria providing a certain measure of protection for aquatic life. When these criteria, e.g., Eq. (65), are based on a well-documented model they are likely to provide adequate protection without being overly restrictive, thus minimizing the cost of pollution abatement.

XI. CONCLUSIONS

We have shown that the univariate weibit transformation provides a better fit to experimental dose-response data for uncoupler test chemicals than probit and logit transformations. This observation is consistent with previous results for other organisms and toxicants, and supports the proposed mechanistic-probabilistic basis of the Weibull model.

The basic shape of isobolograms for two toxicants depends in all cases, except for concentration addition and no addition, on the slopes of the log-linear transformations for the individual toxicants. When one or both of the slopes of the weibit transformations are less than one, portion(s) of the isoboles for response addition and response multiplication lie below the line for concentration addition. The isoboles for response addition are dependent on the chosen standard response level, e.g., whether the nonresponse probability Q is equal to 10, 50, or 90%.

The multiple toxicity index (MTI) introduced by Konemann (1981b) is identical to the overall similarity parameter λ in our noninteractive multiple toxicity model for n toxicants when the following conditions are met. The similarity in action of the n toxicants must be completely characterized by the above λ, there must be full correlation of toxicant tolerances, and MTI must be determined for equal toxic units of the individual toxicants in the mixture. Thus when MTI is determined under these standard conditions, so is λ, and the multiple toxicity model may now be considered to have been calibrated, meaning that it can be used to predict the biological responses for arbitrary combinations of toxicant concentrations.

We have demonstrated that several experimental isobolograms for pairs of organic chemicals from the same or different QSARs (Broderius, 1987) can be fitted to our bivariate toxicity model. The value of the correlation coefficient ρ determined by the fitting is equal to one in all cases. The fitted value of λ determines the position of the isobole. For $0 < \lambda < 1$, the isobole is above the line for concentration addition, and for $\lambda < 1$, it is below this line. When the two toxicants come from the same QSAR, $\lambda \approx 1$ ($0.87 \leq \lambda \leq 1.23$), and otherwise ≤ 0.68. These values of λ are generally consistent with the notion of λ as a similarity parameter describing the degree of overlap of two biological systems affected by the two toxicants. The experimental isobolograms had not previously been fitted to any multiple toxicity model.

An explicit water quality criterion, Eq. (65), has been developed for the above case where there is full correlation of toxicant tolerances and where the similarity in toxic action can be adequately characterized by a single parameter λ. The criterion is very general in that it is independent of the standard response level and of the form of the log-linear transformation for the individual toxicants, i.e., weibit, probit, or logit. When $\lambda = 1$, it reduces to the classical criterion, Eq. (64) valid for concentration addition.

REFERENCES

American Society for Testing and Materials. 1980. *Standard Pracatice for Conducting Acute Toxicity Tests with Fishes, Macroinvertebrates, and Amphibians*, Publication No. E729-80. Philadelphia: ASTM.

Anderson, P. D., and Weber, L. J. 1975a. Toxic response as a quantitative function of body size. *Toxicol. Appl. Pharmacol.* 33:471–483.

Anderson, P. D., and Weber, L. J. 1975b. The toxicity to aquatic populations of mixtures containing certain heavy metals. In *Proceedings of the International Conference on Heavy Metals in the Environment*, pp. 933–954. Toronto: University of Toronto, Institute of Environmental Studies.

Ashford, J. R. 1981. General models for the joint action of mixtures of drugs. *Biometrics* 37:457–474.

Biesinger, K. E., Christensen, G. M., and Fiandt, J. T. 1986. Effects of metal salt mixtures on *Daphnia magna* reproduction. *Ecotoxicol. Environ. Safety* 11:9–14.

Bliss, C. I. 1939. The toxicity of poisons applied jointly. *Ann. Appl. Biol.* 26:585–615.

Broderius, S. J. 1987. Joint aquatic toxicity of chemical mixtures and structure–toxicity relationships. Preliminary and Final Report. Final Report in: *ASA/EPA Conferences on Interpretation of Environmental Data. I. Current Assessment of Combined Toxicant Effects*, May 5–6, 1986, pp. 45–62, EPA-230-03-87-027. Washington, D.C.: U.S. Environmental Protection Agency, Office of Policy, Planning, and Evaluation.

Broderius, S., and Kahl, M. 1985. Acute toxicity of organic chemical mixtures to the fathead minnow. *Aquat. Toxicol.* 6:307–322.

Broderius, S. J., and Smith, L. L., Jr. 1979. Lethal and sublethal effects of binary mixtures of cyanide and hexavalent chromium, zinc or ammonia to the fathead minnow (*Pimephales promelas*) and rainbow trout (*Salmo gairdneri*). *J. Fish. Res. Board Can.* 36:164–172.

Center for Lake Superior Environmental Studies. 1986. *Acute Toxicities of Organic Chemicals to Fathead Minnows*, vols. I–III, eds. L. T. Brooke, D. J. Call, D. L. Geiger, and C. E. Northcott. Superior: University of Wisconsin.

Chen, C.-Y. 1987. The effects of limiting nutrient to algal toxicity assessment—A theoretical approach. Paper presented at the Third International Symposium on Toxicity Testing Using Microbial Systems, Valencia, Spain, May 11–15.

Chen, C.-Y., and Christensen, E. R. 1985. A unified theory for microbial growth under multiple nutrient limitation. *Water Res.* 19(6):791–798.

Chen, C.-Y., and Christensen, E. R. 1986. MULTOX, a computer program for the calculation of response of organisms to multiple toxicants and multiple limiting nutrients. AEEP's computer manual, pp. 166–173, ed. J. C. Crittenden. Houghton, Mich.: Michigan Technological University, Dept. of Civil Engineering.

Christensen, E. R. 1984. Dose-response functions in aquatic toxicity testing and the Weibull model. *Water Res.* 18:213–221.

Christensen, E. R. 1987. Development of models for combined toxicant effects. In *ASA/EPA Conferences on Interpretation of Environmental Data. I. Current Assessment of Combined Toxicant Effects,* May 5–6, 1986, pp. 66–74. EPA-230-03-87-027. Washington, D.C.: U.S. Environmental Protection Agency, Office of Policy, Planning, and Evaluation.

Christensen, E. R., and Chen, C.-Y. 1985. A general noninteractive multiple toxicity model including probit, logit, and Weibull transformations. *Biometrics* 41:711–725.

Christensen, E. R., and Nyholm, N. 1984. Ecotoxicological assays with algae: Weibull dose-response curves. *Environ. Sci. Technol.* 18:713–718.

Christensen, E. R., Scherfig, J., and Dixon, P. S. 1979. Effects of manganese, copper, and lead on *Selenastrum capricornutum* and *Chlorella stigmatophora. Water Res.* 13:79–92.

Christensen, E. R., Chen, C.-Y., and Kannall, J. 1985a. The response of aquatic organisms to mixtures of toxicants. Paper presented at IAWPRC's 12th Biennial Conference, Amsterdam, Holland, 1984. *Water Sci. Technol.* 17:1445–1446.

Christensen, E. R., Chen, C.-Y., and Fisher, N. S. 1985b. Algal growth under multiple toxicant limiting conditions. In *Proceedings of the International Conference on Heavy Metals in the Environment,* held in Athens, Greece, vol. 2, pp. 327–329. Edinburgh, U.K.: CEP Consultants.

Christensen, E. R., Chen, C.-Y., and Fox, D. F. 1986. Computer programs for fitting dose-response curves in aquatic toxicity testing. AEEP's computer manual, pp. 174–180, ed. J. C. Crittenden. Houghton, Mich.: Michigan Technological University, Dept. of Civil Engineering.

Department of Natural Resources, State of Wisconsin, Public Service Commission. 1986. Final Environmental Impact Statement Exxon Minerals Company Zinc-Copper Mine, Crandon, Wisconsin.

Druckrey, H. 1967. Quantitative aspects of chemical carcinogenesis. In *Potential Carcinogenic Hazards from Drugs.* vol. 7, *Evaluation of Risks,* ed. R. Truhaut, UICC Monogr. Ser., p. 60. Berlin: Springer-Verlag.

Durkin, P. R. 1981. Approach to the analysis of toxicant interactions in the aquatic environment. Aquatic Toxicology and Hazard Assessment: Fourth Conference, ASTM STP 737, eds. D. R. Branson and K. L. Dickson, pp. 388–407. Philadelphia: American Society for Testing and Materials.

EIFAC (European Inland Fisheries Advisory Commission). 1980. Report on Combined Effects on Freshwater Fish and Other Aquatic Life of Mixtures of Toxicants in Water. EIFAC Technical Paper No. 37. Rome: Food and Agriculture Organization of the United Nations.

Finney, D. J. 1942. The analysis of toxicity tests on mixtures of poisons. *Ann. Appl. Biol.* 29:82–94.

Finney, D. J. 1971. *Probit Analysis,* 3rd ed. Cambridge: Cambridge University Press.

Gaddum, J. H. 1949. *Pharmacology,* 3rd ed. London: Oxford University Press.

Hermens, J., and Leeuwangh, P. 1982. Joint toxicity of mixtures of 8 and 24 chemicals to the guppy (*Poecilia reticulata*). *Ecotoxicol. Environ. Safety* 6:302–310.

Hermens, J., Canton, H., Janssen, P., and De Jong, R. 1984a. Quantitative structure-activity relationships and toxicity studies of mixtures of chemicals with anaesthetic potency: Acute lethal and sublethal toxicity to *Daphnia magna. Aquat. Toxicol.* 5:143–154.

Hermens, J., Canton, H., Steyger, N., and Wegman, R. 1984b. Joint effects of a mixture of 14 chemicals on mortality and inhibition of reproduction of *Daphnia magna. Aquat. Toxicol.* 5:315–322.

Hermens, J., Broekhuyzen, E., Canton, H., and Wegman, R. 1985. Quantitative structure activity relationships and mixture toxicity studies of alcohols and chlorohydrocarbons: Effects on growth of *Daphnia magna*. *Aquat. Toxicol.* 6:209–217.

Hewlett, P. S. 1969. Measurement of the potencies of drug mixtures. *Biometrics* 25:477–487.

Hewlett, P. S., and Plackett, R. L. 1950. Statistical aspects of the independent joint action of poisons, particularly insecticides. II. Examination of data for agreement with the hypothesis. *Ann. Appl. Biol.* 37:527–552.

Hewlett, P. S., and Plackett, R. L. 1956. The relation between quantal and graded responses to drugs. *Biometrics* 12:72–78.

Hewlett, P. S., and Plackett, R. L. 1959. A unified theory for quantal responses to mixtures of drugs: Non-interactive action. *Biometrics* 15:591–610.

Hewlett, P. S., and Plackett, R. L. 1964. A unified theory for quantal responses to mixtures of drugs: Competitive action. *Biometrics* 20:566–575.

Hewlett, P. S., and Plackett, R. L. 1979. *The Interpretation of Quantal Responses in Biology.* London: Edward Arnold.

Kaiser, K. L. W. (ed.). 1984. *QSAR in Environmental Toxicology.* Dordrecht: Reidel.

Kodell, R. L. 1987. Modeling the joint action of toxicants: Basic concepts and approaches. In *ASA/ EPA Conferences on Interpretation of Environmental Data. I Current Assessment of Combined Toxicant Effects,* May 5–6, 1986, pp. 1–8. EPA-230-03-87-027. Washington, D.C.: U.S. Environmental Protection Agency, Office of Policy, Planning, and Evaluation.

Konemann, H. 1981a. Quantitative structure-toxicity relationships in fish toxicity studies. Part 1: Relationship for 50 industrial pollutants. *Toxicology* 19:209–221.

Konemann, H. 1981b. Fish toxicity tests with mixtures of more than two chemicals: A proposal for a quantitative approach and experimental results. *Toxicology* 19:229–238.

Lewis, M. A., and Perry, R. L. 1981. Acute toxicities of equimolar and equitoxic surfactant mixtures to *Daphnia magna* and *Lepomis macrochirus*. Aquatic Toxicology and Hazard Assessment: Fourth Conference, ASTM STP 737, eds. D. R. Branson and K. L. Dickson, pp. 402–418. Philadelphia: American Society for Testing and Materials.

Loewe, S. 1953. The problem of synergism and antagonism of combined drugs. *Arzneimitt. Forsch.* 3:285–290.

Marking, L. L. 1977. Method for assessing additive toxicity of chemical mixtures, aquatic toxicology and hazard evaluation, ASTM STP 634, eds. F. L. Mayer and J. L. Hameling, pp. 99–108. Philadelphia: American Society for Testing and Materials.

Muska, C. F., and Weber, L. J. 1977. An approach for studying the effects of mixtures of environmental toxicants on whole organism performances. In *Recent Advances in Fish Toxicology, A Symposium,* ed. R. A. Tubb, pp. 71–87. EPA-600/3-77-085, Ecological Research Series. Corvallis, Ore.: Environmental Research Laboratory.

National Academy of Sciences and National Academy of Engineering. 1972. Water Quality Criteria 1972. EPA-R3-73-033, Ecological Research Series. Washington, D.C.: U.S. Environmental Protection Agency.

Plackett, R. L., and Hewlett, P. S. 1948. Statistical aspects of the independent joint action of poisons. I. The toxicity of a mixture of poisons. *Ann. Appl. Biol.* 35:347–358.

Plackett, R. L., and Hewlett, P. S. 1963. A unified theory for quantal responses to mixtures of drugs: The fitting to data of certain models for two non-interactive drugs with complete positive correlation of tolerances. *Biometrics* 19:517–531.

Plackett, R. L., and Hewlett, P. S. 1967. A comparison of two approaches to the construction of models for quantal responses to mixtures of drugs. *Biometrics* 23:27–44.

Reif, A. E. 1984. Synergism in carcinogenesis. *J. Natl. Cancer Inst.* 73(1):25–39.

Sawicki, R. M., Elliott, M., Gower, J. C., Snarey, M., and Thain, E. M. 1962. Insecticidal activity of pyrethrum extract and its four insecticidal constituents against house flies. I. Preparation and relative toxicity of the pure constituents; statistical analysis of the action of mixtures of these components. *J. Sci. Food Agric.* 13:172–185.

Shelton, D. W., and Weber, L. J. 1981. Quantification of the joint effects of mixtures of hepatotoxic agents: Evaluation of a theoretical model in mice. *Environ. Res.* 26:33–41.

Southgate, B. A. 1932. The toxicity of mixtures of poisons. *Quart. J. Pharm. Pharmacol.* 5:639–648.

Spehar, R. L., and Fiandt, J. T. 1986. Acute and chronic effects of water-quality criteria-based metal mixtures on three aquatic species. *Environ. Toxicol. Chem.* 5:917–931.

Spehar, R. L., Leonard, E. N., and De Foe, D. L. 1978. Chronic effects of cadmium and zinc mixtures on flagfish (*Jordanella floridae*). *Trans. Am. Fish. Soc.* 107(2):354–360.

Sprague, J. B. 1986. Multiple aquatic toxicants: Some history and some evidence of the ostrich syndrome. Abstract No. 361, Seventh Annual Meeting, Society of Environmental Toxicology and Chemistry, Alexandria, Va., November 2–5.

Sprague, J. B., and Ramsay, B. A. 1965. Lethal levels of mixed copper-zinc solutions for juvenile salmon. *J. Fish. Res. Bd. Can.* 22:425–432.

U.S. EPA. 1978. The *Selenastrum capricornutum* Printz Algal Assay Bottle Test. Report No. EPA-600/9-78-018. Corvallis, Ore.: U.S. Environmental Protection Agency.

U.S. EPA. 1985. Water quality criteria: Availability of documents. *Fed. Reg.* 50(145):30784–30796.

Veith, G. D., De Foe, D., and Knuth, M. 1985. Structure-activity relationships for screening organic chemicals for potential ecotoxicity effects. *Drug Metab. Rev.* 15(7):1295–1303.

Watson, H. E. 1908. A note on the variation of the rate of disinfection with change in the concentration of the disinfectant. *J. Hyg.* 8:536–542.

Westall, J. C., Zachary, J. L., and Morel, F. M. M. 1976. MINEQL, a computer program for the calculation of chemical equilibrium composition of aqueous systems, Technical Note No. 18. Cambridge, Mass.: Massachusetts Institute of Technology, Dept. of Civil Engineering.

Wong, P. T. S., Chau, Y. K., and Luxon, P. L. 1978. Toxicity of a mixture of metals on freshwater algae. *J. Fish. Res. Bd. Can.* 35:479–481.

Principles and Applications of Surface Water Acidification Models

William D. Schecher and Charles T. Driscoll

Department of Civil Engineering
Syracuse University, New York

I. INTRODUCTION

Recently, interest and concern have developed over the effects of acidic deposition on terrestrial and aquatic systems. However, there is considerable debate over the magnitude and extent of these effects (Krug and Frink, 1983). Although it is evident that anthropogenic activities, such as fossil fuel combustion (Likens et al., 1979) and agricultural practices (van Breemen et al., 1982) have facilitated the release of potentially acidic sulfur and nitrogen compounds into the atmosphere, it is unclear what the long-term consequences will be on the environment and society.

To better understand the impact of acidic deposition on terrestrial and aquatic systems, a number of acidification models have been developed. In general, these models are used (1) as tools to facilitate the integration of research and to demonstrate deficiencies in our current understanding of basic processes and (2) to make predictive assessments of long-term effects and the regional extent of surface water acidification under different loading scenarios of acidic deposition. There is considerable consensus with regard to the most important processes regulating the acid-base status of drainage waters. Variations in model approach can generally be attributed to the objective that a given model is

The subject of acidification modeling is extremely comprehensive and no review can include all contributions to the field. Therefore, we would like to acknowledge the cooperation and work of the following researchers whose models are not discussed in this chapter: Paul A. Arp (Arp, 1983; Arp and Ramnarine, 1983), William G. Booty and Joseph V. DePinto (WAM and ALARM; Scheffe et al., 1986; Booty and Kramer, 1984), M. Hauhs (Hauhs, 1986), and Mary E. Thompson (Cation Denudation Rate model, Thompson, 1982). In addition, we gratefully acknowledge the cooperation of a number of researchers: N. Christophersen, B. J. Cosby, C. S. Cronan, W. de Vries, R. A. Goldstein, J. O. Reuss, J. L. Schnoor, H. M. Seip, M. Small, and R. F. Wright, whose support greatly facilitated an up-to-date review of acidification models. This review was supported through the ALBIOS study with funding provided by the Electric Power Research Institute and the Empire State Electric Energy Research Corporation.

attempting to accomplish and the degree of complexity to which the objective is approached. There is no definitive approach to acidification models and each model has certain advantages and limitations.

The objectives of this chapter are (1) to provide an overview of the important physicochemical processes that are depicted in most acidification models; (2) to examine the role of acidification models in research and predictive assessments; (3) to summarize some prominent acidification models, emphasizing the assumptions and objectives used in the model development; and (4) to suggest structural techniques for possible comparisons between models and for evaluating the quality of model predictions.

II. ACIDIFICATION IN THE ENVIRONMENT

An understanding of hydrological flow paths and major processes involved in the transfer of ionic solutes is critical to an understanding of the acid-base status of surface waters as well as assessments of the sensitivity of these systems to changes in strong acid loading. A forested watershed can be depicted as a chemical reactor in which atmospheric inputs are transported with drainage water through the forest canopy, soil, and the stream-lake environment. A wide variety of biogeochemical processes within watershed-lake systems serve to modify the composition of drainage waters. A critical parameter used in most models to assess the acid-base status of drainage waters is acid neutralizing capacity (ANC), Eqs. (1) and (2):

$$ANC = [HCO_3^-] + 2[CO_3^{2-}] + RCOO_w^- + OH-Al$$
$$+ [OH^-] - [H^+] \tag{1}$$

$$ANC = C_B + [NH_4^+] + 3[Al_T] - [NO_3^-] - [Cl^-]$$
$$- 2[SO_{4T}] - [F_T] - RCOO_s^- \tag{2}$$

where C_B = equivalence of basic cations $(2[Ca^{2+}] + 2[Mg^{2+}] + [Na^+] + [K^+]$, in equivalents per liter (eq/L))

$RCOO_s^-$ = equivalent concentration of organic anions that act as strong acid-base systems, i.e., complete dissociation (in eq/L)

$OH-Al$ = equivalence of OH^- complexed with Al $([AlOH^{2+}] + 2[Al(OH)_2^+] + 4[Al(OH)_4^-]$, in eq/L)

Al_T = total concentration of aqueous Al $([Al^{3+}] + [AlOH^{2+}] + [Al(OH)_2^+] + [Al(OH)_4^-] + [AlF^{2+}] + [AlF_2^+] + [AlF_3^0] + [AlF_4^-] + [AlF_5^{2-}] + [AlSO_4^+] + [Al(SO_4)_2^-]$, in moles per liter (mol/L))

F_T = total concentration of aqueous F (F^- + $[AlF^{2+}]$ + $2[AlF_2^+]$ + $3[AlF_3^0]$ + $4[AlF_4^-]$ + $5[AlF_5^{2-}]$ + $[HF^0]$, in mol/L)

SO_{4T} = total concentration of aqueous SO_4 ($[SO_4^{2-}]$ + $[AlSO_4^+]$ + $2[Al(SO_4)_2^-]$, in mol/L)

$RCOO_w^-$ = equivalence of organic anions that act as weak acid-base systems, i.e., incomplete dissociation (in eq/L)

ANC is the ability of water to neutralize inputs of strong acid to a preselected equivalence point (Stumm and Morgan, 1981). Through electroneutrality, ANC is equivalent to the sum of basic cations (C_B), NH_4^+, and total Al less the sum of NO_3^-, Cl^-, total SO_4, total F, and strong organic anions [Eq. (2)]. As a result, processes that alter the concentration of ionic solutes result in an equivalent change in ANC. For example, processes such as nitrification (net NO_3^- production), assimilation of cationic nutrients by vegetation, production of organic acids or atmospheric inputs of H_2SO_4 all result in a decrease in ANC [Eq. (2)]. Processes such as denitrification (NO_3^- reduction), SO_4^{2-} reduction, or the release of basic cations through cation exchange or chemical weathering reactions all serve to increase ANC (van Breemen et al., 1983).

A. Chemical and Biological Processes in Lake/Watershed Systems

The ultimate source of ANC in most drainage waters is chemical weathering of soil minerals (Schnoor and Stumm, 1985). Weathering reactions increase ANC through the release of basic cations, and surface waters that have been found to be particularly sensitive to acidic inputs exhibit low concentrations of basic cations due to low concentrations of readily available (exchangeable, easily weatherable) basic cations in the surrounding soil (Driscoll et al., 1987a). Factors that influence the rate of chemical weathering in soil and therefore the production of ANC include mineralogy, grain size of particles, contact of water with soil, temperature, pH, CO_2, and concentration of organic ligands in soil solutions (Schnoor and Stumm, 1985). Cation exchange is similar to chemical weathering in that both processes result in the release of basic cations and serve to increase ANC. However, the rate of cation exchange reactions are fast relative to chemical weathering. As a result, there are two distinct pools of basic cations in mineral soil: a large pool of relatively unreactive mineral bases with slow chemical weathering rates and a smaller pool of exchangeable bases that exhibit rapid rates of reaction and are quantitatively depicted as equilibrium reactions.

An important parameter used in many acidification models to assess the acid-base status of soils is percent base saturation (%BS). This parameter is calculated as the amount of exchangeable basic cations on the soil exchange complex divided by the cation exchange capacity (CEC). The difference between CEC and %BS is the quantity of acidic cations (H^+, and Al^{n+}) associated with soil exchange sites. Soils with low %BS are susceptible to acidification due to the low concentrations of basic cations on exchange sites, while systems with high %BS are more enriched in basic cations and therefore inputs of strong acid are completely neutralized by basic cation release.

Cation transfer through vegetation can also be an important process influencing the acid-base chemistry of drainage waters. Vegetation requires cationic nutrients (e.g., Ca^{2+}, Mg^{2+}, K^+), and a net accumulation of cation occurs in biomass and the forest floor of an aggrading forest system (Binkley and Richter, 1987). In most systems, cationic nutrients are assimilated in excess of anionic nutrients, resulting in net acidification of soil and drainage water by vegetation. For example, Driscoll and Likens (1982) reported that 32% of proton sources (815 eq/ha per year) in a northern hardwood ecosystem could be attributed to cation accumulation by biomass and the forest floor.

Soils that do not have large pools of readily available basic cations are characteristically acidic. Acidic soils contain elevated concentrations of Al on cation exchange sites (low %BS) and are susceptible to acidification by strong acid inputs (Reuss and Johnson, 1985). When an acid-sensitive soil is subjected to strong acid leaching, the release of basic cations may not be adequate to neutralize inputs, resulting in soil solutions with elevated concentrations of acidic cations (H^+, Al^{n+}). Aluminum is the most abundant metallic element in the lithosphere, occurring as about 8% by mass (Hem, 1986). Aluminum is also a strongly hydrolyzing metal and therefore is relatively insoluble and immobile in most environments. However, under acidic conditions or in the presence of complexing ligands, such as organic acids, Al is readily mobilized (Driscoll, 1985). Theoretically, the concentration of Al that is in solution in equilibrium with Al-containing minerals increases exponentially with decreases in pH. A number of researchers have reported that acidic waters draining mineral soils exhibit Al concentrations consistent with the pH-dependent solubility of $Al(OH)_3$ (Driscoll, 1985; Johnson et al., 1981). However, organic soils or wetland solutions are generally undersaturated with respect to $Al(OH)_3$ solubility (Cronan et al., 1986).

Elevated concentrations of Al may be ecologically significant. Concentrations of Al above the range of 4–8 μmole/L appear to be toxic to fish (Baker and Schofield, 1982; Muniz and Leivestad, 1980; Schofield and Trojnar, 1980). The speciation of Al is critical to aquatic toxicity, with organisms generally most sensitive to inorganic forms (Driscoll et al., 1980; Baker, 1982). In the terrestrial environment, high concentrations of Al may affect the fine roots of trees due to sorption of Al onto negatively charged root surfaces, resulting in forest dieback (Schnoor and Stumm, 1985). Through direct precipitation or adsorption reactions, Al may alter the cycling of P, organic C (Dickson, 1978), SO_4 (Prenzel, 1983), or trace metals (White and Driscoll, 1985). Aluminum may also act as a coagulant, facilitating the removal of particulates and organic solutes in lakes, and thereby increase transparency and alter the thermal structure of acidic lakes (Effler et al., 1985).

In addition to processes that regulate the release of cations, surface water acidification is also strongly influenced by the transfer of strong acid anions within watershed environments. Atmospheric inputs of sulfur are often in excess of nutritional requirements (Turner et al., 1980; Johnson et al., 1982; Johnson and Reuss, 1984). Therefore, in many watersheds the extent to which acidic deposition causes increased cation leaching (and drainage water acidification if the available cation pools are low) is determined by SO_4^{2-} mobility (e.g., Cronan et al., 1977; Johnson and Cole, 1977; Seip, 1980). Sulfate retention within soils serves to mitigate acidic inputs through decreasing the concentration of strong acid anions in solution and increasing the ANC [Eq. (2)]. The retention of SO_4^{2-} may occur by biological assimilation to organic forms, dissimilatory reduction, or abiotic adsorption-precipitation reactions (Fuller et al., 1985). Sulfate adsorption is thought to largely occur on positively charged iron or aluminum surfaces in soil (Chao et al., 1964). Soils derived in nonglaciated regions generally contain an abundance of free (nonsilicate bound) Fe and Al and therefore readily sorb SO_4^{2-}. In contrast, soils of glaciated regions retain relatively little SO_4^{2-} by adsorption due to small pools of free Fe and Al as well as an abundance of organic matter which competes with SO_4^{2-} for surface adsorption sites (Johnson et al., 1980). Mineral soils express highly pH-dependent behavior with respect to sulfate adsorption (Nodvin et al., 1986). At low pH values, soil particles develop high, positive, surface charge which enhances SO_4^{2-} adsorption (Davis, 1977). Although this phenomenon serves to initially mitigate the acidity of waters due to inputs of SO_4^{2-}, it also postpones the recovery of soil-stream systems due to pH-dependent desorption of SO_4^{2-} under more alkaline conditions (Nodvin et al., 1988) once inputs are reduced (Johnson and Reuss, 1984). In regions with acidic soils experiencing extremely high deposition of SO_4^{2-}, such as central Europe, SO_4^{2-} retention may also occur through precipitation of basic aluminum sulfate minerals such as jurbanite [$Al(OH)SO_4 \cdot H_2O$] (Prenzel, 1983; Johnson, 1984).

While drainage water acidification is most often attributed to elevated inputs of H_2SO_4 relative to the release of basic cations, other mechanisms of acidification may occur. For example, acidification of drainage waters by nitric acid has been reported for (1) systems that experience elevated inputs of NH_4^+ followed by nitrification and NO_3^- leaching (van Breemen et al., 1982), (2) degrading forests (Likens et al., 1970), and (3) stands, such as alder, which assimilate large quantities of nitrogen by nitrogen fixation and exhibit elevated leaching losses of NO_3^- (van Miegroet, 1984).

Reactions involving carbon may also be important in regulating the acid-base status of drainage waters. Microbial respiration in soil environments can produce CO_2 concentrations considerably in excess of atmospheric values (up to 10^{-1} atm; Reuss and Johnson, 1985). Production of CO_2 increases acidity and lowers the pH of soil solutions. The hydration and dissociation of CO_2 to H^+ and HCO_3^- is critical to the leaching of soil cations (van Breemen et al., 1983). Hydrogen ion originating from the dissociation of CO_2 may displace cations from exchange sites or facilitate mineral weathering. Bicarbonate then serves as a counterion in the leaching of metallic cations with drainage water. When soil leachate migrates from the confined soil environment and mixes with surface water, CO_2 degasses as it equilibrates with the atmosphere. Highly acidic solutions (<0 μeq/L ANC) exhibit little change in pH associated with CO_2 degassing due to limited dissociation of H_2CO_3 (CO_2). However, in solutions with positive ANC (>20 μeq/L) CO_2 exolution coincides with marked increases in solution pH. Aluminum mobilized through the soil profile by the dissociation of CO_2 may undergo hydrolysis and deposition on the stream-lake sediment as CO_2 is released to the atmosphere and pH values increase (Norton and Henriksen, 1983; Reuss and Johnson, 1985).

The release of naturally occurring organic acids may influence the chemistry of acid-sensitive drainage waters. Dissolution of strongly acidic functional groups associated with organic acids serves to reduce pH and ANC of drainage waters (Driscoll et al., 1988). For example, Driscoll and Newton (1985) reported several examples of lakes in the Adirondack region of New York State that exhibited elevated concentrations of dissolved organic carbon. Although atmospheric deposition of strong acids contributed to the acidity of these waters, in the absence of strong acid inputs these lakes were probably naturally acidic. Organic acids are also important in the chemistry and transport of Al. Naturally occurring organic ligands may effectively complex Al in solution, facilitating transport through the soil profile (Driscoll et al., 1985). Organic complexation of Al is also thought to mitigate toxicity to fish (Driscoll et al., 1980; Baker and Schofield, 1982).

B. Hydrological Processes

Within a forested watershed, incoming precipitation may undergo a variety of hydrological processes, including interception by the forest canopy, accumulation as snowpack, loss to the atmosphere by evapotranspiration, overland flow, infiltration into the soil, percolation through soil layers, and discharge to stream-lake systems (Fig. 1). A knowledge of hydrological flow paths is critical to understanding the chemical characteristics of surface waters because it provides information concerning the location of water within watershed compartments as well as the hydrological residence time of these compartments (Chen et al., 1984). The residence time of water in soil compartments such as the organic horizon or mineral soil horizon is particularly important in determining

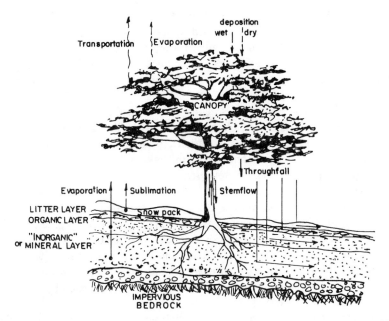

FIG. 1. Common hydrological flow paths which are assumed to be important in many acidification models (after Gherini et al., 1985).

the chemical composition of solutes that are regulated by time-dependent processes, such as chemical weathering, biological mineralization reactions, or biological accumulation by vegetation.

Numerous researchers have demonstrated the role of hydrological flow paths in regulating surface water chemistry (Johnson et al., 1981; Chen et al., 1984; Gherini et al., 1985; Driscoll and Newton, 1985). Surface water composition has been modeled as a mixture of two solutions: groundwater and quick flow (or interflow) (Johnson et al., 1969; Pinder and Jones, 1969). Groundwater is characterized by a long residence time within glacial till and lower mineral soil and, as a result, has near neutral pH values and enriched ANC due to the release of basic cations from weathering reactions (Peters and Driscoll, 1987). Quickflow (or shallow interflow) is water that rapidly moves through the upper acidic soil, probably to a large extent through root channels and macropores, to surface waters during high flow periods. Due to limited contact of this water with the mineral soil, it usually has low pH and low ANC. The relative contribution of groundwater and acidic interflow to surface water flow is primarily controlled by the distribution and thickness of soil and till as well as the hydraulic conductivity (rate at which water is transmitted) of these deposits (Chen et al., 1984). Peters and Murdoch (1985) evaluated the hydrological characteristics of acidic and neutral pH lakes. Runoff from the acidic lake was highly variable, suggesting that water traveled rapidly through the upper acidic soil horizons to the lake. In contrast, runoff from the neutral pH lake was less variable because flow was attenuated by the thick glacial till. Moreover, during dry periods, groundwater from the thick till maintained discharge of a relatively high level.

Hydrological flow paths are also apparently critical to understanding short-term changes in the chemical characteristics of surface waters. Chen et al. (1984) reported that during low flow periods, groundwater, which is enriched in basic cations and ANC from contact with the lower mineral soil and glacial till, comprises a large percentage of total stream discharge (Fig. 2). During high flow periods (e.g., spring snowmelt), the rate of infiltration exceeds the hydraulic conductivity of the soil and the zone of saturation migrates up the soil profile. The hydraulic conductivity of the soil is considerably greater in the upper soil due to macropores and root channels, so when the zone of saturation reaches this depth, water readily migrates laterally to form surface water discharge. During these hydrological events, water bypasses the lower mineral soil, resulting in acidification through dilution of groundwater by inputs of shallow acidic interflow (Fig. 2).

The hydrodynamic characteristics and hydrological retention times of lakes are also important factors influencing their acid-base status. Many lakes that experience surface water acidification are in temperate regions and exhibit dimictic behavior. These lakes are generally completely mixed during fall and spring and are stratified during the summer and the winter ice-cover periods. Drainage water inputs to these systems are generally lowest (and with highest

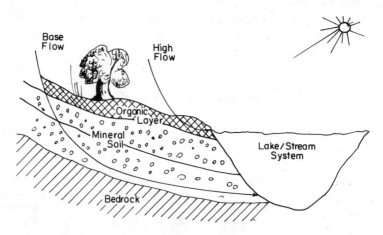

FIG. 2. Variation in hydrologic flow path under high and low flow conditions.

ANC) during the summer due to evapotranspiration by the surrounding forest watershed and during the period of winter snowpack accumulation. During the fall and spring snowmelt periods, discharge is typically high and the water quality acidic (Driscoll et al., 1987a). The quantity and quality of input water and the conditions of stratification all influence lake response. For example, during fall rainfall events, acidic inputs from stream water enter lake systems that are often completely mixed and as a result the entire water column is impacted (Driscoll et al., 1987b). The extent to which fall acid events influence lake water quality depends on the quantity of input water in addition to the lake volume. Smaller lakes which receive elevated discharge of acidic water are most severely impacted. During snowmelt, peak discharge generally coincides with ice cover and inverse thermal stratification of lakes. As a result, this highly acidic water generally does not completely mix with the bulk lake water (Hagen and Langeland, 1973). Rather, temperature stratification causes acidic inputs, which are near $0\,°C$, to migrate across the lake along the ice-water interface. This meltwater generally mixes with only the upper 1–2 m of lake water prior to export via the lake outlet. As a result, during ice cover, most of the lake volume does not experience the effects of episodic acidification (Driscoll et al., 1987b).

Hydraulic retention time also influences the chemical composition of lake water. In-lake processes such as SO_4^{2-} reduction and NO_3^- retention are time-dependent processes. Kelly et al. (1987) have demonstrated that the extent of in-lake retention of SO_4^{2-} and NO_3^- and generation of in-lake ANC is strongly dependent on hydraulic retention time.

III. ACIDIFICATION MODELS

A. The Role of Models

The term model has become popular in the last 25 years, primarily due to the rise of computer-based applications. The definition of model is often assumed a priori, but this assumption has led to a range of uses for the term. For purposes of this discussion, a model is defined as a combination of ideas represented in a quantitative form (such as a computer program or mathematical equation). In addition, a model also includes a human element since computer programs cannot be developed, modified, or verified without human guidance. The development of computer programs for modeling purposes involves a continual feedback of information from calculated results to the programmer. This information is in turn evaluated and considered in future model development. Modeling is an iterative process; both the human ideas and the computer program are modified in each iteration. Modelers are usually intimately linked to the models they develop and this accounts for the bias evident in the literature in descriptions of model development and application. High-quality results are generally obtained from model simulations by their developers while the same program may yield poor results to those who are not intimately knowledgeable about the program.

Comparison and criticism of current acidification models are difficult because these programs were developed for a range of applications and to accomplish different objectives. However, it is possible to review the objective(s) that these models were designed to accomplish and to provide a description of model structure. Generally, these objectives can be categorized into (1) testing research-oriented hypotheses and (2) predictive applications. Many of the research-oriented models tend to also center on soil processes and can be very complicated. Although comprehensive models have been used for predictive purposes, models that have been specifically designed for this purpose are usually simpler in their construction and often empirical.

Output parameters obtained from acidification models vary greatly from simply pH or ANC (Henriksen, 1979, 1980) to a comprehensive summary of concentrations and pools of all major elements within lake-watershed systems (Gherini et al., 1985). Use of output parameters depends on the objective of the user. When models are used as research tools, the output parameters of interest are as varied as the number of research questions. When models are used in policy assessments, output parameters may be considerably more limited in scope. Policy decisions made using acidification models are largely based on water quality criteria. In particular, intensity factors such as pH and concentrations of inorganic Al species are critical to evaluate the response of aquatic organisms to the acid-base chemistry of waters, while the capacity factor, ANC, is used to determine the sensitivity of waters to potential changes in strong acid loading.

B. Empirical Acidification Models

1. Steady-State Models

Simple acidification models have been developed with the philosophy that empirical observations may prove to have great utility. The models were designed to help in classifying which lakes were acidic and to try to predict the effects of future acid loading on lake chemistry. In general, empirical models are abiotic nonhydrological models. While most acidification models employ the concept of electroneutrality, empirical models often simplify ion balance expressions by eliminating chemical constituents that are thought to be insignificant. Consequently, although empirical expressions are useful for many waters, they may not be appropriate for every circumstance. Simple models often assume that cation release is the predominant mechanism for the neutralization of acids and accommodations generally have not been made for processes regulating the retention of strong acid anions.

One of the most prominent empirical models is Henriksen's acidification model (Henriksen, 1979, 1980; Wright and Henriksen, 1983). It should be noted that the Henriksen (1979, 1980) model is similar in its approach to other empirical models (e.g., Dickson, 1980; Dillon et al., 1980; Thompson, 1982). The model originated with the concept that lake acidification can be analogously represented by a beaker which is being titrated with strong acid. It was assumed a priori that freshwater acidification is caused by acid precipitation. Basic cations are assumed to be present in the lake systems either through mineral weathering or cation exchange reactions.

Henriksen's (1979, 1980) model assumes an electroneutrality balance of the form

$$H^+ + Ca^{2+} + Mg^{2+} + Na^+ + K^+ + NH_4^+ + Al^{n+}$$
$$= NO_3^- + SO_4^{2-} + Cl^- + HCO_3^- + RCOO_w^- \qquad (3)$$

of which a fraction of these aqueous species may be from marine sources. After eliminating minor ions (Na^+, K^+, NO_3^-, NH_4^+, and $RCOO_w^-$), Eq. (3) simplifies to

$$H^+ + Ca^{2+} + Mg^{2+} + Al^{n+} = SO_4^{2-} + HCO_3^- \qquad (4)$$

Empirical models generally assume that steady state exists between inputs of strong acid and the production of basic cations. Henriksen (1982) introduced the concept of the F factor, the fraction of acid inputs which are neutralized by the release of basic cations, which is assumed to be constant over time. Equation (4) becomes

$$H^+ + F(Ca^{2+} + Mg^{2+}) + Al^{n+} = SO_4^{2-} + HCO_3^- \qquad (5)$$

where F is estimated at about 0.91 (Wright, 1983). Equation (5) can be segmented into three distinct acidification classes:

Class:	Equation:	pH:	
Acidic	$H^+ + Al^{n+} + F(Ca^{2+} + Mg^{2+}) = SO_4^{2-}$	< 5.0	(6)
Slightly acidic	$F(Ca^{2+} + Mg^{2+}) = SO_4^{2-}$	5.0–5.5	(7)
Neutral	$F(Ca^{2+} + Mg^{2+}) = SO_4^{2-} + HCO_3^-$	> 5.5	(8)

Equations (6)–(8) form the basis of the Henriksen (1979, 1980) nomograph (Fig. 3). The Henriksen (1979, 1980) nomograph provides a simple but useful method for classifying lake resources and assessing the extent of acidification.

Wright (1983) examined 15 data sets from North America and 12 data sets from Europe to test the validity of the Henriksen model. He found that for 90% of the lakes (oligotrophic, clear water, low ionic strength) that were examined, ANC changes could be modeled using the Henriksen equation. Deviations were attributed to analytical errors, road salt, and agricultural pollution. Lakes that have an internal source or sink of sulfur, such as in Florida, were not suitable to analysis by the Henriksen approach.

2. Modifications of Empirical Models

a. Steady-State Trickle-Down Model. The steady-state version of the trickle-down model (Schnoor et al., 1986; Schnoor and Stumm, 1985) is similar to the Henriksen (1982) model but it includes a chemical weathering rate factor which varies nonlinearly with acid inputs. The steady-state version of the trickle-down model (Schnoor et al., 1986) has been used to estimate the percentage of lakes that are at risk from acidic deposition in the upper Midwest of the United States. Schnoor et al. (1986) concluded that most (88%) of the

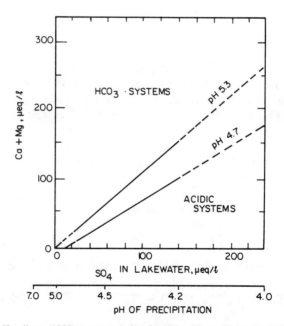

FIG. 3. Henriksen (1980) nomograph for classifying lake acidity (after Wright, 1983).

acidic lakes in this region were seepage lakes that received their water directly from precipitation. They also found that SO_4^{2-} was the dominant anion in the precipitation and chemistry of these lakes and that 90% of the H^+ in the lakes came directly from precipitation. Further use of the model estimated the lake resources that are at risk in the Midwest at various levels of SO_4^{2-} deposition.

   b. *Regional Predictor Model.* Another modification of the empirical approach has been the application of the Henriksen model to regional analysis. The work by Small and Sutton (1986) shows how the empirical model can be modified to address the distribution of watershed responses to strong acid inputs within a region. Although empirical models may be used to predict the chemistry of many lakes in a region, Small and Sutton (1986) believe that variations in weathering rates from one watershed to another limit the use of these models in regional predictions. The approach used by the authors is to represent chemical weathering and lake ANC in a region by a statistical distribution. Rather than apply the empirical equations to each individual lake system, they represent the ANC values of an entire population of lakes by a single set of equations.

3. Limitations

   Although empirical models are effective tools that can be used to help classify lake resources and predict future trends of acidification, they are limited in their ability to help researchers understand mechanisms and environmental pathways for acidic interactions in soils and streams. Moreover, there is always some danger in extrapolating empirical relationships to regions where the processes which regulate the acid-base status of waters are unknown. Under these conditions, the assumptions used in a model may be violated. If processes other than those depicted in the empirical models are important, then regional assessments as well as predictions of surface water quality under different scenarios of strong acid loading are without merit. For example, the Henriksen (1979, 1980) model was developed under the assumption that the acid-base status of waters is regulated by the release of basic cations relative to inputs of H_2SO_4 and that SO_4^{2-} inputs are conservatively transported through a watershed. The application of this model to regions where the ANC of drainage waters is strongly influenced by in-lake processes (Kelly et al., 1987), SO_4^{2-} adsorption (Arp, 1983; Arp and Ramnarine, 1983; Johnson, 1984), or inputs of organic acids (Driscoll et al., 1987a), would result in inaccurate assessments.

   Empirical models are also limited because they cannot address temporal trends in solution chemistry. Important changes in chemistry may be evident over both short- and long-time scales. Episodic acidification of surface waters is evident during hydrological events and may coincide with the most severe water quality conditions over the annual cycle. If organisms are sensitive to episodic rather than chronic acidification, then empirical models may have limited use in biological assessments.

Policy decisions regarding surface water acidification need to be concerned with the response time of systems to changes in strong acid loading. Empirical models provide no information on the nature and length of response of watersheds to increases or decreases in strong acid loading.

C. Simulation Models Based on Lake/Watershed Processes

In order to overcome the shortcomings of empirical models, mechanistic models have been developed which range in complexity from simple equilibrium models (e.g., Reuss and Johnson, 1985; Cosby et al., 1985a, 1985c, 1986; Schecher and Driscoll, 1987) to highly comprehensive solute transport models (e.g., Gherini et al., 1985).

1. In-Lake Neutralization of Acids

Recent studies have demonstrated the importance of acid neutralization from in-lake processes, such as SO_4^{2-} and NO_3^- reduction (Schindler et al., 1980; Kelly et al., 1982; Cook and Schindler, 1983). In some regions, in-lake processes represent an important source of ANC, and assessments that fail to consider this input may seriously overestimate sensitivity to acidic deposition. Despite the importance of in-lake processes, acidification models have generally ignored these sources of ANC. Recently, Kelly et al. (1987) presented a simple model which can be used to predict in-lake retention of SO_4^{2-} and NO_3^- and, therefore, in-lake production of ANC under a wide range of conditions. Application of the model requires only information on hydraulic retention time, mean depth, and mass transfer coefficients of SO_4^{2-} and NO_3^- removal:

$$AN = L \left(\frac{1}{OF/A} - \frac{1}{OF/A + S} \right) \tag{9}$$

where AN = neutralization of acid (μeq/m^3)
   L = area of sulfate of nitrate loading (μeq/m^2 per year)
   OF = lake outflow (m^3/yr)
   A = lake area (m^2)
   S = mass transfer coefficient for SO_4^{2-} or NO_3^-
This model can readily be applied to lakes with oxygenated waters, which is generally characteristic of acid-sensitive, oligotrophic systems.

Kelly et al. (1987) found good agreement for mass transfer coefficients obtained by mass balance calculations and measured rates of sulfate reduction or denitrification. With these independent methods they suggest values of 0.4–0.54

m/yr for SO_4^{2-} and 7.4–9.2 m/yr for NO_3^- mass transfer coefficients to be used in model calculations.

Results of this model indicate that hydraulic retention time is the most critical parameter in predicting in-lake generation of ANC by SO_4^{2-} and NO_3^- retention. Lakes with long hydraulic retention times exhibit considerable removal of SO_4^{2-} and NO_3^- and are therefore capable of effectively neutralizing strong acid inputs. Lakes with short hydraulic retention times, however, demonstrate limited in-lake generation of ANC and therefore are sensitive to acidification. While this model is not comprehensive, it might be effectively used in regional assessments of in-lake generation of ANC to changes in acid loading.

2. Reuss-Johnson Model

Chemical equilibrium is an integral component of many soil process models. The model by Reuss and Johnson (1985) encompasses several of the chemical equilibrium principles that are present in many of the more complicated models. This model calculates the loss of ions from soil as a function of atmospheric deposition, soil cation exchange, partial pressure of soil CO_2, chemical and physical soil properties as well as runoff. The objective of this model is to allow the user to predict the effect of rainfall acidity on leaching of basic cations, Al, and H^+ from noncalcareous soils using established principles of soil chemistry. Leaching of major cations (Ca^{2+}, Mg^{2+}, Na^+, K^+, Al^{n+}) is regulated by exchange reactions using selectivity coefficients. Aluminum inputs to the exchanger are assumed to originate from dissolution of $Al(OH)_3$. Over time steps during a simulation, calculations are performed assuming equilibria with soil CO_2, the soil exchanger, and $Al(OH)_3$ solubility. The input and output parameters used in the model are summarized in Table 1.

The first version of the Reuss-Johnson model (1985) did not consider exchange reactions with Mg^{2+}, Na^+, or K^+ but has since been revised. The model is simple and does not represent processes such as mineral weathering, adsorption, or biological transfer of basic cations, sulfur, nitrogen, or organic acids. Moreover, only a single soil layer is considered as a continuously stirred-tank reactor (CSTR), so variable hydrological flow paths are not depicted in this representation. However, this model could be readily used as a soil module in a more comprehensive ecosystem element cycling model.

The strength of the Reuss-Johnson model (1985) is that it allows the user to explore conditions that result in the production of ANC within the soil profile. The extent of ANC generation is a function of inputs to the soil profile, the distribution of ions on the soil exchange, equilibrium relationships regulating cation release to solution, and the partial pressure of CO_2 within the soil profile. The interplay between weak and strong acids in the release of ANC to drainage water is effectively demonstrated using the model. Reuss and Johnson (1985) suggest that soils with a large percentage of exchange sites occupied by acidic

TABLE 1

Summary of Input and Output Parameters for the Reuss-Johnson Model

Parameter	Unit of measurement
Input parameters	
Precipitation quality	μeq/L
Selectivity coefficients for	
Al-Ca	
Ca-Mg	
Ca-Na	
Ca-K	
Solubility of Al(OH)$_3$	
Soil exchange fractions (0–1)	
CEC	meq/100 g
Soil depth	cm
Bulk density	kg/L
Precipitation input	mm
% evapotranspiration in rain	
Step size through soil	mm
Output interval	
Output parameters displayed through soil profile	
pH	
H$^+$	μeq/L
Al^{3+}, AlOH^{2+}, Al(OH)$_2^+$	μmol/L
Aqueous Ca^{2+}, Mg^{2+}, Na$^+$, K$^+$	μeq/L
Aqueous SO$_4^{2-}$, NO$_3^-$, Cl$^-$	μeq/L
HCO$_3^-$, ANC	μeq/L
% CO$_2$	
Amount of Ca^{2+}, Mg^{2+}, Na$^+$, K$^+$ on exchanger	μeq/L

cations (low percent base saturation) are particularly susceptible to drainage water acidification. Under high percent base saturation of the soil exchanger and/or low inputs of strong acid anions, the partial pressure of CO$_2$ followed by dissociation to H$^+$ and HCO$_3^-$ serves to displace basic cations from the exchanger and produce ANC. As the percent base saturation of the soil decreases, it becomes more difficult to displace basic cations to solution. Moreover, under elevated inputs of strong acids, dissociation of carbonic acid is limited. As a result, acid-sensitive soils (low percent base saturation) experiencing strong acid loading have low pH and high Al concentrations in drainage waters. Reuss and Johnson (1985) concluded that leaching of basic cations will eventually cause serious nutrient depletion in soils while the leaching of Al causes higher biological toxicity within terrestrial ecosystems and to downstream aquatic systems.

3. MAGIC Model

The MAGIC model (Modeling Acidification of Groundwater In Catchments) (Cosby et al., 1985a,c) extends the approach of Reuss and Johnson (1985) by depicting solute transport in a watershed. Cosby et al. (1985a,c) consider additional processes in their acidification model, including a monoprotic organic acid, sulfate adsorption, inorganic complexation of Al, weathering of soil minerals proportional to a fractional power of H^+, and element uptake. A summary of the input and output parameters used in the models is presented in Table 2.

The MAGIC model was developed to predict long-term response of terrestrial and aquatic systems to acidic deposition. To accomplish this objective, Cosby et al. (1985 a,c) felt that a process-oriented, physically based model of catchment water was necessary. However, MAGIC is a relatively simple, lumped parameter model which composites spatially distributed processes. Therefore, spatial variation in biogeochemical processes along a soil profile or within a watershed due to heterogeneity from soil depth or vegetation is not considered. Rather, one set of input parameters that are representative of the watershed are used in simulations to predict the chemical composition of stream water at the outlet.

Using the approach of Reuss and Johnson (1985), selectivity coefficients are used to represent cation release through exchange reactions in soil. Selectivity coefficients are not true thermodynamic constants but vary markedly from soil to soil and with soil conditions (e.g., pH; McBride and Bloom, 1977). Consequently, selectivity coefficients need to be established each time the model is to be calibrated for a watershed. These estimates are generally obtained from watershed observations of the distribution of cations on the exchanger and cation concentrations in soil solutions.

Sulfate adsorption is represented in MAGIC by a Langmuir adsorption isotherm (Cosby et al., 1986). Like cation exchange, the extent of sulfate adsorption varies between soils and with soil solutions/soil characteristics (Johnson et al., 1980). Langmuir adsorption parameters, the half-saturation constant, and sulfate adsorption capacity are generally obtained from SO_4^{2-} adsorption isotherms developed for the soil of interest from batch laboratory experiments.

The distribution of Al on the soil exchanger is assumed to be derived from the dissolution of solid phase $(Al(OH)_3)$. Unlike cation exchange, selectivity coefficients, and Langmuir adsorption parameters, the solubility of $Al(OH)_3$ is reasonably well defined (Schecher and Driscoll, 1987). The user may therefore use literature values or solubility data obtained from laboratory experiments.

The remaining thermodynamic constants involving complexation of Al with OH^-, F^-, and SO_4^{2-}, the solubility of CO_2 and subsequent dissociation of H_2CO_3 to HCO_3^- and CO_3^{2-} are well defined in the literature (Schecher and Driscoll, 1987). These constants can be corrected for variations in temperature

TABLE 2

Summary of Input and Output Parameters for MAGIC Model

Parameter	Unit of measurement
Input parameters	
Soil depth	m
Porosity	
Soil bulk density	kg/m^2
CEC	meq/kg
$1/2$ saturation constant for SO_4 adsorption	meq/m^3
Maximum SO_4 adsorption capacity	meq/kg
SO_4 dry deposition factor	
Al solubility constant for the soil and stream	
Total $RCOO^-$ in the soil and stream	$mmol/m^3$
pK of $RCOO^-$	
Selectivity coefficients for:	
Al-Ca	
Al-Mg	
Al-Na	
Al-K	
% of CO_2 degassing from soil	
Average annual stream flow	m/yr
% of annual stream flow for each month	m/yr
Average annual precipitation volume	m/yr
% of annual precipitation volume for each month	m/yr
Background and present precipitation quality	meq/m^3
Background weathering rates	$meq/m^2/yr$
A proportional weathering factor	power of $[H^+]$
Background and present uptake rates	$meq/m^2/yr$
%BS	
Output parameters for soil and stream for every time step	
pH	
Alkalinity	meq/m^3
Al^{3+}	meq/m^3
Total aqueous Al	meq/m^3
Sum of basic cations	meq/m^3
Sum of acidic anions	meq/m^3
HCO_3^-	meq/m^3
Total aqueous F	meq/m^3
Total aqueous SO_4	meq/m^3

with enthalpy values for the reaction of interest as well as for effects of ionic strength.

Rate-dependent processes such as weathering and element uptake are not explicitly represented in MAGIC. Rather the release of individual solutes from weathering is constant with time and the rate is simply a function of the H^+ to a fractional order. Likewise, watershed uptake of solutes is constant with time. These values are typically obtained through model calibration.

The MAGIC model was developed to include processes that are important in the long-term regulation of stream water quality, while restricting the model complexity and avoiding large data requirements. MAGIC has been used to assess regional responses of surface waters to acidic deposition. Regional analysis is performed primarily to assess the impact of acidic inputs on terrestrial and aquatic resources within regional boundaries. The results of such analyses can be used to aid in policy decisions concerning the extent and timing of air pollution control strategies. The MAGIC model is well suited for regional analysis since input parameters are averaged for an entire watershed. More detailed models require more extensive information and these requirements may limit regional assessments of policy questions. Therefore, this model has been developed primarily as a predictive tool rather than to facilitate research.

An important application of the MAGIC model has been for long-term predictions. Important policy questions include: (1) To what extent have surface waters been acidified by acidic deposition? (2) What are the long-term effects of current loadings of acidic deposition to acid-sensitive systems? (3) How will acid-sensitive watersheds respond to changes in the loading of acidic deposition? Cosby and co-workers (1985b) have attempted to address these questions through the application of the MAGIC model. They calculated that waters draining from soils with low %BS and low SO_4^{2-} adsorption capacity, such as those found in the northeastern United States, would demonstrate increased acidity within one decade of initial acidic inputs. However, in systems where soils have a high %BS and high SO_4^{2-} adsorption capacity, drainage waters would start to show increased acidity in 20–100 years. Cosby et al. (1985b) also calculated that the initial response time to decreased acidic inputs would be rapid relative to the time required for acidification, but that the total time for recovery would be two or more times greater than the acidification period. The validity of this approach to long-term modeling, however, is speculative since long-term data are generally not available to reconstruct historical observations. The alternative seems to wait for responses to occur, which will be in about 25 years (Cosby et al., 1985b, 1986) or to conduct whole-watershed manipulation experiments in order to test predictive acidification models.

As stated previously, the MAGIC model was not designed to be a research model. Although there are advantages to using a lumped parameter approach, averaging input parameters for a region leads to the question of whether spatial and temporal variations in input parameters are significant enough to limit model application. A second limitation is that MAGIC does not simulate detailed hydrological processes that may be important over short time scales. For process oriented applications, a user should turn to the research models that will be discussed.

4. Time-Dependent Trickle-Down Model

A more mechanistic version of the trickle-down model has been developed by Schnoor and co-workers (Schnoor et al., 1984; Lin and Schnoor, 1986) which includes time-dependent processes for seepage lakes. The trickle-down model can be segmented into three submodels: a hydrological model, an ANC/pH model, and an Al speciation model. The hydrological submodel uses the principle of continuity of mass within a watershed compartment including snow, upper soil, unsaturated zone, groundwater, and surface water (Fig. 4). Hydrological processes such as precipitation, evapotranspiration, snowmelt, lateral flow, percolation, and seepage out of the lake are considered. ANC is the conservative state variable for the model and is transported within intercompartment hydrological flow paths. The trickle-down model uses a simplified expression for ANC that includes only species of inorganic carbon, a weak organic acid, and H^+. Solution pH can be calculated from ANC once the partial pressure of CO_2 is established. Chemical weathering reactions involving $CO_{2(aq)}$ and H^+ with minerals (e.g., augite, feldspars, hornblende, biotite, and olivine)

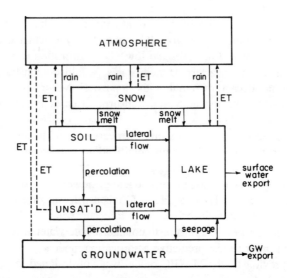

FIG. 4. Schematic representation of compartmentalization in trickle-down model (after Lin and Schnoor, 1986).

are dissolution reactions that release basic cations and consume protons. Strong acid additions originate from atmospheric inputs, soil cation exchange, cation uptake by vegetation, sulfide oxidation, and nitrification by microorganisms. Processes such as sulfate adsorption, sulfur reduction, denitrification, chemical weathering, and nitrate uptake by vegetation contribute to increase the ANC of the drainage water (Schnoor et al., 1984). All of the acid-base processes are lumped into one kinetic formulation for each compartment. Lin and Schnoor (1986) believe that the lumped approach is appropriate since detailed information concerning accurate formulation of nutrient cycling and base cation denudation is not known.

Lin and Schnoor (1986) used the time-dependent version of trickle down to assess 3 years of field data for Vandercook Lake, Wisconsin. They calculated that the steady-state ANC for Vandercook Lake was approximately $10 \ \mu eq \cdot L^{-1}$. Moreover, they determined that if the acidity of precipitation inputs were doubled, the lake would become acidic (ANC < 0) within 5 years. Based on hydrological budgets, 97% of the water entering Vandercook Lake was from precipitation that fell directly on the lake surface. Using the model, Lin and Schnoor (1986) calculated that 99% of the lake acidity came from atmospheric inputs. Neutralization of this acidity came mostly from lake sediments (76%) and from groundwater inputs (24%).

The time-dependent trickle-down model (Schnoor et al., 1984; Lin and Schnoor, 1986) should not be confused with a comprehensive lake process model. The model was specifically developed to address important processes in seepage lakes, and therein lies its strength and limitation. Researchers interested in modeling drainage waters would find the trickle-down model inappropriate due to its limited hydrological detail. However, the trickle-down model has an advantage over more comprehensive models, with regard to seepage lake modeling since it does not require collection of extensive input data.

5. Birkenes Model

The Birkenes model (Christophersen et al., 1982; Rustad et al., 1986; Seip et al., 1985) demonstrates another variation in simple, process-oriented approaches. The model was originally developed to simulate the water quality in streams within the acidified region of Birkenes in southern Norway. The development of the model was based on observations that stream-water chemistry was highly correlated to stream flow. The model consists of a hydrological submodel with two soil reservoirs and a snow reservoir (Fig. 5), a sulfate submodel, which includes sulfur mineralization and a linear sulfate adsorption isotherm, and a submodel, which calculates speciation of H^+, $Ca^{2+} + Mg^{2+}$, Al, Na^+, HCO_3^-, and organic anions. The model considers different hydrological flow patterns during high and low flow periods, cation exchange, weather-

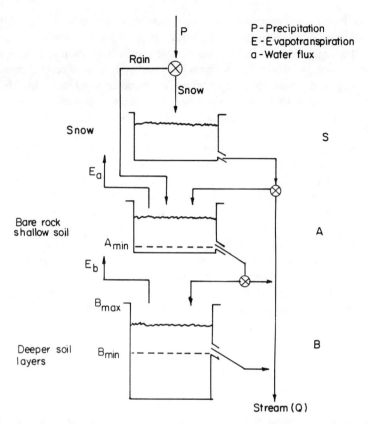

FIG. 5. Representation of flow routing in reservoir system of the Birkenes model (after Rustad et al., 1986).

ing, SO_4^{2-} adsorption/desorption, gibbsite [$Al(OH)_3$] solubility, and CO_2 equilibrium in soils and streams.

The upper soil compartment represents the upper, organic, soil horizon where cation exchange, aluminum solubility, and CO_2 exchange are thought to be important. The lower compartment was designed to represent the lower mineral soil where mineral weathering replaces cation exchange as the source of basic cations and where sulfate adsorption occurs. The most recent version of the Birkenes model includes an organic anion as part of the electroneutrality condition of both soil compartments but assumes that aluminum is present only as Al^{3+}. Nitrate is assumed to be negligible, while in earlier versions Na^+ and Cl^- were assumed to balance each other and were neglected. Although both compartments contribute to the resulting stream-water quality, under high flow conditions, the upper compartment is thought to dominate stream-water chemistry and drainage water quality is calculated utilizing an approach similar to the Reuss-Johnson (1985) equilibrium model. Under low flow conditions, the lower reservoir is in contact with base flow and aqueous chemistry is controlled by chemical weathering. The required input and output parameters for version 4.1 of the Birkenes model are summarized in Table 3.

The authors' objective has been to assess the performance of the model for different catchments. Their philosophy has been that a good fit of observed and calculated results will imply that a limited number of key processes are sufficient in describing trends in stream-water chemistry (Rustad et al., 1986). The model has been successful in simulating the chemistry of stream water in Birkenes and Storgama catchments in Norway (Christophersen et al., 1982) as well as a tributary to Harp Lake in Ontario (Rustad et al., 1986). The model is limited, however, to applications of stream water.

6. ILWAS Model

The Integrated Lake-Watershed Acidification Study (ILWAS) model (Chen et al., 1982, 1983, 1984; Gherini et al., 1985) was developed to predict changes in surface water acidity following changes in inputs of atmospheric deposition to forested ecosystems. Particular emphasis was placed on predictions of sur-

TABLE 3

Summary of Input and Output Parameters for Birkenes Model[a]

Parameter	Unit of measurement
Input parameters	
Precipitation	mm
Average daily temperature	°C
Stage height	mm
Average temperature over 20 days	°C
Concentration of SO_4^{2-} and Cl^- in precipitation	μeq/L
Output parameters for daily time step	
Time step	days
Amount of water in upper and lower compartments	mm
Accumulated precipitation	mm
Accumulated evapotranspiration	mm
Accumulated runoff	mm
Total SO_4^{2-} in and out of the system	μeq/L
Total Cl^- in and out of the system	μeq/L
Concentration of SO_4, H^+, CA^{2+}, Mg^{2+}, Na^+, Al^{3+}, and HCO_3^- in lower compartment and stream	μeq/L

[a]Compiled from Birkenes model version 4.1, which also contains a third soil compartment. All references to the third soil compartment of this version are not explicitly summarized.

face water pH and Al concentrations because those parameters are considered critical to understanding fish response to acidic deposition. Gherini et al. (1985) determined that to accomplish the model objectives, it was necessary to simulate the routing of water through lake-watershed systems, as well as describe all major ANC consuming and producing reactions along the hydrological cycle. The result of this exercise was the development of a theory of surface water acidification that was quantitatively depicted in the ILWAS program. Although ILWAS was developed as a predictive model, it has been extensively used as a research tool. This application has been demonstrated in the integration of several multidisciplinary studies of surface water acidification (Goldstein et al., 1984, 1985, 1987).

To simulate lake-watershed processes, the ILWAS model takes the approach that watershed heterogeneity must be accommodated by subdividing the drainage basin in a series of subcatchments, stream segments and, if appropriate, lake(s). Within a subcatchment, water and solutes are routed through a series of compartments including the canopy, snowpack, organic soil layer, and inorganic soil layers. With the exception of the snowpack, each compartment is treated as a CSTR.

Adequate representation of hydrological flow paths are critical to accurate representation of surface water chemistry in the ILWAS model. The hydrological component of the model is a simple but apparently accurate depiction of the movement of precipitation through the forest system to surface water (Chen et al., 1982). Precipitation enters a watershed as either rain, snow, or a mixture of rain and snow. The fraction of precipitation as snow is calculated as a function of air temperature. The interception of water by the forest canopy is calculated by the leaf area index and precipitation inputs. If precipitation inputs exceed the canopy interception capacity, then the excess moisture passes through the canopy as throughfall.

Under cold conditions, snowpack accumulates above the forest soil; this accumulation is determined through a mass balance on water. Water leaves the snowpack by sublimation and temperature-induced and rain-induced melting. Sublimation is assumed to be constant. Melting rate coefficients and a correction factor for temperature-induced melting are input parameters determined from calibration. Once melting rates are determined, the meltwater may be retained within the snowpack according to its "field capacity." If melting exceeds the field capacity of the snowpack, then meltwater enters the soil.

The ILWAS model uses a one-dimensional approach to describe movement of water through the soil system. Each soil layer is characterized by a volumetric soil moisture content, hydraulic conductivity, field capacity, and saturated moisture content. In model calculations, a mass balance on water is accomplished within each soil layer. This mass balance includes percolation entering the layer from the upper soil horizon, percolation of water leaving to a lower layer, flux of water out of the layer due to evapotranspiration, as well as lateral flow from the layer. Evapotranspiration initially occurs through moisture obtained from canopy interception. If this quantity is not adequate to satisfy the total demand, water is taken from the soil according to the distribution of roots in the soil layers. The ILWAS model assumes that percolation is zero when the soil is at or below field capacity and increases linearly with soil moisture content until saturation is reached. Lateral flow is determined by the hydraulic gradient, the hydraulic conductivity, the width of the soil layer, and the saturated depth.

Stream hydrology is governed by Muskingum routing and lake discharge is calculated by a mass balance of water on the lake including direct precipitation inputs, stream inputs, groundwater inputs, evaporation, lake outflow, and changes in storage. Accurate depiction of the thermal profile is a critical aspect of lake hydrology because stream inflow is routed to the lake level of equivalent density and lake outflow is largely obtained from the surface. Lake hydrodynamics are important in regulating chemical response to hydrological events (e.g., snowmelt). Temperature and therefore water density variations with depth are calculated by a heat budget on the lake.

The ILWAS model includes all major chemical processes that add or consume ANC [Eqs. (1) and (2)] from water as it flows through a basin. Contributions of acid or base are determined from the stoichiometry of reactions occurring within the basin. Chemical processes within the ILWAS model are either represented as equilibrium reactions, if they occur rapidly (90% completion within an hour), or as rate-dependent processes if they occur slowly.

A variety of processes influence the chemical composition of precipitation and dry deposition in the canopy (Chen et al., 1983). The canopy facilitates dry deposition which is determined by the leaf area index, ambient air chemistry, deposition velocity, and collection efficiency. Moreover, solutes are released to solution by foliar exudation which is proportional to the chemical composition of the leaves, a chemical species amplification factor, and a leaf area index. Deposition, in the form of SO_2 and NO_x, rapidly oxidizes to H_2SO_4 and HNO_3 on the leaf surface, while NH_4^+ oxidation to NO_3^- is described by a temperature-dependent rate expression which is proportional to the concentration of NH_4^+.

In the snow compartment, leaching of solutes from snowpack is represented by each volume snowmelt transporting its solute composition plus a fraction of the solutes remaining in the snowpack.

Within a given soil layer, a variety of processes are depicted in the ILWAS model. Leaf litter is deposited to the upper soil layer at rates that vary monthly and with vegetation type. Litter decay is simulated through four stages of decomposition and is represented by rate expressions with first-order dependency on reactant mass or concentration. Each of the rate constants are temperature dependent and the stoichiometry of the mineralization is variable, depending on the type of litter which is decomposed.

In addition to release from decomposition of organic matter, CO_2 can enter the soil from root respiration in the ILWAS model. The CO_2 produced can dissolve in the soil solution according to Henry's law, providing an internal source of acidity. The dissociation of CO_2 may influence cation exchange and weathering reactions, ultimately increasing the ANC of drainage waters. Carbon dioxide that enters the soil layer through respiration processes may be transported out of the soil by advection with soil moisture or by diffusion of soil gas. The model uses a gas-phase diffusion coefficient which is a function of porosity, moisture content, and tortuosity.

Nutrients are assimilated by vegetation from soil at monthly rates that are established by the model's user. Element uptake is used to support bole and foliar growth, as well as exudation from the canopy. Nutrients are taken into vegetation, according to the stoichiometry of biomass for the stand, from soil layers according to the distribution of roots. Element uptake by vegetation results in changes in the composition of soil solutions and the acid-base chemistry of drainage water.

Ammonia that enters the soil is oxidized to NO_3^- by Michaelis-Menten kinetics. Nitrification is an important process consuming ANC in drainage waters. The ILWAS model also simulates the adsorption of SO_4^{2-}, PO_4^{3-}, and organic acids as noncompetitive reversible equilibrium reactions. Sulfate and PO_4^{3-} are partitioned according to a linear isotherm, while the adsorption of organics strongly increases with decreasing pH. Solution cations (Ca^{2+}, Mg^{2+}, Na^+, K^+, NH_4^+, and H^+) may be regulated through cation exchange reactions. The model simulates cation exchange as competitive reversible equilibrium reactions.

The most important source of ANC in most watersheds is mineral weathering reactions. In the ILWAS model, the user may specify individual mineral dissolution according to a given stoichiometry. Mineral weathering proceeds according to a rate expression which is a function of a specific reaction rate constant, the mass of the mineral present, and H^+ concentration with a power dependency (typically 0.3–0.7). In addition to the dissolution of specific minerals, Al may be released by the dissolution of $Al(OH)_3$. Mineral soil solutions and surface waters under low flow conditions are typically close to saturation with the solubility of $Al(OH)_3$. However, organic soil solutions and surface waters under high flow conditions are often highly undersaturated with respect to the solubility of $Al(OH)_3$ (Cronan et al., 1986). To depict these observations, mineral soil horizons are generally assumed to be in equilibrium with $Al(OH)_3$ in the ILWAS model, while in organic horizons, Al is regulated by a rate-dependent mass action expression. Aqueous Al is distributed according to the concentrations of ligands (pH, F^-, SO_4^{2-}, and organic ligands) using thermodynamic equilibrium constants.

In the ILWAS model, calculations can be conceptualized as a series of steps in a CSTR (Fig. 6). Initially, the reactor contains a volume of solution and mass of solute. As the time step proceeds, rate-dependent processes add material to solution. Next, external inputs are added to the reactor. Then materials in solution are allowed to equilibrate and finally outflow is removed from the reactor. Critical to these calculations are solution equilibrium expressions. To determine the solution composition at equilibrium, concentrations of nonconservative solutes (e.g., H^+, HCO_3^-, Al^{3+}) are represented by a series of thermodynamic equilibrium expressions. The solution composition at equilibrium is determined by a numerical procedure which considers electroneutrality and mass balances together with the chemical equilibrium calculations.

The ILWAS model has been used to successfully model the chemistry in three lakes in the Adirondack region of New York: Woods Lake which is acidic (pH = 4.7), Sagamore Lake which is modestly acidic (pH = 5.7), and Panther Lake which is neutral (pH = 6.6). Although these lakes are located in close proximity to one another, their chemical characteristics are different due to differences in geology. The importance of soil type and flow characteristics were shown to be crucial factors in the neutralization of acidic inputs (Chen et

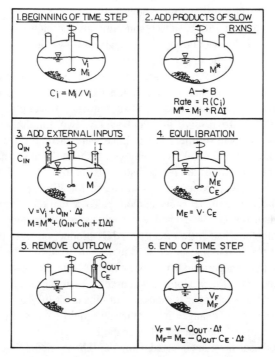

FIG. 6. A series of batch CSTRs and their corresponding changes in water quality which are assumed to be present within the ILWAS model (after Gherini et al., 1985).

al., 1984). Gherini et al. (1985) also calculated that internal production of acidity through soil processes was approximately two-thirds of the amount supplied through atmospheric inputs.

As mentioned previously, the ILWAS model has proven to be a very effective research tool to facilitate a better understanding of the biogeochemistry of acid-sensitive lake-watershed systems. To use the model as a research tool, a scientist typically poses a hypothesis, suggests how the model can be used to test this hypothesis, and indicates the anticipated response of the model to the proposed simulation. For example, it was proposed by the ILWAS researchers that a major factor regulating the chemistry of Woods, Sagamore, and Panther lakes is depth of glacial till (Gherini et al., 1985). To test this hypothesis, the ILWAS model was modified such that the till depth of circumneutral Panther Lake was decreased to values measured for more acidic Sagamore and Woods Lake basins and the simulations using the input parameters were run for 5 years. Model results showed that lake water pH decreased according to till depth, although simulated pH for Sagamore and Woods Lakes were somewhat lower than actual values. The importance of till depth in regulating the surface water chemistry can be attributed to variations in hydrological flow paths. In Panther Lake, the deep till causes surface water flow to be largely derived from deep groundwater flow which has contact with base rich soil minerals and high ANC. Conversely, inflow to Woods Lake was largely from shallow flow which was more acidic due to contact with acidic upper soil horizons.

In another instance, ILWAS researchers were in disagreement over the relative role of mineral weathering and cation exchange in regulating drainage water concentrations of basic cations and subsequent production of ANC (Goldstein et al., 1984). To resolve this controversy, two simulations were made using the ILWAS model on the Panther Lake watershed. First hornblende, an important weatherable mineral, was removed from the model, and second, the cation exchange capacity of the mineral soil was set to zero. Under these conditions, watershed ANC production was simulated for 30 years and compared to an unmanipulated system. Results demonstrated a marked decrease in ANC production when the cation exchange capacity was eliminated, while the removal of hornblende had little effect on ANC production over the simulation period. These manipulations served to illustrate the role of cation exchange in buffering solution concentrations of basic cations. Note, however, that weathering of soil minerals is the predominant source of basic cations to the watershed and the %BS of the exchanger is the result of thousands of years of soil mineral weathering. Therefore, changes in weathering would have a marked effect on the acid-base status of the watershed over the long term.

Many hypotheses have been tested using the ILWAS model that have contributed to researchers' understanding of atmospheric deposition, hydrology, soil geology, and water chemistry in acid-sensitive regions (Goldstein et al., 1984;

Gherini et al., 1985). The ILWAS model has also been used for predictive purposes. Gherini et al. (1985) found that the effect of reductions in sulfur emissions were watershed specific. Following a reduction in the sulfur inputs by one-half, little change was evident in the pH of circumneutral Panther Lake, while marked changes (up to 1 pH unit increase) were calculated for acidic Woods Lake. Gherini et al. (1985) also reported that the lakes responded rapidly (within a few years) to this reduction in loading, with Woods Lake responding more rapidly than Panther Lake.

It would be difficult to use a model that is as complicated as ILWAS for performing regional analysis due to the amount of data required as input parameters. This model functions primarily as an important research tool, giving the degree of compartmentalization that is required in basic research. One of the model's most significant contributions may be in its sensitivity analysis. Modelers who are interested in developing simple models that can describe aqueous systems with a minimum of input parameters may find the sensitivity analysis of the ILWAS model useful in eliminating unnecessary hydrological and chemical processes.

IV. CONFIDENCE ESTIMATION

A major difficulty in designing and implementing any acidification model is understanding its limitations. This chapter has described the key processes utilized in many acidification models and has shown how individual models apply these concepts to address specific questions. It has not been our intent to critically review individual models but rather to provide an understanding of model objectives and the assumptions used in model development that are important in comprehending model limitations. Reuss and co-workers (1986) have written a critical review of acidification models.

An important aspect of understanding model limitations is a knowledge of the degree of certainty which it provides. No model is 100% certain and acidification models are no exception. Young (1983) indicates that any type of sensitivity analysis of environmental models demonstrates that there are no unique solutions; yet it is necessary to utilize mechanistic models if any understanding of the physical system is to be accomplished. Cosby et al. (1985c) also suggest that it may be possible to merge model simplicity and process-oriented approaches to try to minimize the uncertainty in long-term predictions of surface water acidification. Hornberger et al. (1986) also show how a restricted set of calibrated parameter estimates can be established by performing uncertainty and sensitivity analysis with a model, despite the fact that no unique parameter estimate can be found. The lack of a mathematically unique solution has also been demonstrated by Rustad et al. (1986), with the Birkenes model, in which an optimization technique to calibrate the model was used. They found that similar agreement between observed and calculated hydrographs could be obtained using different parameter values.

Uncertainty in measurement or in model calculations is an inherent part of scientific analysis. Analytical thought requires that a system to be examined be logically subdivided and that the interactions between each subdivision be established. The process of subdivision takes unified logical structures (e.g., a forested region) and creates many internal substructures (e.g., subcatchments, lakes, streams, soil horizons). As discussed by Brillouin (1964), the process of increasing the number of internal substructures and their interactions increases the "entropy" of the system and thereby increases uncertainty. If an understanding of environmental systems is to be obtained using analytical and logical thinking, we must also accept uncertainty as a natural consequence.

Structural methods for analyzing uncertainty have been developed for determining the amount of information obtained from measurements and empirical laws (Brillouin, 1962, 1964). However, these methods have not been strictly applied to environmental models. Information in this context is defined mathematically as:

$$I_1 = K \ln \frac{P_0}{P_1} \tag{10}$$

where I_1 = information obtained

K = constant of proportionality

P_0 = number of combinations or interactions between dependent and independent variables that are possible before any observation or calculation is performed

P_1 = number of combinations or interactions between dependent and independent variables that are known to occur after an observation or calculation

This definition of information is simply a quantitative measure of the total information obtained from an observation or calculation and has implications for information that is obtained through experimentation or model simulation. For example, for a stream system, in which there is some information of interest, such as stream discharge over time, data could be collected as a first step toward understanding this system. Initially, assume that limited prior knowledge about the system exists, so that all values of stream discharge within some predetermined bounds are equally probable at any given time. If the upper boundary of stream discharge is b and the upper boundary of time is a, then P_0 can be determined as the product of a and b (Fig. 7). After enough data are

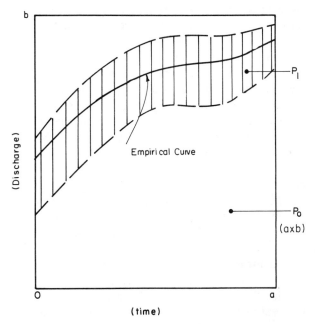

FIG. 7. Schematic representation of the uncertainty which is present before (P_0) and after (P_1) in an experiment or calculation (after Brillouin, 1964).

collected, including replications, confidence limits can be calculated for these data and will define the boundaries for P_1.

When errors in measurement are extremely high (P_1 approaching P_0), the amount of information obtained approaches zero [Eq. (10)]. Conversely, when uncertainty is extremely low (P_1 approaching 0; i.e., observations approaching pure empirical law), the amount of information obtained approaches infinity. According to Brillouin (1962, 1964), the amount of information that can be obtained is related to the natural logarithm of the amount of energy that must be expended to reach that information level and the amount of entropy that must be produced. Consequently, the probability of obtaining infinite information from any experiment or physically based model is zero. For the stream discharge example, complete certainty in the measurements would require the experiment be replicated an infinite number of times, requiring an infinite energy expenditure in the form of money, man hours, and equipment.

Uncertainty in model predictions from acidification models can originate from a number of sources including input parameters such as thermodynamic data, selectivity coefficients, kinetic constants, and hydrological constants. The inherent errors in these parameters arise from experimental and analytical procedures used to determine their values, as well as from numerical techniques used in estimating parameter values when actual data are not available. Uncertainty in model output can also be derived from variability in input variables. Uncertainty in measured constituents, such as solution concentrations, soil cation exchange capacity, or stream discharge, can be due to instrumental or human error, as well as spatial and temporal variability in these constituents.

One of the consequences of information theory on acidification models is that it provides a structural method to compare different acidification models. For the uncertainty of two models to be compared, they must both address the same question, i.e., they must both define the same P_0 domain [Eq. (10)]. However, current acidification models do not all address the same questions, so some of the uncertainties in these models are unique.

Uncertainty analysis needs to be an integral part of the calibration and verification process. Calibration generally means fitting model runs to some observed data while verification takes the model and the fitted parameters to a different system to be tested. Unless the uncertainties in the measured data and the calculated runs are well established, it is unclear whether significant differences exist between two systems. This reiterates the suggestion by Hornberger and co-workers (1986) that a calibrated parameter set can be established through uncertainty analysis during the calibration process.

It is also evident that uncertainty analysis is needed for the realistic prediction of long-term changes in forested and aquatic ecosystems as a result in alterations in atmospheric loadings. Hornberger et al. (1986) indicate that traditional sensitivity analysis of model performance in which parameters are varied about a calibrated value is likely to give a distorted view of the actual uncertainty in the predictions. Long-term predictions lack enough time series data to give highly accurate estimates; therefore, uncertainty analysis is needed for better understanding of models and their predictions.

V. CONCLUSIONS

Acidification of water and land resources is a subject of much controversy. Natural and man-made processes are both thought to contribute to the acidification of surface waters. Processes such as chemical weathering, cation exchange, SO_4^{2-} adsorption, vegetative uptake, nitrification, in-lake retention of NO_3^- and SO_4^{2-}, microbial CO_2 production, and dissolution of Al minerals occur along the hydrological cycle and serve to regulate the acid-base status of aquatic systems. Although there is considerable consensus with regard to the importance of these processes, the complexity of watershed systems requires that acidification models be utilized so that important subtleties in watershed-to-watershed variations can be understood.

Acidification models can be classified into two types: predictive models used for aiding policymaking and research models that examine in-depth questions concerning surface water acidification. Models used for predictive purposes tend to be simple in design and require little input data. Research models tend to be more complicated and need more information to run but can address a wider range of questions. The models discussed in this chapter ranged from extremely simple to very complicated, combining all levels of chemical equilibria, rate dependent processes, biological activity, and hydrology.

All of the models presented have been shown to be useful in certain applications, but assessing the quality of a model requires more quantitative methods. Information theory provides a structural approach to perform uncertainty analysis and assess model performance. It also provides methodology to fairly compare the benefits of each model when applied to specific questions.

REFERENCES

Arp, P. A. 1983. Modelling the effects of acid precipitation on soil leachates: A simple approach. *Ecol. Model.* 19:105–117.

Arp, P. A., and Ramnarine, S. 1983. Verifying the effect of acid precipitation on soil leachates: A comparison between published records and model predictions. *Ecol. Model.* 19:119–138.

Baker, J. P. 1982. Effects on fish of metals associated with acidification. In *Acid Rain/Fisheries*, ed. R. E. Johnson, pp. 165–176. Bethesda, Md.: American Fisheries Soc.

Baker, J. P., and Schofield, C. L. 1982. Aluminum toxicity to fish in acidic waters. *Water Air Soil Pollut.* 18:289–309.

Binkley, D., and Richter, D. 1987. Nutrient cycles and H^+ budgets of forest ecosystems. *Adv. Ecol. Res.* 16:2–51.

Booty, W. G., and Kramer, J. R. 1984. Sensitivity analysis of a watershed acidification model. *Phil. Trans. Roy. Soc. Lond.* B 305:441–449.

Brillouin, L. 1962. *Science and Information Theory.* New York: Academic Press.

Brillouin, L. 1964. *Scientific Uncertainty, and Information.* New York: Academic Press.

Chao, T. T., Harward, M. E., and Fang, S. E. 1964. Iron or aluminum coatings in relation to sulfate adsorption characteristics of soils. *Soil Sci. Soc. Am. Proc.* 28:632–635.

Chen, C. W., Dean, J. D., Gherini, S. A., and Goldstein, R. A. 1982. Acid rain model: Hydrological module. *J. Environ. Eng.* 108:455–472.

Chen, C. W., Hudson, R. J. M., Gherini, S. A., Dean, J. D., and Goldstein, R. A. 1983. Acid rain model: Canopy module. *J. Environ. Eng.* 109:585–603.

Chen, C. W., Gherini, S. A., Dean, J. D., Hudson, R. J. M., and Goldstein, R. A. 1984. Development and calibration of the Integrated Lake-Watershed Acidification Study model. In *Modeling of Total Acid Precipitation Impacts*, ed. J. L. Schnoor, pp. 175–203. Boston: Butterworth.

Christophersen, N., Seip, H. M., and Wright, R. F. 1982. A model for streamwater chemistry at Birkenes, Norway. *Water Resour. Res.* 18:977–996.

Cook, R. B., and Schindler, D. W. 1983. The biogeochemistry of sulphur in an experimentally acidified lake. *Environ. Biogeochem. Ecol. Bull. (Stockholm)* 35:115–127.

Cosby, B. J., Hornberger, G. M., and Galloway, J. N. 1985a. Modeling the effects of acid deposition: Assessment of a lumped parameter model of soil water and streamwater chemistry. *Water Resour. Res.* 21:51–63.

Cosby, B. J., Hornberger, G. M., Galloway, J. N., and Wright, R. F. 1985b. Time scales of catchment acidification. *Environ. Sci. Technol.* 19:1144-1149.

Cosby, B. J., Wright, R. F., Hornberger, G. M., and Galloway, J. N. 1985c. Modeling the effects of acid deposition: Estimation of long-term water quality responses in a small forested catchment. *Water Resour. Res.* 21:1591-1601.

Cosby, B. J., Hornberger, G. M., Wright, R. F., and Galloway, J. N. 1986. Modeling the effects of acid deposition: Control of long-term sulfate dynamics by soil sulfate adsorption. *Water Resour. Res.* 22:1283-1291.

Cronan, C. S., Walker, W. J., and Bloom, P. R. 1986. Predicting aqueous aluminum concentrations in natural waters. *Nature* 324:140-143.

Cronan, C. S., Reiners, W. A., Reynolds, R. C., and Lang, G. E. 1977. Forest floor leaching: Contributions from mineral, organic, and carbonic acids in New Hampshire subalpine forests. *Science* 200:309-311.

Davis, J. A. 1977. *Adsorption of Trace Metals and Complexing Ligands at the Oxide/Water Interface.* Stanford University Ph.D. thesis. Stanford, Calif.

Dickson, W. 1978. Some effects of the acidification of Swedish lakes. *Verh. Inrenat. Verein. Limnol.* 20:851-856.

Dickson, W. 1980. Properties of acidified waters. In *Proceedings of the International Conference on Ecological Impact of Acid Precipitation,* eds. D. Drablos and A. Tollan, pp. 75-83. Oslo, Norway: SNSF Project.

Dillon, P. J., Jeffries, D. S., Scheider, W. A., and Yan, Y. D. 1980. Some aspects of acidification in Southern Ontario. In *Proceedings of the International Conference on Ecological Impact of Acid Precipitation,* eds. D. Drablos and A. Tollan, pp. 212-213. Oslo, Norway: SNSF Project.

Driscoll, C. T. 1985. Aluminum in acidic surface waters: Chemistry, transport, and effects. *Environ. Health Perspect.* 65:93-104.

Driscoll, C. T., and Likens, G. E. 1982. Hydrogen ion budget of an aggrading forested ecosystem. *Tellus* 34:283-292.

Driscoll, C. T., and Newton, R. M. 1985. Chemical characteristics Adirondack lakes. *Environ. Sci. Technol.* 19:1018-1024.

Driscoll, C. T., Baker, J. P., Bisogni, J. J., and Schofield, C. L. 1980. Effect of aluminum speciation on fish in dilute acidified waters. *Nature* 284:161-164.

Driscoll, C. T., N. van Breemen, and Mulder, J. 1985. Aluminum chemistry in a forested Spodosol. *Soil Sci. Am. J.* 49:437-444.

Driscoll, C. T., Fuller, R. D., and Schecher, W. D. 1988. The role of organic acids in the acidification of surface waters in the eastern U.S. *Water Air Soil Pollut.* (in review).

Driscoll, C. T., Yatsko, C. P., and Unangst, F. J. 1987a. Longitudinal and temporal trends in the water chemistry of the North Branch of the Moose River. *Biogeochemistry* 3:37-62.

Driscoll, C. T., Fordham, J. F., Ayling, W. A., and Oliver, L. M. 1987b. The chemical response of acidic lakes to calcium carbonate treatment. *J. Lake Reservoir Manag.* 3:404-411.

Effler, S. W., Schafran, G. C., and Driscoll, C. T. 1985. Partitioning light attenuation in an acidic lake. *Can. J. Fish Aquat. Sci.* 42:1707-1711.

Fuller, R. D., David, M. B., and Driscoll, C. T. 1985. Sulfate adsorption relationships in forested Spodosols of the northeastern USA. *Soil Sci. Am. J.* 49:1034-1040.

Gherini, S. A., Mok, L., Hudson, R. J. M., Davis, G. F., Chen, C. W., and Goldstein, R. A. 1985. The ILWAS model: Formulation and application. *Water Air Soil Pollut.* 26:425-459.

Goldstein, R. A., Gherini, S. A., Chen, C. W., Mok, L., and Hudson, R. J. M. 1984. Integrated acidification study (ILWAS): A mechanistic ecosystem analysis. *Phil. Trans. Roy. Soc. Lond. B* 305:409-425.

Goldstein, R. A., Chen, C. W., and Gherini, S. A. 1985. Integrated lake-watershed acidification study: Summary. *Water Air Soil Pollut.* 26:327-337.

Goldstein, R. A., Gherini, S. A., Driscoll, C. T., April, R., Schofield, C. L., and Chen, C. W. 1987. Lake-watershed acidification in the North Branch of the Moose River: Introduction. *Biochemistry* 3:5-20.

Hagen, A., and Langeland, A. 1973. Polluted snow in southern Norway and the effect of meltwater on freshwater and aquatic organisms. *Environ. Pollut.* 5:45-57.

Hauhs, M. 1986. A model of ion transport through a forested catchment at Lange Bramke, West Germany. *Geoderma* 38:97-113.

Hem, J. D. 1986. Geochemistry and aqueous chemistry of aluminum. In *Kidney International*, vol. 29, suppl.18, eds. J. W. Coburn and A. C. Alfrey, pp. S3-S7. New York: Springer-Verlag.

Henriksen, A. 1979. A simple approach for identifying and measuring acidification of freshwater. *Nature* 278:542-545.

Henriksen, A. 1980. Acidification of freshwaters—A large scale titration. In *Proceedings of the International Conference on Ecological Impact of Acid Precipitation*, eds. D. Drablos and A. Tollan, pp. 68-74. Oslo, Norway: SNSF Project.

Henriksen, A. 1982. Acid Rain Report. *Preacidification pH-Values in Norwegian Rivers and Lakes.* Oslo, Norway: Norwegian Inst. Water Research.

Hornberger, G. M., Cosby, B. J., and Galloway, J. N. 1986. Modeling the effects of acid deposition: Uncertainty and spacial variability in estimation of long-term sulfate dynamics in a region. *Water Resour. Res.* 22:1293-1302.

Johnson, D. W. 1984. Sulfur cycling in forests. *Biogeochemistry* 1:29-43.

Johnson, D. W., and Cole, D. W. 1977. Sulfate mobility in an outwash soil in western Washington. *Water Air Soil Pollut.* 7:489-495.

Johnson, D. W., and Reuss, R. O. 1984. Soil-mediated effects of atmospherically deposited sulphur and nitrogen. *Phil. Trans. Roy. Soc. London B* 305:383-392.

Johnson, D. W., Hornbeck, J. W., Kelly, J. M., Swank, W. T., and Todd, D. E. 1980. Regional patterns of soil sulfate accumulation: Relevance to ecosystem sulfur budgets. In *Atmospheric Sulfur Deposition: Environmental Impact and Health Effects*, eds. D. S. Shriner, C. R. Richmond, and S. E. Lindberg, pp. 507-520. Ann Arbor, Mich.: Ann Arbor Science.

Johnson, D. W., Henderson, G. S., Huff, D. D., Lindberg, S. E., Richter, D. D., Shriner, D. S., and Turner, J. 1982. Cycling of organic and inorganic sulfur in a chestnut oak forest. *Oecolgia* 54:141-148.

Johnson, N. M., Likens, G. E., Bormann, F. H., Fisher, D. W., and Pierce, R. S. 1969. A working model for the variation in stream water chemistry at the Hubbard Brook Experimental Forest, New Hampshire. *Water Resour. Res.* 5:1353-1363.

Johnson, N. M., Driscoll, C. T., Eaton, J. S., Likens, G. E., and McDowell, W. H. 1981. Acid rain, dissolved aluminum and chemical weathering at the Hubbard Brook Experimental Forest, New Hampshire. *Geochim. Cosmochim. Acta* 45:1421-1437.

Kelly, C. A., Rudd, J. W. M., Cook, R. B., and Schindler, D. W. 1982. The potential importance of bacterial processes in regulating rate of lake acidification. *Limnol. Oceanogr.* 27:868-882.

Kelly, C. A., Rudd, J. W. M., Hesslein, R. H., Schindler, D. W., Dillon, P. J., Driscoll, C. T., Gherini, S. A. and Hecky, R. E. 1987. Prediction of biological acid neutralization in acid-sensitive lakes. *Biogeochemistry* 3:129-140.

Krug, E. C., and Frink, C. R. 1983. Acid rain and acid soil: A new perspective. *Science* 221:520.

Likens, G. E., Bormann, F. H., Johnson, N. M., Fisher, D. W., and Pierce, R. S. 1970. Effects of forest cutting and herbicide treatment on nutrient budgets in the Hubbard Brook watershed-ecosystem. *Ecol. Monogr.* 40:23-47.

Likens, G. E., Wright, R. F., Galloway, J. N., and Butler, T. J. 1979. Acid rain. *Sci. Am.* 241:43-51.

Lin, J. C., and Schnoor, J. L. 1986. Acid precipitation model for seepage lakes. *J. Environ. Engr.* 112:677-694.

McBride, M. B., and Bloom, P. R. 1977. Adsorption of aluminum by a smectite, II, An $Al^{3+} - Ca^{2+}$ exchange model. *Soil Soc. Am. J.* 41:1073–1077.

Muniz, I. P., and Leivestad, H. 1980. Toxic effects of aluminum on the brown trout *Salmo trutta L.* In *Ecological Impact of Acid Precipitation,* eds. D. Drablos and A. Tollan, pp. 268–269. Oslo, Norway: SNSF Project.

Nodvin, S. C., Driscoll, C. T., and Likens, G. E. 1986. The effect of pH on sulfate adsorption by a forest soil. *Soil Sci.* 142:69–75.

Nodvin, S. C., Driscoll, C. T., and Likens, G. E. 1988. Soil processes and sulfate loss at the Hubbard Brook Experimental Forest. *Biogeochemistry* 5:185–199.

Norton, S. A., and Henriksen, A. 1983. The importance of CO_2 in evaluation of effects of acidic deposition. *Vatten* 4:346–354.

Peters, N. E., and Driscoll, C. T. 1987. Hydrologic controls of surface-water chemistry in the Adirondack region of New York State. *Biogeochemistry* 3:163–180.

Peters, N. E., and Murdoch, P. S. 1985. Hydrogeologic comparison of an acidic-lake basin with a neutral-lake basin in the west-central Adirondack mountains, New York. *Water Air Soil Pollut.* 26:387–402.

Pinder, G. F., and Jones, J. F. 1969. Determination of the ground-water component of peak discharge from the chemistry of total runoff. *Water Resour. Res.* 5:438–445.

Prenzel, J. 1983. A mechanism for the storage and retrieval of acid in acid soils. In *Effects of Accumulation of Air Pollutants in Forest Ecosystems,* eds. B. Ulrich and J. Pankrath, pp. 157–170. Dordrecht, Holland: D. Reidel.

Reuss, J. O., and Johnson, D. W. 1985. Effect of soil processes on the acidification of water by acid deposition. *J. Environ. Qual.* 14:26–31.

Reuss, J. O., Christophersen, N., and Seip, H. M. 1986. A critique of models for freshwater and soil acidification. *Water Air Soil Pollut.* 30:909–930.

Rustad, S., Christophersen, N., Seip, H. M., and Dillon, P. J. 1986. Model for streamwater chemistry of a tributary to Harp Lake, Ontario. *Can. J. Fish. Aquat. Sci.* 43:625–633.

Schecher, W. D., and Driscoll, C. T. 1987. An evaluation of uncertainty associated with aluminum equilibrium calculations. *Water Resour. Res.* 23:525–534.

Scheffe, R. D., Booty, W. G., and DePinto, J. V. 1986. Development of methodology for predicting reacidification of calcium carbonate treated lakes. *Water Air Soil Pollut.* 31:857–864.

Schindler, D. W., Wagemann, R., Cook, R. B., Ruszcyunski, T., and Prokopowich, J. 1980. Experimental acidification of Lake 223, Experimental Lakes Area: Background data and the first three years of acidification. *Can. J. Fish. Aquat. Sci.* 37:342–354.

Schnoor, J. L., and Stumm, W. 1985. Acidification of aquatic and terrestrial systems. In *Chemical Processes in Lakes,* ed. W. Stumm, pp. 311–338. New York: Wiley-Interscience.

Schnoor, J. L., Palmer, W. D., and Glass, G. E. 1984. Modeling impacts of acid precipitation in northern Minnesota. In *Modeling of Total Acid Precipitation Impacts,* ed. J. L. Schnoor, pp. 155–174. Boston: Butterworth.

Schnoor, J. L., Nikolaidis, N. P., and Glass, G. E. 1986. Lake resources at risk to acidic deposition in the Upper Midwest. *J. Water Pollut. Cont. Fed.* 58:139–148.

Schofield, C. L., and Trojnar, J. R. 1980. Aluminum toxicity to fish in acidified waters. In *Polluted Rain,* eds. T. Y. Toribara, M. W. Miller, and P. E. Morrow, pp. 347–366. New York: Plenum Press.

Seip, H. M. 1980. Acidification of freshwater—Sources and mechanisms. In *Proceedings of the International Conference on Ecological Impact of Acid Precipitation,* eds. D. Drablos and A. Tollan, pp. 358–366. Oslo, Norway: SNSF Project.

Seip, H. M., Seip, R., Dillon, P. J., and de Grosbois, E. 1985. Model of sulfate concentration in a small stream in the Harp Lake catchment, Ontario. *Can. J. Fish. Aquat. Sci.* 42:927–937.

Small, M. J., and Sutton, M. C. 1986. A direct distribution model for regional aquatic acidification. *Water Resour. Res.* 22:1749-1758.

Stumm, W., and Morgan, J. J. 1981. *Aquatic Chemistry.* New York: Wiley-Interscience.

Thompson, M. E. 1982. The cation denudation rate as a quantitative index of sensitivity of eastern Canadian rivers to acidic atmospheric precipitation. *Water Air Soil Pollut.* 18:215-226.

Turner, J., Johnson, D. W., and Lambert, M. J. 1980. Sulfur cycling in a Douglas-fir forest and its modification by nitrogen applications. *Oecol. Plant* 15:27-35.

van Breemen, N., Burrough, P. A., Velthorst, E. J., van Dobben, H. F., de Wit, T., Ridder, T. B., and Reijnders, H. F. R. 1982. Soil acidification from atmospheric ammonium sulfate in forest canopy throughfall. *Nature* 299:548-550.

van Breemen, N., Mulder, J., and Driscoll, C. T. 1983. Acidification and alkalinization of soils. *Plant Soil* 75:283-308.

van Miegroet. 1984. The impact of nitrification on soil acidification and cation leaching in a Red Alder ecosystem. *Soil Sci. Am. J.* 49:1274-1279.

White, J. R., and Driscoll, C. T. 1985. Lead cycling in an acidic Adirondack lake. *Environ. Sci. Technol.* 19:1182-1187.

Wright, R. F. 1983. *Acid Rain Research. Predicting Acidification of North American Lakes.* Oslo: Norwegian Institute for Water Research.

Wright, R. F., and Henriksen, A. 1983. Restoration of Norwegian lakes by reduction in sulfur deposition. *Nature* 305:422-424.

Young, P. C. 1983. The validity and credibility of models for badly defined systems. In *Uncertainty and Forecasting of Water Quality,* eds. M. B. Beck and G. van Straten, pp. 69-98. New York: Springer-Verlag.

Fate and Transport of Sediment-Associated Contaminants

Allen J. Medine

Water Science, Boulder, Colorado

Steve C. McCutcheon

Environmental Research Laboratory
U.S. Environmental Protection Agency, Athens, Georgia

I. INTRODUCTION

A. Concerns about Contaminated Sediments

Contaminated sediments are an important aspect of hazard assessments in aquatic systems. While is is generally felt that sediments provide beneficial effects in terms of water quality improvement through the "sorption" of pollutants from the water column, contaminated sediments may also result in adverse effects. There is increasing evidence that illustrates the importance of sediment-associated contaminants to direct toxicity and bioaccumulation in benthic organisms (Knezovich et al., 1987; Krantzberg, 1985; Salomons, 1985) and to continued water quality problems following source control due to release of previously bound contaminants from sediments (Salomons, 1985; Reece, et al., 1978). Stream and lake sediment reconnaissance for chemicals from various sources has long been used for understanding the historical nature of chemical loadings (Hamilton-Taylor, 1983; Nriagu et al., 1983; Thomas et al., 1984), integrating water quality over specific time periods, monitoring mobility from chemical sources (Davison et al., 1985; Abernathy et al., 1984), and tracking the effectiveness of remedial actions imposed at a hazardous waste site to reduce or eliminate releases of contaminants (Fleischhauer and Engelder,

Wu and Gschwend developed the sorption kinetics model described herein and Madsen developed the sediment transport framework in cooperative work with the U.S. EPA-Athens. Eric Weber and Fred Fong of U.S. EPA-Athens were kind enough to review and comment on the work. The authors wish to recognize the importance of discussions with Philip Gschwend, Ole Madsen, and S-C. Wu of MIT and Samuel Karickhoff, Robert Ambrose, Jr., David Brown, and Sue Wolfe of U.S. EPA-Athens in formulating these ideas.

1985; Sly, 1982). Attention is also presently being focused on the direct benthic toxicity of contaminants bound to sediments.

The environmental hazard of chemical contaminants present in the aquatic environment is largely related to exposure to bioavailable forms of the contaminant. The bioavailable forms will be determined by the net result of interacting phenomena governing the environmental partitioning (fate). For example, sorption, hydrolysis, photolysis, biodegradation, and volatilization processes may be important for an organic contaminant while adsorption, complexation, hydrolysis, chemical precipitation, and oxidation-reduction processes may be important in the environmental fate of toxic metals (e.g., As, Tl, Zn, Cu, Cr, Hg, Se). The extent to which each process influences the exposure concentration will largely determine the actual toxicity of the chemical in the receiving water. For the majority of chemicals of interest with respect to aquatic hazards, the interactions of dissolved forms of the chemicals with solids (suspended or bed region) are major fate-influencing processes.

To provide for the appropriate level of protection for aquatic life and other uses of the resource (e.g., drinking and irrigation), it is important to be able to predict the environmental distribution of important chemicals on spatial and temporal scales and to do so with particular emphasis on the water column concentrations. Regulatory levels reflected in water quality criteria or standards usually are based on water column concentrations. Predicting water column concentrations requires a consideration of the interactions of water column contaminants with both bed sediments and suspended particulates as a critical component in the assessment.

B. Fate of Contaminated Sediments

Contaminants which have been associated with freshwater sediments have a variety of both temporary and more permanent resting points, or fates, in streams, rivers, lakes, reservoirs, and estuaries. It is generally known that contaminated sediments are located throughout a stream reach which is below chemical loadings from particular sources, such as industrial discharges, domestic waste waters, mining and milling, metal finishing, landfill leachates, or contaminated groundwater discharge. Reversing currents in lakes and estuaries can disperse contaminated sediments throughout the water body.

The concentration of a contaminant observed at a particular point in response to these chemical loadings reflects not only the loading of the chemical relative to the water residence time and important processes that cause association of the contaminant with the sediments but also the dynamics of the natural environment and sediment transport processes (movement and deposition). Consider a contaminated river entering a lake system, for example. The lake water quality will reflect the loadings from the river (relative to other loads) and the movement of contaminated sediment and dissolved contaminant in response to hydrodynamics and transformation processes. Sediment concentrations throughout the lake will, in most cases of organic and inorganic chemicals, reflect the loading and also show areal patchiness, vertical patchiness, laminated distributions (seasonal or long-term), and effects of river mouth sedimentation as well as wind/wave action.

The concentration of contaminants in the sediments will be highly site specific and dependent on the physical, chemical, and biological factors affecting sediment-water exchange. While it is difficult to describe sediment mass and contaminant concentration in dynamic systems in a very detailed manner with current models, it is possible to describe the major processes taking place and the environmental response under the more steady conditions that may exist. As our understanding of sediment transport and fate is coupled with understanding of chemical, physical, and biological factors affecting contaminant fate and transport, we will be able to describe, in more detail, the behavior of contaminants in complex systems under a variety of conditions.

A scientific basis to aquatic hazard assessments requires an understanding and description of the important fate-influencing processes and their relative contribution to controlling exposure from water, sediments, interstitial water, and food. Current efforts to model the fate and transport of contaminants in the aquatic environment are directed at mathematically representing the major controlling processes through both process-oriented (mechanistic) as well as empirical (or statistical) approaches. In this chapter the important processes that affect contaminant associations in sediments and aspects of physical transport of contaminated sediments are examined with respect to the current understanding.

II. MODELING CONTAMINANT TRANSPORT AND FATE: OVERVIEW

A number of physical, chemical, and biological processes can be incorporated into digital computer models that describe sedimentary contaminant fate and transport. These generally include

- Adsorption and desorption
- Ionization and acid-base speciation
- Hydrolysis
- Oxidation-reduction (redox) reactions
- Complexation
- Dissolution and precipitation
- Volatilization

- Biological degradation
- Photolysis
- Diffusion in particle aggregates and bed sediment
- Advective exchange with groundwater systems
- Bioturbation (biological mixing)
- Suspended and near-bed sediment transport
- Advective-dispersive transport in the water column

The physicochemical properties of the contaminant and the physical, biological, and chemical characteristics of the sediment-water system through which the contaminant is moving governs the relative importance of these processes. In developing mechanistic models of sedimentary contaminants, it is useful to categorize processes according to whether or not the process involves transformation or fate of the contaminant or whether it involves transport of the contaminant. From this arises the often used terminology of fate and transport modeling.

Relative importance determines which processes are typically included in models that are, by definition, limited in extent and approximate in nature. The distinction between processes included in models and those deemed unimportant for a model or certain application is usually based on dimensionless ratios that take into account energy and time requirements to define critical or rate-controlling steps. Table 1 defines the important dimensionless parameters for building models of this type.

In terms of energy requirements, we expect sorption, hydrolysis, redox reactions, complexation, dissolution and precipitation, volatilization and diffusion to be the least energetic, while biological degradation and photolysis have significantly greater energy requirements. Greater still are the energy requirements for groundwater, exchange, bioturbation, advective-diffusive transport and sediment transport. Because the process energy requirements vary, fate and transport in some aquatic systems, such as lakes, will be controlled by the lack of process energy. In certain other cases, critical processes will control fate and transport. For example, recent work (Gschwend et al., 1986) indicates that intraparticle diffusion can control sorption rates while bioturbation and sediment transport can limit the flux of contaminant between the bed and the water column.

In general, biogeochemical and physicochemical reactions are involved in the transformation of chemical contaminants. Biogeochemical reactions include biological degradation of organic molecules and in some cases mediation of photolysis or redox reactions. Physicochemical reactions include sorption, hydrolysis, redox reactions, complexation, volatilization, and precipitation and dissolution.

Physical transport of a contaminant is generally thought to include diffusive,

TABLE 1

Dimensionless Parameters Useful in Defining Important Model Processes for Sedimentary Contaminants

Parameter	Definition	Description
Reynolds number	$Re = \dfrac{Ud}{\nu}$	Ratio of kinematic to viscous forces of flowing fluids that defines the importance of turbulence or laminar flow in water column, bed pores, or groundwaters. U is the average velocity, d is the length scale of the flow (depth for wide surface water flows or pore diameter for flows in porous media), and ν is the kinematic viscosity for water or sediment-water mixtures.
Roughness Reynolds number	$RE^* = u_f D/\nu$	Defines whether or not turbulent or viscous forces are important at the bed. Rough turbulent flow exists for Re* > 800 (Gschwend et al., 1986). u_f is the fluid velocity near the bed and D is the length scale (bed particle diameter or roughness height of bed protrusions into the flow).
Shields parameter	$\psi = \dfrac{\tau_0}{(s-1)\rho_w gD}$	Dimensionless bed shear stress (τ_0 that determines the importance of sediment transport for a sediment class. If $\psi > \psi_c$, sediment transport occurs. Y_c is the critical shear stress at which the movement and entrainment begins, s is the specific gravity of sediment particles, ρ_w is the density of water, and g is the acceleration of gravity.
Froude number	$Fr = \dfrac{U^2}{gD}$	Ratio of inertia forces to weight that determines whether or not the flow is rapid or tranquil and is useful in predicting bed forms in streams and rivers which in turn influence flow characteristics in the water column and exchange with the bed (Streeter and Wylie, 1975).
Péclet number	$Pe = \dfrac{UD}{E}$	Measure of the relative importance of advective transport compared to diffusive or dispersive transport. For diffusive transport, E is the molecular diffusivity or sorptive retarded diffusivity. For dispersive transport, E is the dispersion coefficient or eddy diffusivity for turbulent mixing including turbulent mixing caused by bioturbation.
Ratio of molecular diffusion to turbulent mixing	$D_x = E_x$	Measure of the importance of diffusive transport including retardation by sorption compared to transport by turbulent diffusion and dispersion including mixing by bioturbation. This is usually redundant with the Reynolds number especially when the Prandtl number $(\nu/D_x) \approx 1$. Bowie et al. (1985) give typical ranges of mixing coefficients.

TABLE 1 (Continued)

Parameter	Definition	Description
Ratio of the mean residence time to the contaminant half-life ($1/k$)	tk	Describes the maximum possible relative importance of first-order transformation processes in portions of water bodies. k is a first-order rate constant for volatilization, hydrolysis, biodegradation, and sorption. We generally assume other transformations occur very rapidly and judge the significance of these reactions by the ratio of mass transformed to the total mass of contaminant.
Ratio of change in mass in a specified time to total mass	$\Delta C/C$	Generally used to judge the relative significance of biogeochemical reactions but especially those assumed to occur extremely rapidly compared to mean residence time such as ionization, chemisorption, redox reactions, complexation, dissolution, and precipitation.

dispersive, and advective movement of the dissolved species and transport by attachment to solids and as solid particles in the case of some chemical precipitates. Molecular diffusive transport is due to diffusion of the dissolved species and is rarely important outside quiescent areas such as those found in bed sediments and in the interior of natural aggregates. As it turns out, however, molecular diffusion can be a critical controlling process despite its low energy requirements and the limited extent of quiescent laminar flows in surface waters. Turbulent dispersive transport arises from residual decaying turbulence where mean shear is zero. Examples include turbulence arising from lake seiches (regular, resonating waves which pass from one end to the other) and other stirring events not associated with net advective flow through the water body. Dispersion, in contrast to diffusion, is the turbulent mixing of the dissolved phase caused by nonuniform velocity distributions of flowing waters (Fig. 1).

Advective transport is generally considered to include transport of the contaminant by water movement including movement of the dissolved species by laminar and turbulent water movements and movement of the particulate-sorbed species by the movement of solid particles. In formulating models, we typically invoke the advective-diffusive or advective-dispersive equation derivable from first principles (namely, that mass is conserved for a specified control volume). Considering particulate forms, for example the sorbed species, we write an analogous mass balance equation for sediments and solid particles and use empirical or semiempirical methods to keep an accounting of the exchange of contaminant between the dissolved and sorbed phases.

As implied above, the basis for most mechanistic modeling is the mass

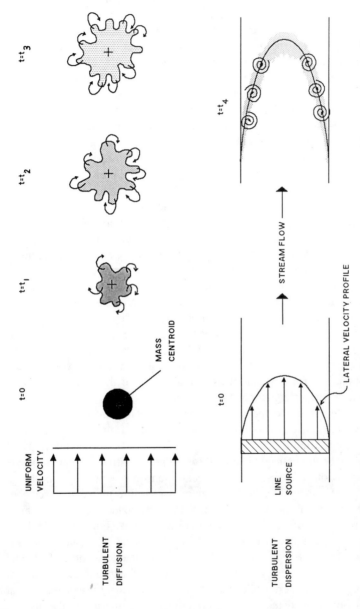

FIG. 1. Mixing of contaminants due to turbulent diffusion or turbulent dispersion. Dispersion is mixing due to nonuniform velocity gradients.

balance for discrete control volumes. For surface waters there are at least two quite similar approaches for contaminant modeling. These involve mass balances for the important components in the water column and specification of the benthic flux in one of two ways. In the simplest case, the benthic flux is measured or estimated independent of the water column mass balances. In more elaborate simulations, the benthic flux can be specified with a separate mass balance for the benthic area. In either case, but especially for the latter case, it is important to determine which processes are important for describing or estimating benthic exchange. These benthic exchange processes are a subset of the overall set of processes that affect contaminant fate and transport.

In describing the exchange or movement of contaminant (exclusive of transformation) between the water column and benthos, at least one physicochemical process and five transport processes have been deemed important for a general-purpose modeling framework (Gschwend et al., 1986). These processes include direct sorption, diffusion, advection, bioturbation (biological mixing), and sediment transport of bed and suspended loads (Fig. 2).

In the remainder of this chapter, we describe the important processes and review the algorithms that have proven useful for modeling. Without a doubt, the most important processes governing the association of contaminants and sediments in response to industrial discharges or other relatively large mass loadings are adsorption and desorption. In certain cases, the processes of chemical precipitation and dissolution are important in controlling the fate and transport of toxic metals, either through direct precipitation of toxic metal solids ($ZnSiO_3$, CdS, $CuCO_3$, for example) or through coprecipitation where a major ion precipitate is formed ($CaCO_3$, $FeOOH$, and others) which binds toxic metals in the process. Other processes, such as ionization, hydrolysis, and complexation, control the molecular or ionic form of the contaminant in the dissolved form (water column and interstitial waters). This will affect both the dissolved, interstitial and the sediment-bound contaminant concentrations by altering equilibrium conditions or the kinetics of adsorption/desorption and chemical precipitation/dissolution processes. Toxicity as well as reactivity of a pollutant may be dramatically altered through these relatively simple processes.

Redox reactions affect sediment concentrations in several ways. In the case of organic contaminants, reaction with an oxidant will result in decreased chemical concentrations in the water column with corresponding changes in the kinetics of processes driven by dissolved concentrations. For toxic metals, not only will redox conditions control the form of the metal in both the water column and the bed region, but will also affect the formation of chemical precipitates (oxides, sulfides, carbonates) of iron and manganese. Volatilization and photolysis processes, while important processes in terms of the fate and transport of organic contaminants, exert influences primarily on the dissolved contaminant and may indirectly affect the sediment concentrations. Inclusion of

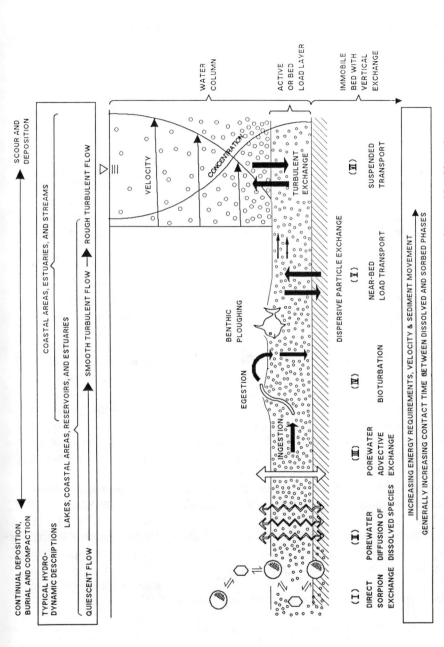

FIG. 2. Important processes that regulate exchange of contaminant between water column and benthic region.

233

these processes, as appropriate, in an overall model of the fate and transport of contaminants is necessary but they will not be discussed here relative to processes directly affecting sediment associations. In certain cases, photolysis of sorbed chemicals may be important.

III. CHEMICAL AND BIOLOGICAL TRANSFORMATIONS

A. Adsorption and Desorption

Solutes in water rarely, if ever, have a perfect affinity for dissolution. As a result, there is usually a tendency for some of the solute to attach itself to solid surfaces. Many radionucleides, metals, and organic chemicals have a strong affinity for solid surfaces. In surface and groundwaters, the most readily available surfaces for contaminants to adsorb to are the surfaces of sediment and soil particles.

Adsorption is not fully understood, but essentially it arises because of the fluid movement and random molecular motions that carry dissolved molecules to the surface of the particle and the chemical and thermodynamic properties of the water, contaminant, and solids that cause attachment. The amount of hydrophobic material attached to particles depends on the particle concentration and surface area and the amount of organic carbon associated with the particles. The chemical properties of the sorbate (hydrophobicity) and water determine the affinity for attachment to the surface. The attachment of polar compounds and metal ions depends on the charges of the dissolved species and active sorption sites on the solid. Density and orientation of sorption sites as well as competitive ion effects may be important.

We presently assume in most cases that an equilibrium is soon reached between the dissolved and sorbed phases where an equal amount of molecules are simultaneously being adsorbed and desorbed. This may be generally useful for metals, although the equilibrium attainment time may be as long as 24 h in certain water quality conditions as shown for Zn and Cu in the White River (Medine and Bicknell, 1986). As a result of observations like those of Medine and Bicknell for certain metals, and a number of investigators studying organic chemicals, we are beginning to recognize that in many cases equilibrium cannot be reached before transport processes change the sediment-water system in which sorption is occurring. As a result, new sorption kinetics models have been developed that will be explored later in this section.

1. Organic Contaminants

For the purpose of predicting sorption of dilute organic contaminants, we generally recognize that distinct processes are possible (Karickhoff, 1984).

These include both hydrophobic and nonhydrophobic bonding. Hydrophobic bonding, primarily important for neutral organics and perhaps some polar compounds, results when the contaminant lacks an affinity for water and, in effect, dissolves into the organic carbon films on natural aggregates in a way that is conceived of as a solvent extraction process (Dzomabak and Morel, 1987). Nonhydrophobic bonding may contribute to or dominate sorption when the sorbate is highly polar or has highly polar or ionizable functional groups that readily bind to specific sites on the sorbent surface consisting of polar groups or charged sites, or when the organic content in the sorbent is low; especially when the clay content is also large (Karickhoff, 1984). The ionic bonding or chemisorption is especially important for some polar organic compounds and metal ions.

In the cases where it is appropriate to assume that equilibrium has been achieved, Karickhoff (1984) has shown, by examining contaminant fugacities (see Mackay, 1979) for dilute concentrations, that the relationship between the dissolved and sorbed contaminant can be approximated by a linear isotherm (plot of concentration of the sorbed phase C_s vs. the concentration of the dissolved concentration C)

$$C_s = K_p C \tag{1}$$

in which K_p is the equilibrium partition coefficient and is equal to ϕ^w/ϕ^s, or for composite sorbents,

$$K_p = \phi^w \sum_i \frac{\psi_i}{\phi_i^s} \tag{2}$$

where ϕ^w = fugacity coefficient for the dissolved phase
ϕ^s = fugacity coefficient for the sorbed phase
ψ_i = weight fraction of each component of the sorbent

To avoid confusion, note that we typically work in concentration units for the dissolved phase of mass of contaminant per volume of the water (milligrams of contaminant per liter of water, mg_c/L_w) and for the sorbed phase of mass of contaminant per unit mass of sediment (milligrams of contaminant per kilogram of sediment, mg_c/kg_s) giving K_p units of L_w/kg_s, or ml/g (volume of water per mass of sediment).

The fact that K_p is a weighted average of the fugacity coefficients for the sorbed phase of components of the sorbent permits the often-used description of multiphase partitioning for various fractions of the water-solids system as follows:

$$f_0 = \frac{1}{1 + \sum_i (K_{pi} S_i / n)} \tag{3}$$

$$f_1 = \frac{K_{p1} S_1 / n}{1 + \sum_i (K_{pi} S_i / n)} \tag{4}$$

$$f_2 = \frac{K_{p2} S_2 / n}{1 + \sum_i (K_{pi} S_i / n)} \tag{5}$$

$$f_3 = \frac{K_{p3} S_3 / n}{1 + \sum_i (K_{pi} S_i / n)} \tag{6}$$

where f_0, f_1, f_2, and f_3 are the fractions dissolved, sorbed to mineral particles with partial or complete organic carbon coatings, sorbed to humic and/or colloidal particles, and sorbed to biological solids such as bacteria and algae cells and aggregated cell masses, respectively; K_{p1}, K_{p2}, and K_{p3} are partitioning coefficients (L_w/kg_s) for mineral sediments, colloids, and biological solids, respectively; S_1, S_2, and S_3 are concentrations (kg_s/L) of mineral sediments, colloids, and biological solids, respectively; and n is porosity (L_w/L, volume of water per unit volume of water-sediment mixture). Note that S in Eqs. (3)–(6) should be multiplied by 10^6 if units of mg_s/L are used. The reader is referred to WASP (Ambrose et al., 1986) and TOXIWASP (Ambrose et al., 1983) for discussion of partitioning to biological and mineral solids and to Gschwend et al. (1986) for partitioning to natural aggregates and colloids.

As a simple example, consider the equilibrium-phase partitioning between the dissolved and one solid phase for 4,4'-dichlorobiphenyl having $K_p = 214$ L_w/kg_s. Table 2 shows the expected distribution for a wide range of sediment concentrations. This makes it clear that, in a water column with typical amounts of suspended sediment (10–100 mg_s/L), the dissolved phase is the most important component despite indications that the sorbed phase is always the most significant (Blachford and Ongley, 1984). At high concentrations, it is true that the sorbed phase is the most important. Such high concentrations (10^4–10^6 mg_s/L) are typical of bed and near-bed concentrations for some stream-type flows (Einstein and Chien, 1955); benthic areas of lakes, reservoirs, and estuaries; and dredged sediments. As a result, it is clear that the sorbed phase is only dominant when high sediment concentrations are well mixed in the water

TABLE 2.

Equilibrium Partitioning for 4,4′-Dichlorobiphenyl with K_p = 214 L_w/kg$_s$ and Sediment Specific Gravity of 2.65

Sediment concentration			Fraction dissolved	Fraction sorbed
mg$_s$/L	kg$_s$/L	kg$_s$/L$_w$		
10	10^{-5}	0.00001	0.998	0.002
10^2	10^{-4}	0.00010	0.979	0.021
10^3	10^{-3}	0.00100	0.824	0.176
10^4	10^{-2}	0.01004	0.318	0.682
10^5	10^{-1}	0.10392	0.043	0.956
10^6	10^0	1.60606	0.003	0.997

column to facilitate equilibrium or when water is only slowly advected through the bed to allow equilibrium.

Equations (3)–(6) are written for application with a mass balance for the water column or the bed. Under limited circumstances, however, those equations may be applicable for a mass balance of the water column that also includes some surficial sediments (to a depth of the order of 15 cm; see Thibodeaux and Boyle, 1987, for examples of the depths to which advective exchange with the water column can occur). In such a case, the analyst would have to estimate the depth of sediment that could quickly come to equilibrium with the water column using experiments like those of Thibodeaux and Boyle or other estimates from observations of bioturbation (1–20 cm; see Gschwend et al., 1986). Overall, such estimates may vary from a depth of one-half the grain diameter on the bed to the order of 20 cm. Such a method is useful for screening but forgoes a kinetic description of exchange with the bed, which may be critical for some spills in which the water column concentration varies dramatically with time.

To illustrate the method of looking at the potential effect of surficial sediments as a contaminant source or sink, examine Table 3 to determine when average sediment concentrations may be significant. The average sediment concentrations in Table 3 were computed by averaging the surficial sediments in 0.01-, 0.1-, 1-, 10-, and 100-mm layers over the water column. The water column is assumed to have negligible amounts of suspended sediments and the surficial sediments are assumed to have a specific gravity of 2.65 (adequate for clays, silts, and sands) and a bed porosity of 0.5. Comparison of the results in Tables 2 and 3 shows that there is a very real potential for the bed sediments to serve as significant source or sink for 4,4′-dichlorobiphenyl if the upper 1–10 cm of the bed sediments come to equilibrium with the dissolved concentrations

TABLE 3

Average Sediment Concentration (in kg_s/L_w) over Water Column and Surficial Bed Depths with a Bed Porosity of 0.5 and Sediment Specific Gravity of 2.65

Waterbody type	Water column depth (m)	Surficial bed depth in contact with the water column (mm)				
		0.01	0.1	1	10	100
Streams	0.1	0.0001	0.0013	0.0132	0.1324	1.3243
Streams, ponds, and shallow lakes	1	0.0	0.0001	0.0013	0.0132	0.1324
Rivers, reservoirs, lakes, and estuaries	10	0.0	0.0	0.0001	0.0013	0.0132
Lakes	100	0.0	0.0	0.0	0.0001	0.0013

in the water column before the concentrations change. This is likely for the deeper water bodies if bioturbation provides rapid sediment turnover compared to water column residence time. This is also more probable for shallower waters where bioturbation or advective pumping caused by bed dunes (see Thibodeaux and Boyle, 1987) is significant.

Table 3 also indicates that it is likely that surficial sediments represent a significant source or sink for hydrophobic chemicals in most shallow waters since only the swiftest streams would be moving too fast to establish equilibrium with the topmost 0.01 mm of sediment. The moderately fast streams are not likely to fully approach equilibrium, but this is offset somewhat by the fact that these streams are likely to have irregular beds (dunes) that lead to greater depths of advective pumping of water through the beds.

In Eqs. (3)–(6), we have written general expressions that allow the specification of different values of K_p for mineral particles with organic coatings vs. humic and colloidal materials vs. biological solids. In addition, most general models would solve two sets of these equations for the water column and bed sediments and assume that the three K_p values could differ between the sediments and the water column (Gschwend et al., 1986). However, we expect that K_{p1}, K_{p2}, and K_{p3} are approximately the same for hydrophobic bonding when the organic carbon fractions are the same for the three types of solids. We believe this to be especially true if the organic coating on particles and colloidal humics derive from a common source. Karickhoff (1984) notes that hydrophobic bonding referenced to organic carbon content should be thermodynamically amenable to the same treatment for organic coated minerals, colloidal humics, and biological solids. Brownawell (1986) and Chiou et al. (1986) provide some evidence that $K_{p1} \approx K_{p2}$ (mineral and colloid partitioning), but Karickhoff notes that sorption to the "prepared" humics from soil and sediment systems used in many earlier studies appear to be sensitive to the source and method of preparation. Sorption to bacterial biomass is better understood. Steen and Karickhoff (1981) and Baughman and Paris (1981) show that if sorption is referenced to organic carbon in contrast to bacterial biomass, the partitioning coefficients are approximately equal to those for partitioning to sediment. At present, Karickhoff (personal communication, 1986) notes that it is not clear what relation may exist for hydrophobic bonding to algae cells and plant tissue. Swackhorne (1985) indicates that some information has been collected for sorption to algae cells and the Athens Environmental Research Laboratory (U.S. EPA) is undertaking some study of sorption to vascular plants at present (S. Wolfe, personal communication, 1987). Furthermore, DePinto of Clarkston College has evidently investigated this question as well.

Some controversy remains regarding what relationship exists between hydrophobic partitioning coefficients and sediment concentrations. As a result, contaminant fate and transport models such as WASP (Ambrose et al., 1986) are formulated so that different values of K_p can be specified for the water column and bed where sediment concentrations are greatly different. In other cases, models are being developed with empirical relationships between sediment concentration and K_p (O'Connor and Connolly, 1980). However, recent work fully indicated that K_p, when properly normalized to organic carbon content, did not vary with sediment concentration (Karickhoff, 1984; Bowman and Sans, 1985; Gschwend and Wu, 1985; Gschwend et al., 1986).

Karickhoff (1984) gives an explanation of why the varied relationship between K_p and sediment concentration has been observed. Generally, the controversy arises because of poorly conceived experimental methods and semantic differences regarding what is dissolved and what is particulate matter. Specifically, the primary causes of the controversy are:

1. A considerable amount of work has not properly taken into account the interference of colloids on sorption and, hence, a "third phase" needs to be invoked to explain adsorption/desorption experimental results.

2. Some experiments have not achieved equilibrium for adsorption phases.

3. Sinks of contaminant at low sediment concentrations have not been fully taken into account.

In contrast to the experimental evidence noted above, Di Toro (1985) proposed a particle interaction model (mechanism uncertain) with an additional desorption reaction which occurred as the particles interacted at high particle concentrations and enhanced desorption due to increased particle collisions. Mackay and Powers (1987) also developed a conceptual model for particle-induced collision effects on the partition coefficient. It is consistent with Di Toro's model but included a mass transfer approach to determine when particle-induced desorption would become important. It is clear, from these nascent opinions concerning particle concentration effects on K_p, that there is some need to further investigate the basic nature of sorption of hydrophobic chemicals in aquatic systems.

Earlier we mentioned the limiting behavior that the sorption isotherm approached linearity for dilute contaminant concentrations. Karickhoff (1984) notes that hydrophobic bonding follows a linear isotherm if the dissolved concentration is less than the lower of 10^{-5} molar concentration or one-half the solute solubility. As a result, we expect that linearity may be assumed for most surface water and groundwater contamination problems. Nonlinear isotherms may need to be defined if concentrated organic wastes are involved, like those expected to occur in some landfill leachates and spills into surface waters before sufficient dilution occurs. Karickhoff (1984) also notes that the linear nonhydrophobic bonding has also been observed. However, sorbent and solution phase chemistry may also be important, and this prevents a simple definition of the limits of applicability.

Karickhoff (1984) has found that, for neutral organics of limited solubility ($<10^{-3}$ molar), which are not susceptible to speciation changes or complexation, partitioning can be predicted from the behavior of partitioning in carbon referenced systems such as those for octanol-water systems. As a first approximation

$$K_p = \frac{\phi^w f_{oc}}{\phi^{oc}} = K_{oc} f_{oc} \tag{7}$$

where $K_{oc} = \phi^w \phi^{oc}$. The organic carbon fraction f_{oc} is easily measured. Furthermore, K_{oc}, the partition coefficient normalized to the organic carbon content, has been related to chemical properties (K_{ow}, octanol-water partitioning coefficient) for a number of classes of compounds. Table 4 gives a compilation of the empirical relationships between K_{oc} and K_{ow}, which are of the general form

$$\log K_{oc} = a \log K_{ow} + B \tag{8}$$

where a and B are empirical coefficients.

Karickhoff (1984) also notes that K_{oc} can be estimated a priori from molecu-

TABLE 4

Relationship between K_{ow} and $_{oc}$ for Selected Compounds (Karickhoff, 1984; Gschwend et al., 1986)

Contaminants	Relationship	Reference
13 Halogenated benzenes	$\log K_{oc} = 0.72 \log K_{ow} + 0.49$	Schwarzenbach and Westall, 1981
13 Pesticides	$\log K_{oc} = 1.029 \log K_{ow} - 0.18$	Rao et al., 1982
5 Polynuclear aromatics	$\log K_{oc} = 0.989 \log K_{ow} - 0.346$	Karickhoff, 1981
22 Polynuclear aromatics	$\log K_{oc} = \log K_{ow} - 0.317$	Hassett et al., 1980b
Aromatic hydrocarbons	$\log K_{oc} = \log K_{ow} - 0.21$	Karickhoff, 1981

lar fragments for some compounds. This approach assumes that extrapolation from the thermodynamic properties of components of a molecule is possible. When the addition of ring fragments are involved, such as unsubstituted condensed ring aromatics, the ring fragment constant for attachment at the α-carbon location on the ring can be written as

$$f^{\alpha}_{C_4H_2} = \log k_{oc} \text{ (naphthalene)} - \log K_{oc} \text{ (benzene)} \approx 1.16 \qquad (9)$$

This is the specific relationship for naphthalene and benzene where the ring fragment constant is about 1.16. The practical importance of this is that if we know the $\log K_{oc}$ of naphthalene or benzene we can compute the other. Karickhoff (1984) reviews applications of this form to three-ring systems (phenanthrene and anthracene), four-ring systems (tetracene), and five-ring systems (dibenz[a,h]anthracene) and notes how six-membered ring aromatic hydrocarbons may be handled. This approach also works well for functional groups (denoted by X) replacing a hydrogen atom, H, on a hydrocarbon matrix, R—H. An experimentally determined substituent constant is defined as

$$\Pi^s_x = \log K_{oc}(\text{R—H}) - \log K_{oc} \text{ (R—H)} \qquad (10)$$

These substituent constants are relatively constant for similar organic molecules. This method seems to work well except for (1) the addition of polar groups —NH_2 and —OH to molecules, (2) addition of electron releasing or removing groups to compounds that can potentially ionize in sediment-water systems, and (3) multiple substitution of functional groups on a hydrocarbon matrix.

Karickhoff (1984) notes that this method is not readily adaptable to estimating K_{oc} for most pesticides but it does offer considerable predictive potential for mono- and di-substituted compounds. As a result, tremendous advantage can be obtained from relationships between K_{oc} and K_{ow} in Table 4. Karickhoff notes that because K_{oc} and K_{ow} are linearly related, substituent constants for K_{oc} approximate those for K_{ow} which are readily available (Karickhoff refers to Leo, 1975, and Leo et al., 1971).

While we are able to forecast a priori K_p for some nonpolar organic compounds where hydrophobic bonding is important, we have not had equal success in describing nonhydrophobic bonding. We expect that both hydrophobic and nonhydrophobic bonding is simultaneously occurring and that nonhydrophobic bonding is important when low organic carbon contents are observed and highly polar sorbent or sorbates are involved.

One complicated feature of nonhydrophobic sorption is that bonding may be occurring to different components of the sediment simultaneously or may involve different binding mechanisms for bonding to the same component. When bonding to different components are involved, Karickhoff (1984) recommends that Eq. (2) be used if care is exercised. As a result, it would be possible to separate the effect of bonding to organic carbon and mineral surfaces as

$$K_p = f_{oc}K_{oc} + f_m K_m \tag{11}$$

where f_m is the weight fraction of mineral sediment and K_m is the partitioning coefficient for the mineral surface. Where different binding mechanisms are involved in bonding to the same sediment component, Karickhoff (1984) suggests that these contributions to K_p should be log additive. Karickhoff also mentions that these approaches are tentative because of the absence of definitive, quantitative work to determine actual distribution of components of the sediment mass and multiple sorptive processes.

Karickhoff (1984) notes that sorption of pesticides to cleaned clays has been explored, but these surfaces bear little resemblance to the oxide- and organic-covered surfaces occurring in the natural environment. Other experiments have attempted to remove various components, such as the organic carbon films, but it is not clear that the remaining surfaces are unaffected by the removal processes.

Karickhoff (1984) notes that Eq. (11) should also take into account the interference of one component with the sorption of another. For example, organic carbon coatings will reduce the mineral surface area and thus reduce the nonhydrophobic bonding that would otherwise occur. In effect, Eq. (11) should be based on surface area and not mass fraction as assumed. However, the surface area of different components cannot be easily measured or predicted. As a result, we attempt to use a modified form of Eq. (11) that is based on measurements of mass fractions and estimated activity coefficients and indirect compensation for surface area coverage such that

$$K_p = f_{oc}a_{oc}K_{oc} + f_m a_m K_m \tag{12}$$

where a_{oc} and a_m are active available fractions of the organic carbon and mineral sediment that vary between 0 and 1. Because organic carbon coats natural particles and hydrophobic bonding seems to involve solution of organic contaminants into the organic material, we generally expect that $a_{oc} = 1$ unless organic carbon is isolated in the interior of a mineral particle. Isolation may involve complete isolation from water and contaminants or simple isolation of the organic film in the smaller pores of a particle that will exclude diffusion of larger molecules. Generally, we expect that $a_m < 1$ because the organic film may cover the mineral surface and nonhydrophobic bonding is considered to be a surface phenomenon.

Because sorption of organic contaminants to mineral surface is a surface phenomena and because swelling clays, such as montmorillonite and hecorite, are the most active binding materials, Karickhoff suggests that mineral sorption indexed to swelling clay content may be useful. This follows from observations of Hassett et al. (1980b) that failure of organic carbon referenced sorption was related to existence of swelling clays. Karickhoff (1984) derives this functional dependence from Eq. (12) by referencing K_p to organic carbon and assuming that $a_{oc} = 1$

$$K_{p:oc} = K_{oc} + \frac{f_m a_m K_m}{f_{oc}} \tag{13}$$

where $f_m a_m K_m / f_{oc}$ shows the nonhydrophobic contribution of mineral bonding. In reviewing the work available, Karickhoff (1984) finds that nonhydrophobic sorption is generally not important unless $f_m / f_{oc} > 30$ but gives somewhat better guidance in Table 5.

Where polar functional groups affect sorption, the change in sorption compared to nonpolar sorption involves increased K_p if the polar groups are positively

TABLE 5

Semiquantitative Guide for Determining the Mineral Contribution to Sorption

	Neutral organic contaminants with polar functional groups	Nonpolar organic contaminants	Large nonpolar organic contaminants
K_{oc}/K_m	10–50	50–100	>100
Threshold f_m/f_{oc} for mineral contributions	25–60	>60	Insignificant in natural sediments

Source: Karickhoff, 1984.

charged and a decrease in K_p if the functional groups are negatively charged (Karickhoff, 1984). Evidently this derives from an interaction of negative electrostatic charges on the organic matrix. Where multiple binding mechanisms seem to be at work, a log-additive formulation is proposed by Karickhoff (1984):

$$\log K_{oc} = \log K_{oc}^h + \sigma_i \tag{14}$$

where K_{oc}^h and σ_i are the hydrophobic and nonhydrophobic components, respectively. Except for $\sigma_{NH_2} \approx 0.65$, no other quantitative information seems to be available. However, Karickhoff (1984) conjectures that $-NH_2$ and $-SO_3^-$ groups represent the positive and negative polar group extremes and that modifications of K_p values to account for nonhydrophobic bonding are only as large as an order of magnitude. Generally, changes in sorption by a factor of two are anticipated.

2. Toxic Metals

The sorptive interactions of metals with particulate matter is the major process affecting the fate of toxic metals in the natural environment. However, the interactions in the environment are complicated due to the heterogeneous nature of aquatic sediment surfaces and the presence of inorganic and organic complexing ligands. Şalomons and Förstner (1984) summarized the results of a number of investigations of the role of particulates in regulating the fate and transport of metals in aquatic environments. They noted that metal speciation can increase or decrease sorption depending on the ability of the ligands to be sorbed, that coatings on solid phases of hydrous iron and manganese oxide or organics can enhance metal removal, that stronger bonding can often result from aging and other diagenic processes (reduced availability), and that dissolved organic matter in the sediments can lead to transformations of metals (e.g., methylation) and increased solubility through complexation. The substrates available for metal removal include detrital minerals, organic residues, reactive organic matter, metal precipitates, iron and manganese oxyhydroxides, and calcium carbonate and sulfidic precipitates.

The bulk of the research on the adsorption of metals to surfaces has been directed at oxide surfaces, predominantly iron, aluminum, manganese, and silica (Morel, 1983; Stumm and Morgan, 1981; Anderson and Rubin, 1981), and has led to considerable advances in our knowledge. For example, it has been observed that adsorption of metals onto oxide surfaces increases from essentially 0% to nearly 100% over a very narrow pH range of about two pH units. Generally, the maximum adsorption occurs at a pH $<$ 6.5-7.0. High concentrations of adsorbate and low concentration of surfaces tend to shift the "adsorption edge" to slightly higher pH values. In a recent study, McIlroy et al. (1986) observed these characteristic sharp adsorption edges in Cu and Zn experiments using natural sediments and the pH range was found to be 4–5.5 and 6–7 for Cu and Zn, respectively. It should be noted that the effects of particle concentration on reducing the partition coefficient were noted for metals (Zn, Cu) by McIlroy et al. (1986) in Flint River (Michigan) experiments and for Cu in White River (Utah) experiments by Medine et al. (1983). Metal sorption is governed by attachment to specific surface sites which vary nonuniformly in natural streams, generally decreasing on a unit weight basis as flow and solids carrying capacity increase. A decreased partition coefficient would be expected. However, these experiments attempted to eliminate the variability in surface sites per unit weight and preserve the relative ratios between suspended sediment types as suspended solids concentration was varied. As a result, experimental procedures, surface bonding characteristics of natural sediments, and the particle interaction theory of Di Toro need further investigation with respect to metal sorption to natural sediments and bed sediments.

While oxides are a component of aquatic sediments and are a convenient laboratory model surface, fresh oxide surfaces may not constitute the principal sorption sites for toxic metals (Morel, 1983). Organic surfaces and organic coatings on inorganic surfaces are believed to provide the majority of the functional groups for metal adsorption. According to Morel (1983), organic particles containing carboxyl, phenolic, or sulfhydryl groups will interact with metals in a way that is conceptually similar to the solution process of complexation with additional interactions of long-range coulombic forces. Organic coatings, in contrast, can yield four surface reactions: metal adsorption, ligand adsorption, metal complexation by the adsorbed ligand, and ligand coordination to the adsorbed metal ion.

The modeling of metal adsorption to surfaces (suspended particulates and bed sediments) is presently receiving much interest due to its importance in regulating metal movement in aquatic systems. The increased attention is being directed at determining an adsorption model for complex natural systems with a number of adsorbants and adsorbates, including the effects of supporting electrolyte ions, complexation, and other effects. Continued laboratory research on the adsorption of metals to natural sediments using complex water chemistry will be necessary if we are to develop realistic models to describe metal fate and transport in complex systems.

When using literature data or other information not specifically collected on the natural system being modeled, it is important that the data reflect similar water chemistry and sediment characteristics. Natural sediments and suspended solids can show very different metal removal as a function of pH, adsorbable metal concentration, and solids concentration (Medine and Bicknell, 1986) than laboratory oxides. The difference is due to the integrated characteristics of natural sediments which are a mixed population of solid types (oxides, clays, detritus, hydroxides, silicates, oxides with organic coatings, etc.).

Sorption can be considered to be physical adsorption on the external particulate surface due to Van der Waals forces; chemical adsorption due to chemical associations between dissolved ions or molecules and particle surfaces; or ion exchange where an exchange of ions of like charge occurs between the bulk solution and the surface. Metal adsorption to oxide surfaces has been described with a number of theories which are classified as physical or chemical. Physical theories are based on electrostatic interactions between the ions and available surface charges and ion–solvent interactions (Gouy-Chapman-Stern-Graham model, adsorption-hydrolysis model and the ion-solvent interaction model) while chemical approaches involve reactions of chemical entities on the surface with dissolved species (ion exchange model and surface complexation models). The chemical approach is presently favored by most investigators (Salomons and Förstner, 1984), although both concepts can explain experimental data. Davis and Leckie (1979) developed an adsorption model based on surface complexation but corrected the mass-law expressions for the effect of electrostatics, and thus, included both physical and chemical interactions.

Available sorption models for metals have been incorporated into the geochemical equilibria model, MINTEQA1, to describe the chemical speciation and solid-phase partitioning of toxic metals. MINTEQ was developed at Battelle's Pacific Northwest Laboratories (Felmy et al., 1984a). Subsequent changes were made at the U.S. EPA Environmental Research Laboratory located in Athens, Georgia, and the current model is referred to as MINTEQA1 (Brown et al., 1986).

Six different sorption algorithms are available in the MINTEQA1 code which are treated as being analogous to aquatic speciation and include

1. Activity Kd
2. Activity Langmuir
3. Activity Freundlich
4. Ion exchange
5. Constant capacitance surface complexation
6. Triple-layer surface complexation

3. Distribution Coefficient Model

The standard Kd relationship between the contaminant adsorbed to the surface [MeS] to the total dissolved contaminant [Me] as affected by solids concentration [S] is obtained from the basic mass law for the reaction as follows

$$S + Me \leftrightarrow MeS \tag{15}$$

The standard distribution coefficient is therefore

$$Kd \text{ (std)} = \frac{[\text{MeS}]}{[\text{Me}][\text{S}]} \tag{16}$$

The equilibrium Kd adsorption model in MINTEQA1 assumes that only the *uncomplexed* metal ion adsorbs to surfaces as opposed to all dissolved aqueous species (replacing total dissolved metal concentration with free ion activity, $\{\text{Me}^{+2}\}$, that the solid concentration is in excess and that the activity or concentration of the solid is one giving the relationship below for a divalent metal, such as Cd, Pb, Zn:

$$Kd \text{ (act)} = \frac{[\text{MeS}]}{\{\text{Me}^{+2}\}} \tag{17}$$

The units for concentrations or activities are moles per liter except for S which is kilograms per liter. If it is assumed that $[\text{Me}] = \{\text{Me}^{+2}\}$, then Kd (act) = Kd (std) S.

4. Activity Langmuir Isotherm Adsorption

The activity Langmuir isotherm is preferable to the activity Kd model because it considers the effects of variable metal concentrations, and it considers a mass balance on surface sites (solids concentration). However, more detailed experimental data are required to implement the activity Langmuir isotherm. The standard Langmuir isotherm is based on the simplest reaction between a metal and a surface site, Me + S ↔ MeS, where MeS represents the metal adsorbed to the surface. Using the mass law for this reaction, and noting that the total sorption sites are limited, gives the standard Langmuir adsorption isotherm

$$[\text{MeS}] = \frac{K_L S_T [\text{Me}]}{1 + K_L [\text{Me}]} \tag{18}$$

where [MeS] is the amount of metal adsorbed in moles per liter, [Me] is the equilibrium concentration of total dissolved metal in moles per liter, K_L is the Langmuir adsorption constant (L/mole), S_T is the maximum adsorption in similar units as [MeS]. The isotherm, in a plot of [Me] vs. [MeS], is a rectangular hyperbola passing through the origin and possessing a definite adsorption maximum.

The activity Langmuir isotherm model is based on the replacement of the total dissolved metal with the activity of the free ion, $\{\text{Me}^{+2}\}$, requiring the simulation of the water chemistry (metal speciation) used in the laboratory determination of the standard Langmuir isotherm to obtain the activity of the uncomplexed metal ion. The isotherm is then redrawn to obtain the "activity" Langmuir K and the total surface sites. Finally, the new isotherm, along with the appropriate stoichiometric information for the adsorption reaction, is used to represent Langmuir adsorption for a divalent metal:

$$[\text{MeS}] = \frac{K_L S_T \{\text{Me}^{+2}\}}{1 + K_L \{\text{Me}^{+2}\}} \tag{19}$$

The "activity" forms of the previous three models and the activity Freundlich isotherm, $\text{MeS} = K_F^{\text{act}} \{\text{Me}^{+2}\}^{1/n}$, where K_F^{act} and n are Freundlich isotherm constants, have a greater range of applicability over their standard counterparts. The models have limitations as they do not consider a charge balance on surface sites, the interactions of other bulk solution ions with sites, the effect of pH and electrolyte composition on surface hydroxyl group distribution, or electrostatic interactions between the surface and the adsorbing ion. The surface complexation models include these effects and may be generally more applicable.

5. Surface Complexation Models

Where the adsorbing surface is composed of primarily single oxides or a mixture of oxides, the surface complexation models offer realistic simulation of metal sorption. The development of the models and data for application to specific metals can be found in Huang and Stumm (1973), Schindler et al. (1976), Davis et al. (1978), Davis and Leckie (1978, 1979), Benjamin and Leckie (1982), and Balistrieri and Murray (1981, 1982). The application of the surface complexation models require data specifically collected for model application. The constant capacitance surface complexation model requires, in addition to the general water quality data, the analytically determined surface site density (N_s: sites/m^2), the specific surface area (SA: m^2/g), the concentration of the solid in suspension (C_{SS}: g/L), estimates of the electrostatic potentials (E_o: volts), and the inner layer capacitance (farads/m^2). The triple-layer surface complexation model also requires the above data but also requires the outer layer capacitance (farads/m^2) to be specified. Equilibrium constants and stoichiometry for the surface reactions are also needed. Use of these models for natural soils or sediments which are not primarily oxides may require additional field and laboratory data collection and interpretation, although they have been used with clays (James and Parks, 1982).

The surface complexation models consider the oxide surface to contain an array of hydroxyl groups (sorption sites) which can interact with major ions, metals, or other ion associations in a manner analogous to complexation reactions among dissolved constituents. In addition, surface complexation models include the effects of the variable electrostatic interaction energy between the charged adsorbing metal and the solid surface charge. The details of the model formulations are found in Felmy et al. (1984a).

6. Sorption Kinetics

A comparison of the time available for contact between the dissolved and sorbed phases (controlled by water and sediment transport processes), and the time required to approach equilibrium is used to determine whether or not equilibrium partitioning is an appropriate assumption. The time required to approach equilibrium is a function of sorbate hydrophobicity and the characteristics of the particles to which the chemical is sorbed. The time required to achieve half of the equilibrium concentrations ranges from a few minutes to 30 days or more for typical sediment and chemical characteristics. The contact time between the dissolved phase and sediment varies over a wide range of seconds to years. Therefore, equilibrium partitioning for a specific substance is an appropriate assumption for certain flow and sediment transport conditions but not necessarily for all transport conditions.

It seems appropriate to assume that equilibrium is approached between organic chemicals and slowly settling particles in water bodies and between bed and aquifer solids when pore water velocities are slow (on the order of meters per year or less). Metal equilibrium, in many cases, seems to be achieved rapidly even when compared to the smallest typical contact times between the dissolved and sorbed phases (i.e., contact time between a stream and its bed).

In general, however, one expects contact times to be relatively short and the time to achieve equilibrium relatively long. Several studies indicate that sorptive transfer does not proceed to completion before the incompletely equilibrated fluid surrounding the particles in the flow and bed is swept away and replaced by new fluid. Kay and Elrick (1967), van Genuchten and Wierenga (1976), Rao et al. (1979), and Schwarzenbach and Westall (1981) found both asymmetric distributions of chemical concentrations vs. depth and nonsigmoid or tailing breakthrough curves in evaluating the transport of organic compounds through leached soil columns. These observations are best explained by recognizing that the sorptive exchange "reactions" or diffusive mass transfer are slow with respect to advective flow of the pore fluids. Investigations into the release of organic pollutants from contaminated bed sediments also provide evidence that sorptive exchange is slow compared to transport. For example, transfer of phthalates, chlorinated benzenes, and chlorinated biphenyls from natural sediments, especially those that have been aggregated into fecal pellets by infauna and those that have been exposed to the pollutant for extended times (in excess of months), has been seen to occur on time scales of days to months, even when good solution-solid contact is maintained (Karickhoff, 1984; Karickhoff and Morris, 1985). Since bioturbation (Rhoads, 1963; Fisher et al., 1980; Lee and Swartz, 1980; Karickhoff and Morris, 1985) and sediment-produced gas release (Medine et al., 1980) can "renew" the surface layers of bottom sediments on a time scale of days to weeks, it is obvious that the rate of desorptive transfer to or from exposed surface sediments may limit bed-water column transfers before these solids are reburied.

This conclusion seems to be supported by the field data of Turk and Troutman (1981), who found that the dissolved polychlorinated biphenyls (PCB) concentration in the Hudson River was inversely proportional to the river discharge. The PCB load originated primarily from contaminated sediments located upstream. Bopp and his colleagues (1981) have measured very high concentrations of PCBs in exposed surface sediments from various sites along the Hudson River. Assuming other processes, such as interaction with groundwater, are unimportant compared to sorptive exchange, the finding that the PCB concentration was inversely proportional to discharge is a very good indication that sorption equilibrium was not obtained before the water adjacent to the bed was swept downstream.

7. Modeling Sorption Kinetics

Sorption kinetics have been described in a number of ways. The mass transfer or chemical reaction viewpoint is typically adopted along with a conceptual definition of one or more boxes or compartments (Fig. 3) describing sorption sites or reservoirs (collection of sites). Karickhoff and Morris (1985) note that an infinite depth, planar source diffusion model is useful in examining laboratory data, but the boundary conditions that must be specified are too restrictive for use in a general, chemical fate model. Di Toro and Horzempa (1982) distinguish between reversible and resistant sorption in such a way that Karickhoff and Morris (1985) interpret the model as being quasikinetic, except that Di Toro and Horzempa did not address the temporal variation of the resistant component. The resistant component is evidently assumed to be the result of sorption into sites or a reservoir that does not desorb or readily desorb the chemical following a decrease in the ambient or bulk dissolved concentration of the solution. Evidently, it is assumed that if the resistant component desorbs, it only desorbs the chemical very slowly compared to the sorptive rate for material into the sites or compared to the desorption rate for material released from the reversible sites. This model, viewed as a mass transfer from solution to two solid compartments arranged in series (see Fig. 3), can be represented as

One-box model

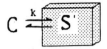

Two-box model

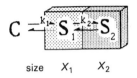

size X_1 X_2

Diffusion model

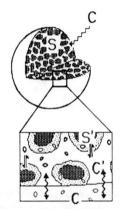

FIG. 3. The conceptual diagram for mass transfer or chemical reaction used for describing sorption kinetics, showing one-box (compartment), two-box, and the diffusion model (Wu and Gschwend, 1986). Copyright © 1986 American Chemical Society.

$$\frac{dC}{dt} = X_1 \frac{s}{n} k_1 (C_{s1} - K_p C) \tag{20}$$

$$\frac{dC_{s1}}{dt} = k_1 (K_p C - C_{s1}) + k_2 (C_{s2} - C_{s1}) \tag{21}$$

$$\frac{dC_{s2}}{dt} = \frac{X_1}{X_2} k_2 (C_{s1} - C_{s2}) \tag{22}$$

where s/n is the solids to water ratio (kg$_s$/L$_w$), C_{s1} is the concentration of sorbate in the rapidly exchanging solid phase compartment (mg/kg$_s$), C_{s2} is the concentration of the sorbate in the slowly exchanging solid-phase compartment (mg/kg$_s$), X_1 is the fraction of solid mass rapidly equilibrating, and X_2 is the fraction of solid mass slowly exchanging. We can demonstrate the solution of Eqs. (20)–(22) for a closed system such as a lake with negligible inflows where initially there is a dissolved concentration, C and no sorbed contaminant, $C_{s1} = C_{s2} = 0$ (i.e., just after a storm pulse loading). The mass conservation condition for closed system is

$$(C)_t + \frac{s}{n}X_1C_{s1} + \frac{s}{n}X_2C_{s2} = (C)_{t=0} \tag{23}$$

and the analytical solution is the widely recognized form

$$(C)_t = Ae^{-b_1t} + Be^{-b_2t} + (C)_{t=\infty} \tag{24}$$

in which A, B, b_1, and b_2 are coefficients derived from K_1, K_2, X_1, s/n and K_p (see Gschwend et al., 1986, for the form) and $(C)_{t=\infty}$ is the equilibrium concentration after sorption is completed given by

$$\frac{(C)_{t=0}}{1 + (s/n)K_p}$$

Therefore, this model description of sorptive uptake can be more accurately fit to the observed data with an early phase of sorptive exchange occurring on a time scale of minutes (i.e., k_1 on the order of 0.1 min^{-1}) and a second, slower step continuing for hours to months (i.e., k_2 on the order of 0.1 h^{-1} to 0.1 per month). Karickhoff (1980) has demonstrated the utility of this two-box model by adjusting the coefficients to fit sorption data for three polycyclic aromatic hydrocarbons. However, Karickhoff (1984) notes that the description of sorption perturbations is limited to those perturbations occurring over times comparable to the time scale of the quicker, early phase of sorptive exchange. Similarly, Gschwend and Wu (1985) have calibrated this model for chlorinated benzenes.

By comparison, the two-box model is similar to the supposition of Di Toro and Horzempa (1982) that sorption sites are of two kinds. However, Di Toro and Horzempa seem to view their two compartments of reversible and irreversible sorption as parallel processes. Furthermore, the two-box model described in Eqs. (15)–(17) is completely reversible. Experimental problems with the work of Di Toro and Horzempa evidently led to this divergent view.

These involve the time required to approach equilibrium and sorption onto colloidal or nonsettling particles. Equilibrium was assumed to occur within 3 h, whereas experiments on PCBs using natural sediments indicate that equilibrium is approached only after 30 or more days (Coates, 1985). Desorption experiments were run after sorption experiments and involved washing out colloidal and nonsettling particles and the material sorbed to these particles. By contrast, experimental procedures using gas purges measure the true dissolved concentration (Karickhoff and Morris, 1985; Coates, 1985; Wu and Gschwend, 1986). As a result, sorption hysteresis and irreversibility should be viewed as operationally oriented explanations rather than sound physical and chemical arguments.

The simplest sorption kinetics model is the one-box, first-order kinetics model (see Fig. 3) where

$$\frac{dC_s}{dt} = k_f C - k_b C_s = K(CK_p - C_s) \tag{25}$$

or

$$(C_s)_t - (C_s)_{t=\infty} = [(C_s)_{t=0} - (C_s)_{t=\infty} [1 - \exp(-Kt)] \tag{26}$$

in which k_f is the sorption uptake rate constant in $L_w/(kg_s\text{-s})$, k_b is desorption rate constant in s^{-1}, and $K = k_f/K_p = k_b$ when sorption and desorption are equivalent processes. The HSPF (Johanson et al., 1984) and SERATRA models (Onishi and Wise, 1982) are examples in which first-order kinetics are used. Unfortunately, K must be chosen by trial and error unless K is derived from less approximate models. Furthermore, van Genuchten and Wierenga (1976), Karickhoff (1980), and Leenheer and Ahlrichs (1971) have conducted experiments for sorption of organic chemicals into natural soil and sediment that show a rapid initial uptake followed by a slow approach to equilibrium that does not lend itself to a first-order description. As a result, it is clear that a multibox approach is needed.

The simplest multibox model conceptually involves subdividing the sorbent into two types of reaction sites for sorption (Fig. 3). The first compartment rapidly reaches equilibrium or is immediately accessible. The second compartment reacts or exchanges more slowly (van Genuchten and Wierenga, 1976).

While the two-box model may better explain sorption kinetics than the one-box model, it is empirical and relies on experimental definition of the empirical coefficients k_1, k_2, and X_1. The determination of the coefficients requires a tedious experimental study of each chemical and sediment system of interest. Furthermore, it is evident, from the review of Hendricks and Duratti (1982), that mass transport approaches are also empirical forms requiring calibration. However, Gschwend et al. (1986) have proposed and validated a physically based model that relies exclusively on known physical measurements and thus not only describes but predicts sorption kinetics.

Gschwend et al. (1986) have developed an infinite-box model based on rate limiting diffusion into natural aggregate particles retarded by microscale partitioning that is consistent with observations by Karickhoff and Morris (1985), Leenheer and Ahlrichs (1971), Weber and Rumer (1965), and Rao et al. (1979)—see Fig. 3. Such a model is conceptually appealing given the aggregate nature of suspended solids (Chase, 1979; Zabawa, 1978), sediments (Johnson, 1974), and soils (Stevenson, 1982; Brady, 1974; Black et al., 1965). Also see Gibbs (1983) regarding the occurrence of aggregates.

Chemical engineers have long held intraparticle diffusion as one of the likely mechanisms limiting sorption of organics to activated carbon, synthetic resins, and some porous catalysts (Mathews and Weber, 1976; Sudo et al., 1978; Weber and Liu, 1980). In addition, soil scientists have introduced intraparticle diffusion to explain transport of chemicals through certain types of soils (Leenheer and Ahlrichs, 1971; Rao et al., 1979, 1980). Separation scientists have used intraparticle diffusion models to explain chromatographic phenomena (Karger et al., 1973). Given the two well-understood mechanisms, intraparticle diffusion retarded by local equilibrium partitioning, and the assumptions that natural aggregates can be approximately described as porous spheres, an expression for sorption kinetics can be mathematically derived (Gschwend et al., 1986).

At any point in a porous sphere, the total concentration will be a function of the radial location

$$C_T(r) = (1 - n)\rho_s C_S(r) + nC(r) = [(1 - n)\rho_s K_p + n]C(r) \quad (27)$$

where ρ_s = density of solid phase (kg/L), n is intraparticle porosity, and $C(r)$ is the dissolved concentration at a distance r along the particle radius. The Fickian diffusion equation for dissolved contaminants in the pore fluid is

$$\frac{\partial nC(r)}{\partial t} = D_m n\psi \left[\frac{\partial^2 C(r)}{\partial r^2} + \frac{2}{r} \frac{\partial C(r)}{\partial r} \right] - \frac{\partial(1 - n)\rho_s C_s(r)}{\partial t} \quad (28)$$

The first term on the right describes diffusion in radial coordinates in which D_m is molecular diffusivity and ψ is a geometry correction factor which is a function of intraaggregate porosity and tortuosity. The second term describes sorption or desorption to local surfaces which can also be expressed by Eq. (1).

As such, Eq. (28) assumes no diffusion of solid bound contaminant. In terms of the local total concentration

$$\frac{\partial C_T(r)}{\partial t} = \frac{D_m n \psi}{(1 - n)\rho_s K_p + n} \left[\frac{\partial^2 C_T(r)}{\partial r^2} + \frac{2}{r} \frac{\partial C_T(r)}{\partial r} \right] \quad (29)$$

when the effective diffusivity is

$$D_{\text{eff}} = \frac{D_m n \psi}{(1 - n)\rho_s K_p + n} \quad (30)$$

In studies of diffusion in benthic sediments, Ullman and Aller (1982) found that ψ could be empirically related to porosity, $\psi = n^i$, where i varies between 1 and 2 for a wide variety of sediments. If we treat intraparticle diffusion similarly and assume $i = 1$, then

$$D_{\text{eff}} = \frac{D_m n^2}{(1 - n)\rho_s K_p + n} = \frac{D_m n^2}{(1 - n)\rho_s K_p} \quad (31)$$

where for hydrophobic solutions with high sediment concentrations $(1 - n)\rho_s K_p \gg n$. Table 6 shows estimated interaggregate porosities which yield the observed D_{eff} for $i = 1$.

As a result, Eq. (29) predicts sorption within particles from five physical parameters. Earlier we reviewed methods to estimate K_p from the organic carbon fraction and the octanol-water partitioning coefficient for nonpolar organic chemicals. Molecular diffusivity can be determined from any of several compilations (see, for example, Weast, 1986) or predictive formulas. For example, Hayduk and Laudie (1974) give

$$D_m = \frac{13.26 \times 10^{-5}}{\nu^{1.14} V_m^{0.589}} \quad (32)$$

in which ν is the kinematic viscosity of water (centipoise) and V_m is solute molar volume (cm^3/mole). Lyman et al. (1982) better describe methods to estimate molecular diffusivity and give methods to estimate molar volume. For context, Gschwend et al. (1986) note that values of molecular diffusivity are of the order of 0.6×10^{-5} cm^2/s. As examples, $D_m = 0.73 \times 10^{-5}$ cm^2/s for dichlorobenzene and $D_m = 0.45 \times 10^{-5}$ cm^2/s for hexachlorobiphenyl.

Dry solid density ρ_s can be easily measured (Lambe, 1951). Alternatively, ρ_s can be estimated from the specific gravity of the materials involved [$\rho_s = 2.65$

TABLE 6

Interaggregate Porosities which Yield Observed D_{eff} for $i = 1$ in $\psi = n^i$

Sediment source	Compound	Estimated particle porosity
Sorption experiments		
Charles River sediments	Pentachlorobenzene	$(0.32)^a$
Charles River sediments	1,2,4-Trichlorobenzene	0.17
Charles River sediments	1,4-Dichlorobenzene	0.17
Charles River sediments	1,2,3,4-Tetrachlorobenzene	$(0.26)^a$
Charles River sediments	1,2,3,4-Tetrachlorobenzene	$(0.33)^a$
Charles River sediments	1,2,3,4-Tetrachlorobenzene	$(0.39)^a$
Iowa soils	1,2,3,4-Tetrachlorobenzene	0.15
Iowa soils	Pentachlorobenzene	0.15
North River sediments	1,2,3,4-Tetrachlorobenzene	0.09
North River sediments	Pentachlorobenzene	0.07
Range Point[b]	Kepone	0.11
	Average	0.13 ± 0.04
Desorption experiments		
Charles River sediments	1,2,3,4-Tetrachlorobenzene	0.18
North River sediments	1,2,3,4-Tetrachlorobenzene	0.14

Source: Gschwend et al., 1986.
[a]These values were not used in averaging because of changing particle size distributions due to particle disaggregation.
[b]Data from Connolly, 1980.

for clay, silts, and sands (Terzaghi and Peck, 1967) and $\rho_s = 1$ for organic matter (Gschwend et al., 1986)].

The only remaining parameter necessary to estimate D_{eff} is the intraparticle porosity n. This is a great deal more difficult to measure and, as a result, this parameter is estimated by fitting forms of Eq. (29) to experimental data assuming that $i = 1$. For the sediments and soils studied, Gschwend et al. (1986) determined (in Table 7) that $n = 0.13 \pm 0.04$. The sediments included those from the Charles River and North River in Massachusetts, from an Iowa soil, and from Range Point marsh sediments collected by Connolly (1980). These data do not cover an extensive range but do seem sufficient to indicate that for silt and clay aggregates n is approximately 0.13 when i is assumed to be 1. The few measurements or estimates of intraparticle porosity are of about the same magnitude but not fully consistent with the estimate that $n = 0.13 \pm 0.04$. These values are given in Table 7 for comparison. Perhaps the assumption that $i = 1$ should be reexamined based on the limited available data.

If we examine sources of uncertainty in D_{eff}, we see that estimates of K_{oc} and

n are the least certain, while estimates of ρ_s, f_{oc}, and D_m are the most certain. For strongly hydrophobic chemicals and large sediment concentrations where $(1 - n)\rho_s K_p \gg n$, errors in K_{oc}, f_{oc}, ϕ_s, and D_m are reflected in D_{eff} in a linear fashion such that, for example, an error of α in any of these four parameters causes an error of α in D_{eff} and an error of α in all four simultaneously causes an error of 4α in D_{eff}. Errors in n, expressed as a fraction of n, are different in that these errors translate into roughly twice as much error in D_{eff} expressed as a fraction of D_{eff}. As a result, uncertainty in K_p of an order of magnitude will result in an order of magnitude uncertainty in D_{eff}. Typically, we expect order of magnitude errors in estimates of K_p unless chemical specific correlations like those in Table 4 are available; then errors on the order of a factor of 2-3 are expected (Gschwend et al., 1986). At most, we expect uncertainties in n of a factor of 5 that yields approximate uncertainties in D_{eff} of an order of magnitude. Obviously, we have to develop methods to precisely measure n and also continue to refine methods to estimate K_p.

Equation (29) can be solved analytically for special cases to determine the flux of contaminant to and from natural aggregates, but these solutions are too limited for a general modeling framework. Analytical solutions can be posed for systems where the ambient bulk solution concentration is constant or the system is infinite and the natural aggregates are of a constant size. In addition, Crittenden et al. (1986) lists other such special solutions.

Gschwend et al. (1986) have developed a finite difference method for multisize sediments and varying contaminant concentrations, as have others

TABLE 7

Measured Intraparticle Porosities

Source	Material	Mean particle radius (μm)	Measured intraparticle porosity
Passioura and Rose, 1971	Cununurra clay soil	362	0.53
		720	0.52
	Sepiolite clay	362	0.33
		720	0.34
	Crushed porcelain	362	0.21
		720	0.23
Crittenden et al., 1986	Keweenaw 7, B horizon, sandy loam	75[a]	0.33[b]

[a]Median aggregate radius.
[b]Estimated by assuming microporosity and macroporosity are equal. Generally this is not correct and the true value is expected to be lower based on measurements by Passioura and Rose (1971).

(Crittenden et al., 1986). Gschwend et al. write Eq. (29) in a dimensionless form and integrate for average concentration in a particle. A finite difference form of the governing equation is formulated and an explicit Euler method for integration in each time step is used. The integration procedure proceeds over each of the n' particle classes and steps through each of the m concentric spherical surfaces equally spaced along the radial axis for each particle class. As a result, we can write for a finite system that the dissolved concentration at time step $k + 1$ will be

$$
C_w^{k+1} = \frac{1}{(1 + a)} \Bigg[C_w^k - \sum_{i=1}^{n'} \frac{f_i V_s (\Delta x_i)^2}{V} \Bigg(\bigg\{ \sum_{j=2}^{m_i-1} [3 + (-1)^j]
$$
$$
(j - 1)(u_{i,j}^{k+1} - u_{i,j}^k) \bigg\} - (m_i - 1)u_{i,mi}^k \Bigg) \Bigg]
\tag{33}
$$

where

$$
a = \sum_{i=1}^{n'} \frac{f_i V_s (\Delta x)^2}{V} (m_i - 1)K_p (1 - n)\rho_s
\tag{34}
$$

and, in addition, f_i is the mass fraction of each size group, i, n' is the number of size groups; m is the grid spacing for each particle class; V_s is the volume of the solids; V is the volume of the aqueous phase; $u = (r/R)C_s$; R is the particle radius; r is the radial distance to each grid point along the radius; and $x = r/R$. We can further write the averaged sorbed concentration as

$$
C_{sw}^{-k+1} = \sum_{i=1}^{n'} (\Delta x_i)^2 \bigg\{ \sum_{j=2}^{m_i-1} [3 + (-1)^j](j - 1)u_{i,j}^{k+1} + (m_i - 1)u_{i,mi}^{k+1} \bigg\}
\tag{35}
$$

For completeness, please note that Gschwend et al. (1986) derive these equations in some detail and illustrate their use. From Gschwend et al., we can note that these equations are relatively accurate when spatial discretization involves 50 or more grid points in a particle class. However, this seems to be relatively inefficient, and it is suspected that more accurate and efficient schemes may be possible. Short time steps are required when sorption is rapid. Long time steps are necessary when sorption proceeds slowly because the solution can otherwise require a great deal of computing time.

For convenience, we can report the progress of sorption in terms of the fraction of completion in a closed system (total chemical and water mass constant),

$$
\begin{aligned}
\frac{M_t}{M_\infty} &= \frac{C_w(t = 0) - C_w(t)}{C_w(t = 0) - C_w(t = \infty)} \\
&= \frac{\overline{C}_{sw}(t = 0) - \overline{C}_{sw}(t)}{\overline{C}_{sw}(t = 0) - \overline{C}_{sw}(t = \infty)} \\
&= \frac{\overline{C}_s(t = 0) - \overline{C}_s(t)}{\overline{C}_s(t = 0) - \overline{C}_s(t = \infty)}
\end{aligned}
\tag{36}
$$

in which the bar indicates an average particle aggregate property. This allows us to track the progress of sorption or desorption in a consistent fashion.

Being able to solve Eqs. (33) and (35) allows us to explore several issues in modeling of contaminant transport. For example, we can investigate how many particle classes are necessary to adequately reproduce observed behavior. Furthermore, we can investigate if equilibrium partitioning is appropriate (and when it is not) and whether or not first-order approximations can be used to simplify the solution.

Equilibrium models of contaminant fate and transport usually include specifications of one to three sediment sizes [WASP (Ambrose et al., 1986) includes three sizes, and TOXIWASP (Ambrose et al., 1983) includes one sediment class]. For nonequilibrium sorption, we can explore the question of how many size classes are necessary. In this regard, Gschwend et al. (1986) illustrate the effect of particle size and distribution on sorption kinetics. Figure 4 illustrates that particles with a diameter of 40 μm will require on the order of 100 min to approach equilibrium. This is important since the operational definition of dissolved material is 0.45 μm and indicates that typical experiments of 2–3 h barely allow the colloidal material to reach equilibrium. Larger particles on the order of 800 μm in diameter will require over 2 months to reach equilibrium. This illustrates the difficulty in interpretation of most experiments conducted over very short periods.

Figure 4 also illustrates that the breakup of natural aggregates will decrease the time required to achieve equilibrium. For example, if dredging or storm events suspend and disaggregate bed sediments, desorption can more quickly reach an equilibrium than was required in sorbing the material. Figures 5 and 6 show that a limited distribution of particles over an order of magnitude can be represented quite well as a single class within a geometric mean. Wider distributions over three orders of magnitude are less adequately represented as a single class.

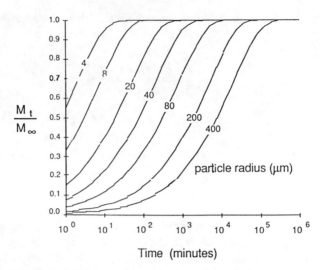

FIG. 4. Effects of particle size on the rate of sorption. Simulation parameters are $n = 0.13$, $\rho_s = 2.5$ g/cm^3, $D_m = 6 \times 10^{-6}$ cm^2/s, $f_{oc} = 0.02$, and $K_{ow} = 10^5$ cm^3/g. K_{oc} is calculated by using $\log K_{oc} = \log K_{ow} - 0.21$ and solid concentration S is adjusted so that $K_p S = 1$ (Gschwend et al., 1986).

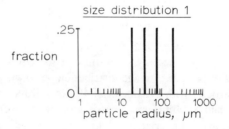

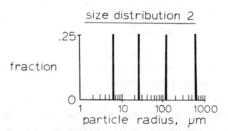

FIG. 5. Two hypothetical size distributions (with same geometric mean size of 63 μm) used in the model simulations to represent narrow and wide particle size range.

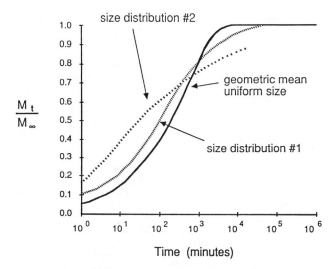

FIG. 6. An analytical solution using geometric-mean single size to represent polydispersed particles is compared with solutions obtained by using numerical method and knowledge of size distribution. The simulation parameters are the same as those in Figure 4. The geometric mean uniform size is 63 μm.

Figure 7 is based on an analytical solution that describes when a closed system approaches equilibrium. Gschwend et al. (1986) note that a water column suspension of 10^{-5} g/ml for all sorbates with $K_p < 10^5$ g/ml will approach equilibrium by about $0.3R^2/D_{eff}$. Given the particle radius, R, and D_{eff} computed from Eq. (31), we can solve for the time required to approach equilibrium. For typical bed conditions with $S = 1$ g/ml and $K_p > 10^2$ cm^3/g, the time to achieve equilibrium between the pore waters and sediments will be $0.003R^2/D_{eff}$.

Gschwend et al. (1986) also note that when at least an approximation of sorption kinetics is important enough to include in a model, that the radial diffusion model can be used to derive an approximate first-order model. The first-order rate coefficient can be expressed in terms of physically measurable parameters. The first-order model is an approximation because it contains only the first term of the infinite series of exponential terms that the radial diffusion model represents. Figure 8 illustrates the nature of the approximation.

To define the first-order rate constant k in terms of physically based parameters, Gschwend et al. (1986) note that it is useful to assume that k is proportional to D_{eff}/R^2. The proportionality constant is evaluated by assuming that the first-order and radial diffusion models are equivalent at one-half of the time required to approach equilibrium ($M_t/M_\infty = 0.5$). As a result we see that

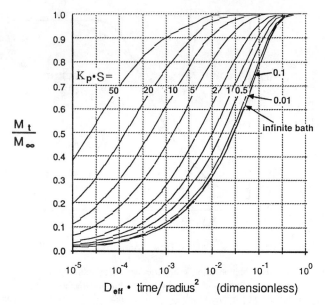

FIG. 7. Analytical solutions for the radial diffusive uptake or release by spherical particles suspended in a closed system. The numbers on curves show the final ratio of the mass sorbed on solids to the mass dissolved in the solution.

$$k = (10.56\, K_p S + 22.7)\, \frac{D_{\text{eff}}}{R^2} \qquad (37)$$

Therefore, this form can be used to simplify treatment of sorption kinetics when an approximate calculation (i.e., HSPF and SERATRA models) is adequate. We also note that a properly chosen first-order approximation [i.e., Eq. (37)] will predict the initial sorption or desorption and overpredict the time required to approach equilibrium.

B. Ionization and Acid-Base Reactions

Ionization is a process whereby a molecule is separated or dissociated into particles of opposite charge. Generally, ionization is a straightforward, extremely fast reaction which has a large impact on the behavior of a chemical substance and may be very important in mixing zones and where pH changes rapidly. The process is especially important for metals and may be important for organics when volatilization of the neutral species is important (Mills et al., 1985) and for positively charged organic compounds (E. Weber, U.S. EPA,

Athens, Georgia, personal communication, 1988) in which ionization may control sorption via the cation exchange mechanism.

Since the reaction is quite rapid, the equilibrium description is most adequate for modeling. The distribution of ionic forms ($[H^+]$ and $[A^-]$, concentration of hydrogen ions and weak organic acid, respectively, in moles per liter and neutral species ($[HA]$ in moles per liter) is described by the acid dissociation constant.

$$K_a = \frac{[H^+] \, [A^-]}{[HA]} \tag{38}$$

The base dissociation constant relates the concentration of hydroxyl ions $[OH^-]$ and weak base $[B^+]$ to the neutral species $[BOH]$ (all concentrations are in moles per liter).

$$K_b = \frac{[OH^-] \, [B^+]}{[BOH]} \tag{39}$$

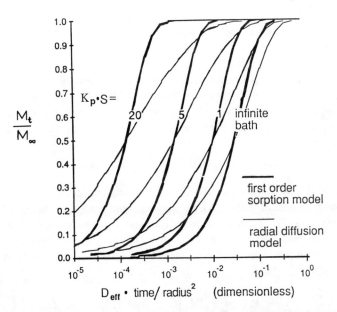

FIG. 8. Comparison of the solution of the first-order sorption model with the analytical solution of the radial diffusion model. Numbers on curves show final ratio of mass sorbed on solids to mass dissolved in solution.

Values of K_a and K_b have been experimentally defined for a number of contaminants (Mills et al., 1985; Donigian et al., 1983) and are reported as $pK_a = -\log_{10}(K_a)$ or $pK_b = -\log_{10}(K_b)$. The fraction of total contaminant ($HA + A^-$ or $BOH + B^-$) that is in neutral form and thus subject to volatilization is for acids

$$f_a = \frac{1}{1 + 10^{pH - pK_a}} \tag{40}$$

and for bases

$$f_b = \frac{1}{1 + 10^{pK_w - pK_b - pH}} \tag{41}$$

where pK_w is the log base 10 of $K_w = [H^+][OH^-]$. Mills et al. (1985) give example calculations of f_a and f_b. Figure 9 shows the relationship between pK_a and f_a and pK_b and f_b. It should be noted that it is not clear exactly how important ionization is for organic compounds.

C. Hydrolysis

Hydrolysis reactions lead to transformations of pollutants from a cleavage of a chemical bond and subsequent formation of a new bond. It is usually promoted by an acid or base although it can occur in neutral environments. Hydrolysis of metal ions is essentially a complexation process and is described in a following section. With organic pollutants, hydrolysis proceeds through the reaction of an organic compound, RX, with water; the formation of a new carbon-oxygen bond and the elimination of the RX bond in the compound:

$$RX + H_2O \leftrightarrow ROH + HX \ (or \ H^+, \ X^-) \tag{42}$$

The rate of the hydrolysis reaction for various organic compounds is pH and temperature dependent with half-lives varying from a few seconds to 10^6 years (Bonzountas, 1983). The acid and base reactions are generally modeled using a second-order rate expression (Callahan et al., 1979) containing the concentration of the hydrolyzed substance, C, and the hydrogen ion or hydroxyl ion concentration,

$$\frac{dC}{dt} = k_h C = k_A[H^+]C + k_B[OH^-]C + k_N C \tag{43}$$

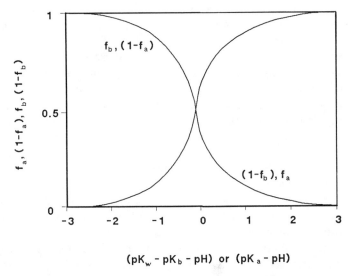

$$(pK_w - pK_b - pH) \text{ or } (pK_a - pH)$$

FIG. 9. Relationship between the fraction of contaminant in various forms and the difference between pK_a and pK_b and the system pH.

where k_h is the first-order rate constant at a specific pH, k_A and k_B are second-order rate constants for acid- or base-mediated processes, and k_N is the first-order rate constant for the neutral hydrolysis, a first-order reaction.

D. Complexation of Metals

Complexation of metals is basically defined as a reaction of two species to form a third species. The metal (Zn, Cu, Cd, Hg, Ag, and others) reacts with a ligand (OH^-, CO_3^{2-}, SO_4^{2-}, Cl^-, F^-, NH_3, S^{2-}, CN^-, amino acids, EDTA, humates, fulvates, and others) to form various species in relation to the formation reaction generally represented as

$$mM + nL + hH \leftrightarrow M_mL_nH_h \qquad (44)$$

where M represents a metal or metal complex, L is the ligand, and H is the proton, and m, n, and h represent stoichiometric coefficients. Charges on the reactants or products have been omitted. The extent to which a given complex formation reaction proceeds is given by the simultaneous advancement of all possible chemical reactions, complete with interactions among all species and individual rate laws governing the advancement of each reaction. In spite of the apparent complexity of defining or applying a kinetic approach to chemical problems, we have found that the equilibrium approach to chemical complexation is a good approximation for natural systems. Chemical thermodynamics provides the means for defining the composition of the system and the extent to which complexation reactions will proceed. The discussion of complexation reactions is covered in great detail in Stumm and Morgan (1981) and Morel (1983).

Calculation of the complexation reactions for metals in natural environments is best left to computer programs such as MINTEQA1 [MINTEQ (Felmy et al., 1984a) code and documentation available from U.S. EPA, Athens, Georgia]. MINTEQA1 will compute the equilibrium composition in a multimetal, multiligand system based on mass balance and mass action equations and applicable thermodynamic constants. MINTEQA1 can also include the effects of redox, chemical precipitation, and sorption in calculating the equilibrium composition of an aqueous system. As previously mentioned, sorption routines are based on the free-metal ion activity (i.e., "activity" Langmuir, "activity" K_d). Complex equilibrium models have been incorporated into environmental fate and transport models [MEXAMS using MINTEQ for metal chemistry, and RIVEQL using MINEQL (Westall et al., 1976) to perform aqueous speciation]. The details of MEXAMS are given by Felmy et al. (1984b) and RIVEQL by Chapman et al. (1982).

The net effect of complexation reactions is to increase the soluble concentration of a given metal (given a solid source) or to redistribute the metal between various complexed forms. The net result of complexation can be significant effects on the mobility of toxic metals and the corresponding distribution between the water column and suspended and bed sediments (Saar and Weber, 1982; Förstner and Salomons, 1983). The concentration of the free metal (unbound) can be affected very substantially through interactions with ligands with subsequent effects on processes driven by the free ion concentration (or more precisely, activity) of a chemical species, such as sorption, chemical precipitation, and biological uptake (as well as toxicity). Determining the concentration (or activity) of the species that is participating in a given reaction will provide a greater ability to model behavior in complex systems.

It is generally assumed that equilibrium with respect to complexation is the rule rather than the exception. This assumption will be especially valid when the kinetics of complexation are compared to dispersive flow phenomena and the processes of sorption, chemical precipitation and redox, which can be very slow. Complexation of metals by ligands is relatively rapid, and the equilibrium assumption provides a realistic approach to describing the metal forms existing in a given environment. If it is necessary to describe short-duration effects on the order of minutes or several hours, complexation kinetics may need to be taken into account in a pseudo-equilibrium approach (Morel, 1983). There are two competing factors that may affect complexation reaction kinetics. First, a multitude of chemical species for metals provide many potential reaction pathways and accelerate the reactions. Second, the low concentrations of metals and ligands in surface waters often compensate for a very high second-order reaction rate constant that results in a nonequilibrium condition for specific metal complex-formation reactions.

E. Chemical Precipitation and Dissolution

The full importance of chemical precipitation and dissolution reactions in aquatic systems to the fate and transport of toxic metals is not fully understood. For toxic metals, the formation of metal hydroxides, carbonates, and sulfides is generally limited to specific systems rather than being a common process in all natural systems. Direct toxic metal precipitation is used quite frequently in industrial waste treatment for removing metals, such as Zn, Cr, Cu, Pb, and Ag. The concentrations of the free metals and anionic species involved in the precipitation reaction are manipulated to result in the substantial exceedance of appropriate solubility products to force precipitation of a solid in a treatment process. However, the concentrations of reactants usually found in receiving waters are much lower, and direct precipitation may not be the dominant process whereby metals are removed from solution. The direct metal precipitation in surface waters occurs as a result of changes in water chemistry, often being found in response to changes in pH, oxidation reduction conditions (redox), or in the concentrations of reacting species. These changes occur from changes in sediment-water interactions, mixing of two waters and from the exchange of gases (Förstner and Salomons, 1983). The reader is referred to Stumm and Morgan (1981) for discussion of the principles of chemical precipitation/dissolution.

Direct chemical precipitation of toxic metals is difficult to observe directly due to the interactions of removal processes in natural aquatic systems and the heterogeneity of sediments. For example, the aqueous concentrations of a particular metal might be explained equally well by Langmuir adsorption and chemical precipitation of a metal carbonate. Often, calculations of the degree to which the solubility product is exceeded are used to infer the precipitation of a pure metal phase without actually obtaining the direct evidence to suggest precipitation controls on the metal concentrations measured in the aquatic system (Carigan and Tessier, 1985; Carigan and Nriagu, 1985; Medine and Porcella, 1980; Hem, 1972, 1977; Dhillon et al., 1981; Jenne et al., 1979; Benjamin and Bloom, 1981; Chapman et al., 1983; see also Rai and Zachara, 1984).

Chemical equilibrium models, such as MINTEQA1, can examine the process of precipitation of pure metal forms in aqueous systems. However, the calculations of controlling solid, saturation indices, and equilibration concentrations are based on equilibrium. According to Morel (1983), solid-phase reactions can be exceedingly slow but are mediated by high temperature, nucleating surfaces, and biological activity. In addition, metastable forms can be initially precipitated that are less stable, more soluble, and often in nonstoichiometric proportions for a particular metal. The application of equilibrium in a water with a particular solid phase is largely governed by the knowledge of the process and the surface water chemistry.

Chemical precipitation/dissolution for the macro metals (Ca, Mg, Fe, Mn, and others) can exert a controlling influence on water quality and be the dominant substrate affecting the fate and transport of toxic metals. The formation of iron and manganese oxides, sulfates, and carbonates (of Ba, Sr, and Ca) in surface waters has been shown to exert considerable impact on toxic metal distributions through processes of coprecipitation, sorption, and ion exchange. Whether the observed coprecipitation process is the direct formation of a solid form ($CdCO_3$ in association with $CaCO_3$) or the result of "sorptive" bonding of the toxic metal to the bulk precipitate is not well understood. Mechanistically, there appears to be little difference between sorption and coprecipitation. As a result, available models treat observed removal of metals in the presence of suitable substrate (such as iron oxides or calcite) as sorptive removal rather than attempt to describe coprecipitation vs. sorption separately. There have been few approaches to describing the kinetics of sorption and chemical precipitation processes in a natural environment. Rate expressions have been developed for simulating precipitation/dissolution processes, such as the dissolution of silicate minerals (White and Claassen, 1979) or calcite formation (Plummer et al., 1979; Morel, 1983). However, they have not been incorporated into solute transport models.

Chapman et al. (1982) used a pseudokinetic approach to incorporating kinetics of dissolution of suspended and bed sediment precipitates by allowing only a portion of the total mass of available precipitate to dissolve. This modeling approach can be relatively straightforward if there is only one solid to consider (such as in acid mine drainages in which metal behavior will be dominated by iron transformations) but becomes increasingly complex if more than one solid can exert controls on general water chemistry and the distribution of trace substances. The limited knowledge of the time dependency of chemical processes, whether it be chemical precipitation, complexation, or sorption, makes it difficult to ascertain the limits of equilibrium approaches. This is especially disconcerting when describing systems with large flows of matter and energy. Increasing our understanding of the occurrence and kinetics of precipitation/dissolution processes for the majority of important solid phases will be necessary prior to development of complex chemical speciation/transport models to predict the effects of chemical precipitation/dissolution of major ions and trace metals and the important interactions between species.

F. Redox Reactions

Oxidation-reduction reactions exert significant controls on the chemistry of both major ions and trace metals. The most significant effects of redox reactions in natural systems are on the transfers of chemicals between solids (suspended sediments and bed sediments) during formation and dissolution of chemical precipitates. Chemical precipitates formed during oxidation of iron and manganese provide substrates for subsequent sorption of toxic metals. This may result in control over the dissolved metal concentration. Reduction reactions, such as the conversion of sulfate to sulfide, can also result in the formation of metal sulfides (FeS_2, CdS, ZnS) and control the dissolved toxic metal concentrations. These reactions are generally recognized as being important in the mobility of metals and are included in most complex equilibrium chemistry models, such as MINTEQA1, to describe the chemistry of toxic metals.

Redox reactions can be exceedingly slow, however, and the application of equilibrium to all species may not be an appropriate assumption for describing metal behavior. The same limitations to modeling chemical precipitation exist for redox chemistry. Calculations for systems with characteristic water residence times shorter than the half-life of a particular redox reaction may not realistically simulate metal transport and fate. Unfortunately, very little kinetic information is available for most of the important redox processes. This results in limited ability to incorporate a kinetic approach into fate and transport modeling.

Redox kinetics of iron, manganese, and sulfur are the most important in regulating the fate and transport of toxic metals. The rate laws for the oxidation of Fe(II), Mn(II), and sulfide have been studied extensively (Stumm and Morgan, 1981; Morel, 1983). Iron oxidation was found to be dependent on Fe(II) concentrations, pH, and the partial pressure of oxygen. Manganese kinetics appeared to be first order with an autocatalytic term reflecting the precipitated MnO_2. If sufficient information is available to determine whether equilibrium is a valid assumption for a particular system, the modeling of chemical behavior is greatly simplified. In highly dynamic environment with active redox processes, it may not be possible to realistically model the chemical movement at the present time.

Synthetic organic contaminants may be lost from water by reaction with an oxidant (Callahan et al., 1979). Generally, the reaction can be represented as a second-order reaction as follows:

$$\frac{d\text{C}}{dT} = k_{\text{Ox}}[\text{Ox}]\text{C} \tag{45}$$

where K_{Ox} is the second-order rate constant for the reaction of the chemical, with the oxidant Ox, and C and [Ox] are the respective concentrations. In addition to oxidation of organics, reduction has also been observed (Tinsley, 1979). Uncertainty still remains concerning the kinetics of redox processes for most pollutants.

G. Biological Degradation

The representation of microbial degradation of organic contaminants is based on the well-known Monod analogy to Michaelis-Menton enzyme kinetics (Burns, 1983; Callahan, et al., 1979). For many pollutants in the environment, the chemical concentration is much less than the saturation constant K_s, and the Monod equation simplifies to an expression that is first-order with respect to chemical concentration and microbial biomass and second-order overall:

$$\frac{d\text{C}}{dT} = k_{b2}\,\text{CB} \tag{46}$$

where k_{b2} is the second-order rate constant and C and B represent the concentrations of the chemical and the biomass, respectively. In natural environments, the cell concentration is relatively high while the pollutant concentration is low, leading to a first-order formulation. If the microbial biomass does not increase significantly as a result of the decomposition of an organic contaminant, then a first-order formulation is appropriate for modeling

$$\frac{d\text{C}}{dT} = k'_b\,\text{C} \tag{47}$$

where k'_b is the pseudo-first-order rate constant in which B is incorporated into the second-order rate constant k_{b2}. Degradation of organics in sediments can exert influence on the total contaminant concentration and subsequent chemical transport.

Evaluation of the microbial population size B, required in Eq. (46), is described by Burns (1983). The rate constants k_{b2} and k'_b have been determined for a number of contaminants using pure and mixed cultures. However, the inability to exactly reproduce natural populations of bacteria and the failure to fully investigate the effects of pH, nutrients, and other geochemical influences makes it difficult to forecast biological kinetics. Generally, models are calibrated with in situ measurement and thus are not reliably predictive.

IV. PHYSICAL TRANSPORT OF CONTAMINATED SEDIMENTS

A. General Knowledge of the Process

Transport of contaminants is described by solving the conservation of mass and momentum equations for water and the conservation of mass equations for contaminants and solids. The conservation of mass and momentum equations for water are descriptions of hydrodynamic behavior. In this regard, Rodi (1980) gives the best description of methods. At the present time, a number of model types are being used to examine contaminant transport and fate. Most promising is the three-dimensional model of Sheng (1986) that uses second-order turbulence closure. Sheng's models are receiving widespread use by the Army Corps of Engineers and the U.S. Geological Survey, among others, to model estuaries, coastal regions, and some lakes. Laterally averaged two-dimensional models (Buchak and Edingers, 1984) are also equally useful for some estuaries and more frequently used in reservoirs and lakes as well as vertically averaged hydrodynamic models.

CE-QUAL-R2 is an extension of Edingers' laterally averaged hydrodynamics model (Buchak and Edingers, 1984) that solves a contaminant mass balance as well. This model has been used in a number of model applications and while it does not specifically address all phases of metal transport, it can be used to describe the behavior of a surface for sorption and/or coprecipitation of metals. An appropriate application of the model to contaminant transport would be in a system that is heavily dominated by iron dynamics such as acid mine waters. In this regard, the model was used by Medine and Lamarra (unpublished data) to describe iron dynamics in a hypothetical reservoir in which acid mine waters would be neutralized, reacted, and allowed sufficient sedimentation time. The processes incorporated in the model would include the oxidation-reduction behavior of iron and chemical precipitation. Because this particular situation involved a system heavily dominated by iron, the toxic metal levels in the neutralized acid waters could be adequately simulated. In this hypothetical reservoir, the behavior of zinc, copper, and lead would be primarily governed by the formation of insoluble iron oxides, sorption of metals to the surface, and subsequent sedimentation of iron solid phases. Sorption of metals to iron precipitates has been described by Davis and Leckie (1978), Benjamin and Leckie (1981a, 1982), Huang et al. (1977), Kinniburgh and Jackson (1981), Laxen (1985), and Swallow et al. (1980).

One-dimensional stream hydraulics are modeled by a number of different methods involving various approximations. The routing model of Fread (1974) seems to be most widely used. The innovative solution of DeLong (1986) involving finite elements seems to be the most stable and useful approach. Some additional work is needed to incorporate our latest understanding of hydraulic routing into contaminant transport modeling.

The mass balance or advective-dispersive equation has the form

$$\frac{\partial C}{\partial t} = \frac{1}{A_x} \frac{\partial}{\partial x} \left(A_x E_x \frac{\partial C}{\partial x} \right) + \frac{1}{A_y} \frac{\partial}{\partial y} \left(A_y E_y \frac{\partial C}{\partial y} \right) + \frac{1}{A_z} \frac{\partial}{\partial z} \left(A_z E_z \frac{\partial C}{\partial z} \right)$$

$$- \frac{\partial(u_x C)}{\partial x} - \frac{\partial(u_y C)}{\partial y} - \frac{\partial(u_z C)}{\partial z} + S\,(C,x,y,z,t) \tag{48}$$

in which C is sediment or total contaminant concentrations (a mass balance is written for both; and written for individual particle classes); A_x, A_y, and A_z are areas of a control volume in yz, xz, and xy planes, respectively; E is the eddy diffusivity or mass transport coefficient; u_x, u_y, and u_z are velocity components in the x, y, and z directions, as determined by solving the hydrodynamic equations; and S is source or sink term sediment or contaminant. For the contaminant, S would quantify the effects of biodegradation, hydrolysis, and transformation processes. For sediment, S would quantify the effects of erosion and deposition from the water column and settling, resuspension, and flow into the water body and locations above the bed.

The most frequently used method for solving the advective-dispersive equation is the box model approach, where a one-dimensional form of the mass balance equation is written in x, y, and z directions for a control volume segment of the water body. Ambrose et al. (1986) illustrate the method in detail for sedimentary and other contaminants and provide a good purpose model, WASP4. WASP4, available from the U.S. EPA, Athens, Georgia, allows the user to specify up to 20 constituents and the appropriate kinetics for modeling.

An innovative approach to solving the mass balance equation involves Lagrangian solution. This method is highly accurate because the advective terms, such as $\partial u_x c / \partial x = 0$, allow the use of numerical integration rather than differentiation in the solution. The Lagrangian solution has been successfully applied to mass transport in one-dimensional streams, but not to transport in two and three dimensions because of the difficulty in solving for E_y and E_z. However, David Schoell-Hamer (U.S. Geological Survey) is completing a laterally averaged two-dimensional model and Ralph Cheng (U.S. Geological Survey) has worked extensively with Eulerian-Lagrangian models for estuaries.

The most significant research issue in solving the hydrodynamic and mass balance equation seems to evolve around the question of whether or not water quality models can solve the mass balance equations over longer time steps and with more coarse discretization of the water body compared to the hydrodynamics model. In highly advective systems like estuaries that have rapidly changing and highly variable velocities, it is not an easy task to link hydrodynamic and water quality models. Chesapeake Bay is an example where care is needed in matching model space and time scales.

B. Bed Diffusion and Advection

Diffusion and advection between the water column and bed, as well as within the bed, involve the transport of the dissolved contaminant and colloid-bound contaminant for some cases where the colloid diameter is less than the pore diameter. The flux through the bed surface or any horizontal plane in the bed is written as

$$\text{Flux} = n^{i+1} D_m \frac{\partial C_w}{\partial z} - n^{i+1} D_c \frac{\partial S_c C_c}{\partial z} + w C_w + w S_c C_c \tag{49}$$

where D_m and D_c are molecular diffusivities of the contaminant and colloids, respectively; n^{i+1} is porosity to the $i + 1$ power to correct for the effects of tortuosity in the bed; S_c is concentration of colloids; C_c is concentration of contaminant sorbed to colloids; and w is the advective velocity in the bed. For surface-water and groundwater exchange, w can be sufficiently high to be important but difficult to estimate or measure. Even when net exchange is zero, Thibodeaux and Boyle (1987) show that pumping through porous stream bed dunes can be important.

C. Bioturbation

Bioturbation is important for several reasons. Tubificid worms ingest buried contaminated sediment and egest bound fecal pellets or mucous-covered particles that tend to armour the bed and inhibit erosion. This conveyor belt process slowly mixes the contaminated sediments and exposes sediments to the water column. Other epifauna species burrow into surficial sediments, mixing the surface layer and entraining sediments. The ploughing organisms tend to loosen sediments and make them more readily erodable by hydrodynamic processes.

At present, the most useful approach seems to involve the eddy diffusion approach (Schink and Guinasso, 1978; Berner, 1980; Fisher et al., 1980)

$$\frac{\partial C}{\partial t} + \frac{\partial w_b(z)C}{\partial z} = \frac{\partial}{\partial z} \left(E_b \frac{\partial C}{\partial z} \right) + C \frac{\partial w_b(z)}{\partial z} \tag{50}$$

where $w_b(z)$ is a bioturbation-induced sediment velocity related to feeding and egestion rates of conveyor belts types, and E_b is due to biologically induced mixing primarily occurring at the bed surface. Gschwend et al. (1986) expand this model and supply a compilation of values for E_b, mixed-layer depth, and feeding rates for important species. As a result, it is possible to quantify benthic fluxes when population densities can be estimated or measured.

D. Near Bed Sediment Transport

Bed load transport is perhaps the most empirical component of our knowledge of sediment transport. Simons and Sentürk (1977) discuss at least 18 different formulas for bed load to illustrate the proliferation of the empirical approaches.

Generally, the past approach to describing transport near the bed is limited to describing existing measured conditions and has limited predictive capability. However, at least one of the more useful forms, the Meyer-Peter and Mueller form, bears close resemblance to a theoretically derivable form.

Theoretically, we can derive a bed load formulation from a horizontal force balance for grains sliding along the bed. Figure 10 illustrates the forces involved.

Starting with an immobile uniform grain, the fluid drag force can be written as (Gschwend et al., 1986)

$$F_1 = \frac{1}{2} \rho_f C_D A_x U_f^2 \tag{51}$$

where ρ_f = fluid density
  C_D = drag coefficient
  A_x = particle cross-sectional area
  U_f = fluid velocity at the particle
The balancing frictional resistance force is

$$F_2 = (s - 1)\rho_f g V_p \tan \phi_s \tag{52}$$

where $(s - 1)\rho_f g V_p$ = immersed weight of the particle
  s = relative density
  V_p = volume of particle
  g = acceleration of gravity

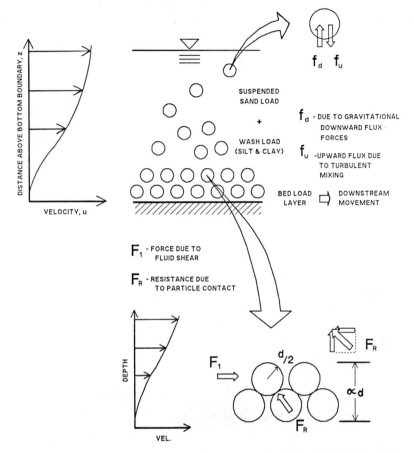

FIG. 10. Diagram showing elements of the force balance for deriving bed load formulation.

ϕ_s = stationary friction angle

The force balance for a spherical particle with diameter d at the initiation of motion yields

$$\frac{U_{fc}^2}{(s-1)gd} = \frac{4 \tan \phi_s}{3C_D} \tag{53}$$

where U_{fc} = fluid velocity of the particle at the initiation of motion.

For the rough turbulent flow distinguished by the absence of a viscous sublayer (Fig. 11), we would expect, for most streams but not necessarily for all other flows, such as lakes and estuaries, that the fluid velocity profile near the bottom will be described as

$$u = \frac{u'_*}{K} \ln \frac{30z}{d} \tag{54}$$

where $u'_* = (\tau'/\rho_f)^{1/2}$ = sheer velocity. If we arbitrarily but reasonably assume that the fluid drag exerted on the particle results from the velocity exerted at a height of $z = 0.8d$, which is the approximate value of z at the mid-level height of exposure to the flow, then the appropriate velocity is expressed as

$$U_f = \frac{u'_*}{K} \ln [30(0.8)] = 8 u'_* \tag{55}$$

ROUGH   TURBULENT   FLOW , $u_* d/v > \approx 70$

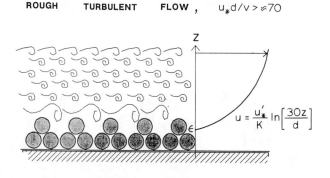

VERSUS SMOOTH TURBULENT FLOW, $u_* d/v < \approx 5$

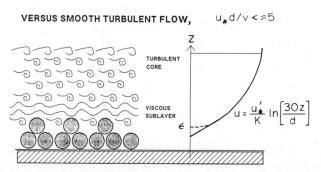

FIG. 11. Fluid velocity profile for rough turbulent flow and smooth turbulent flow.

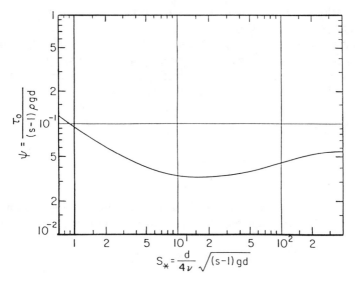

FIG. 12. Modified Shields diagram (Madsen and Grant, 1976) showing relationship to particle properties.

Using this result, Eq. (53) may be expressed in terms of the critical value of the Shields parameter

$$\psi_c = \frac{\tau_c}{\rho_f(s - 1)gd} = \frac{u_{*c}^2}{(s - 1)gd} = \frac{1}{64}\left(\frac{4 \tan \phi_s}{3C_D}\right) \tag{56}$$

The critical Shields parameter, a dimensionless expression of the critical shear stress at which initiation of motion occurs, allows considerable flexibility in taking into account the effect of differing fluid and particulate densities. ψ_c should not be confused with the porosity parameter $\psi = n^i$ used earlier.

If we assume that movement results in grains sliding over the grains underneath and that grains remain in contact with the bed, we can relate the critical sheer stress acting on particles to the drag force on all the moving particles as

$$\tau' - \tau_c = N_2 F_{DM} \tag{57}$$

where N_2 is the number of moving exposed grains. It has been found that the drag coefficient is constant at $C_D = 0.4$ when $u_f d/v > 800$. This corresponds to rough turbulent flow. When $u_f d/v < 800$, C_D is a function of $u_f d/v$. Figure 12 from Gschwend et al. (1986) shows the general relationship between ψ_c and fluid and particle properties.

Once sediment movement begins, the bed sediments deform into well-understood bed forms. These forms, shown in Fig. 13, increase the resistance to flow and lead to a distribution of the total shear stress into that acting on bed forms τ'' and that acting on individual particles τ' such that $\tau_0 = \tau' + \tau''$. The empirical results of Engelund and Fredsoe (1982) related dimensionless forms of τ' and τ_0.

$$\psi = \frac{\tau_0}{\rho(s - 1)gd} \quad \text{and} \quad \psi' = \frac{\tau'}{(s - 1)gd} \tag{58}$$

For the two common bed form types that include, first, dunes in flows having a Froude number less than about 0.7, the Shields parameter is

$$\psi' = 0.06 + 0.3\psi^{1.5} \tag{59}$$

while for plane beds, the Shields parameter is

$$\psi' = \psi \tag{60}$$

Graphic relationships exist for the less common bed forms (Gschwend et al., 1986).

Drag on a moving particle is analogous to that on a stationary particle. If we repeat the force balance for a moving particle and solve for the average velocity of sediment particles, the average velocity in the layer we will later label 2 in Figure 16 is

$$u_2 = 8(u_*' - 0.7\, u_{*c}) \tag{61}$$

where $0.7 = (C_{DS} \tan \phi_m / C_{DM} \tan \phi_s)^{1/2}$. Then we can derive the nondimensional transport rate as

$$\Phi_B = \frac{q_B}{[(s - 1)gd^3]^{1/2}} = 8\,(\psi - \psi_c)\frac{\sqrt{\psi'} - 0.7\sqrt{\psi_c}}{\tan \phi_m} \tag{62}$$

If we assume that the particles are roughly spherical such that $\phi_m = 45°$, we note that when $\psi' \gg \psi_c$, the bed load transport equation is virtually identical to the widely used Meyer-Peter and Mueller empirical formulation. This is illustrated in Fig. 14.

If we adopt the suggestion of Einstein (1950) that the bed load layer

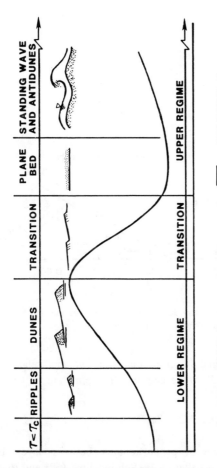

FIG. 13. Typical bed forms under different sediment movement power (stream power, Froude number, velocity).

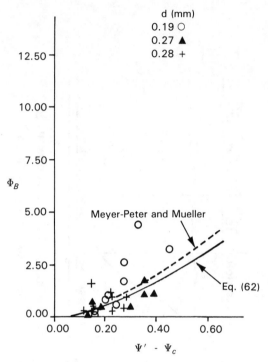

FIG. 14. Nondimensional bed load transport rate, Φ_B, as a function of nondimensional excess shear stress, $\psi' - \psi_c$ (from Gschwend et al., 1986).

thickness is proportional to grain diameter, we can relate a mean volume concentration in layer S_* to the bed load transport rate

$$q_B = \alpha_2 dS_* u_2 \tag{63}$$

where α_2 is a constant of proportionality between layer thickness and d (Einstein recommends a value of 2). Solving Eqs. (62) and (63) for S_* yields

$$S_* = \frac{1}{\alpha_2(\psi' - \psi_c)} \tag{64}$$

Equation (64) gives the boundary condition for suspended sediment profiles in the water column. Madsen (in Gschwend et al., 1986) compares predicted values of S_* with measurements by Guy et al. (1966) in Fig. 15 and shows that Eq. (64) is reasonably accurate.

E. Advective Dispersive Transport in the Water Column

In estuaries, coastal areas, lakes, reservoirs, and streams, the velocity field is in general quite dynamic. The velocity components u_x, u_y, and u_z vary with time and spatial location to the point that highly dynamic systems can only be investigated with a coupled hydrodynamic-sediment model.

The specific advective-dispersive equation for sediment transport in the water column is

$$\frac{\partial S}{\partial t} = \frac{1}{A_x}\frac{\partial}{\partial x}\left(A_x E_x \frac{\partial S}{\partial x}\right) + \frac{1}{A_y}\frac{\partial}{\partial y}\left(A_y E_y \frac{\partial S}{\partial y}\right) + \frac{1}{A_z}\frac{\partial}{\partial z}\left(A_z E_z \frac{\partial S}{\partial z}\right)$$

$$- \frac{\partial(u_x S)}{\partial x} - \frac{\partial(u_y S)}{\partial y} - \frac{\partial(u_z S - WS)}{\partial z} \pm L \pm E \pm P \qquad (65)$$

where W is the settling velocity of particles, L is the inflow or outflow of sediment from the modeled system, E describes entrainment or deposition of

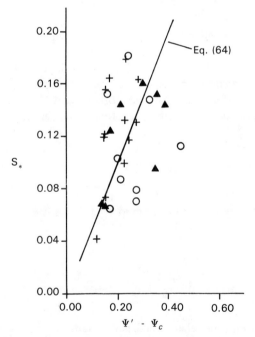

FIG. 15. Reference concentration S_* as a function of nondimensional excess shear stress, $\psi' - \psi_c$ (from Gschwend et al., 1986).

sediment from the bed, and P describes the production or loss of particulates by biological or geochemical processes.

To better understand sediment transport we can look at the simplified condition where there is a steady-state flux into the water body ($\partial S/\partial t = 0$ and $L = 0$), concentrations are laterally and longitudinally constant ($\partial S/\partial x = \partial S/\partial z = 0$), the suspended concentration profile is in dynamic equilibrium with the bed ($E = 0$), the production and destruction of solids are unimportant in the short term ($P = 0$), and it is assumed that the net settling velocity $W_f = u_y - W$. The result is that

$$E_y \frac{\partial S}{\partial y} = - W_f S \tag{66}$$

The restrictions involved generally limit Eq. (65) to steady-state, unidirectional boundary-layer flows like those in streams and in the benthic boundary layers of lakes and oceans.

If we define the bottom boundary condition as $S = S_*$ at $y = a = \alpha_2 d$, then

$$\ln \frac{S}{S_a} = - W_f \int_a^y \frac{dz}{E_y} \tag{67}$$

As a first approximation, E_y has been taken as equal to the turbulent eddy viscosity, $E_y = k(\tau'/\rho)^{1/2}y$, in the area near the bed ($y < 0.2D$) and constant over the rest of the depth. This is an approximation because it ignores the effect of stratification caused by a nonuniform concentration profile. In general, $E_y = f(\partial S/\partial y)$. However, for small total loads, Bradley and McCutcheon (1987) note that Eq. (66) provides a reasonable approximation, at least for cohesiveless particles.

If we use these assumptions in Eq. (67), we can solve for the suspended sediment profile.

$$S(y) = S_* \left(\frac{\alpha_2 d}{0.2D}\right)^G \exp\left[G\left(1 - \frac{y}{0.2D}\right)\right] \qquad \text{for } 0.2 \, D \leq y \leq D \tag{68}$$

and

$$S(y) = S_* \left(\frac{\alpha_2 d}{y}\right)^G \qquad \text{for } 0 < y \leq 0.2$$

where $G = W_f/ku_*'$, the relative tendency of particles to fall out of suspension (W_f) compared to the tendency of the flow to resuspend (ku_*'). From Eq. (68)

and the velocity profile derived from the assumed eddy viscosity profile, we can derive a suspended load equation (see Gschwend et al., 1986).

$$
q_s = \int_{\alpha_2 d}^{D} Suy \, dy \approx \frac{S_* \alpha_2 D u_*}{(1 - G)k} \left[\left(\frac{0.2D}{\alpha_2 d} \right)^{1-G} \right.
$$

$$
G^{-1} \left[G \left(\ln \frac{0.2D}{D} - \frac{1}{1 - G} + 0.8 \right) \right.
$$

$$
\left. + \left(\ln \frac{30D}{d} - 0.8 \right) \left[1 - (1 - G) \exp \left[G \left(1 - \frac{D}{0.2D} \right) \right] \right] \right]
$$

$$
\left. - \left(\ln \frac{30 \alpha_2 d}{d} - \frac{1}{1 - G} \right) \right]
$$
(69)

This is a rather complex result even for a simplified version of the full equation, but it does allow us to understand the limitations of our knowledge and to formulate a method of tracking contaminated sediments. For tracking contaminated sediments, Madsen (in Gschwend et al., 1986) formulates a sediment exchange model shown in Fig. 16. The parameters p_{21}, p_{12}, p_{32}, and p_{23} represent exchange between arbitrarily chosen layers. Equations (56) and (63) define the horizontal flux of sediment.

The vertical fluxes can be derived (see Gschwend et al., 1986) as

$$
p_{23} = p_{32} \frac{N_3}{N_2} = \frac{W_f}{\alpha_2 d} = p_{12} \frac{N_1}{N_2} = p_{21}
$$
(70)

where N_1, N_2, and N_3 are the number of particles in layers, 1, 2, and 3. As a result, if the number of particles per unit area is known for each layer, along with the settling velocity and particle diameter, we can compute the absolute vertical fluxes. From the vertical fluxes and the horizontal transport equations

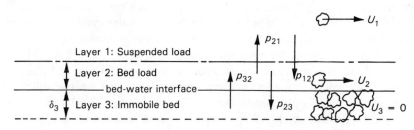

FIG. 16. Sediment transport system showing the three layers used to describe sediment interactions; the suspended load, the bed load, and the immobile bed regions.

we can estimate the dispersion of contaminated sediments. See Gschwend et al. (1986) for example calculations to illustrate the usefulness of this method.

At this time, we have not fully incorporated the effects of cohesiveness and differences in particle size into the crude mechanistic models like the conceptual model described in this chapter. We have relied on empirical derivations to describe cohesive effects on erosion and to describe consolidation in the bed. It remains a difficult task to adequately measure settling velocities in the water column, especially when high concentrations occur in the near bed region. In addition, we are limited to describing aggregation and disaggregation with Krone's aggregation classes. Given our present understanding of sorption kinetics, it is crucial that we be able to mechanistically simulate aggregation and bed consolidation. If we are to precisely track contaminated sediments, we must also know the effects of particle size distribution on the critical shear stress and material stress in the bed to adequately quantify entrainment of sediment. These research needs for cohesive sediments are especially crucial since we have observed that these fine grain sediments are more important than sands and gravels in transporting pollutants.

V. SUMMARY

The fate and transport of both metals and organic contaminants in aquatic systems is affected to a significant degree by the interactions with suspended and bed sediments. At present there are a number of processes that are important in assessing the hazards associated with contaminated sediments of which adsorption-desorption are clearly the most dominant. Additional research is needed to formulate mechanistic models which accurately reflect the adsorption-desorption process for both metals and organics in natural systems containing a mixture of sediment types. Other processes are also important in regulating the exchange of contaminants between the sediment region and need to be considered in modeling frameworks to describe not only the benthic concentration of toxic substances but the extent to which contaminated sediments can retard water quality improvement or improve water quality through removal processes. We have attempted to describe the mathematical formalisms that are used to describe the environmental behavior of organic and metal pollutants and indicate how these are incorporated into complex fate and transport models. Continued research and investigations into the bioavailability of sediment-associated contaminants and mobility is necessary to fully understand the effects of contaminant loading to aquatic systems as a whole, as opposed to just examining surface water quality.

REFERENCES

Abernathy, A. R., Larson, G. L., and Mathews, R. C., Jr. 1984. Heavy metals in the surficial sediments of Fontana Lake, North Carolina. *Water Res.* 18:351-354.

Ambrose, R. A., Jr., Hill, S. S., and Mulkey, L. A. 1983. *User's Manual for the Chemical Transport and Fate Model TOXIWASP, Version 1,* EPA-600/3-83-005. Athens, Ga.: U.S. EPA, Environmental Research Laboratory.

Ambrose, R. A., Jr., Vandergrift, S. B., and Wool, T. A. 1986. WASP3, *A Hydrodynamic and Water Quality Model—Theory, User's Manual, and Programmer's Guide.* EPA/600/3-861 034. Athens, Ga.: U.S. EPA, Environmental Research Laboratory.

Anderson, M. A., and Rubin, A. J. 1981. Adsorption of inorganics at solid-liquid interfaces. Ann Arbor, Mich.: Ann Arbor Science.

Balistrieri, L. S., and Murray, J. W. 1981. The surface chemistry of goethite in major ion seawater. *Am. J. Sci.* 281:788-806.

Baughman, G. L., and Paris, D. F. 1981. Microbial bioconcentration of organic pollutants from aquatic systems—A critical review. *CRC Crit. Rev. Microbiol.,* pp. 205-228.

Benjamin, M. M., and Bloom, N. S. 1981. Effects of strong binding of anionic adsorbates on adsorption of trace metals on amorphous iron oxyhydroxide. In *Adsorption from Aqueous Solutions,* ed. P. H. Tenari, pp. 41-60. New York: Plenum.

Benjamin, M. M., and Leckie, J. O. 1981a. Multiple-site adsorption of Cd, Cu, Zn, and Pb on amorphous iron oxyhydroxide. *J. Colloid Interface Sci.* 79:209-211.

Benjamin, M. M., and Leckie, J. O. 1982. Effects of Complexation by Cl, SO_4, and S_2O_3 on adsorption behavior of Cd on oxide surfaces. *Environ. Sci. Technol.* 16:162-170.

Berner, R. A. 1980. *Early Diagenesis: A Theoretical Approach.* Princeton, N.J.: Princeton University Press.

Blachford, D. P., and Ongley, E. D. 1984. Biogeochemical pathways of phosphorus, heavy metals and organochlorine residues in the Bow and Old Man rivers, Alberta 1980-1981. *Sci. Ser. no. 138,* Ottawa, Canada Inland Water Directorate, Water Quality Branch.

Black, C. A., Evans, D. D., White, J. L., Ensminger, L. E., and Clarke, F. E. 1965. Methods of Soil Analysis, No. 9 in the Ser. Agronomy. Madison, Wisc.: American Society of Agronomy.

Bonzountas, M. 1983. Soil and groundwater fate modeling. In *Fate of Chemicals in the Environment,* eds. R. L. Swann and A. Eschenroeder, pp. 41-65. Washington, D.C.: American Chemical Society, Symp. Ser 225.

Bopp, R. F., Simpson, H. J., Olsen, C. R., and Kostyk, N. 1981. Polychlorinated biphenyls in sediments of the tidal Hudson River, New York. *Environ. Sci. Technol.* 15:210-216.

Bowie, G. L., Mills, W. B., Porcella, D. B., Campbell, C. L., Pagenkopf, J. R., Rupp, G. L., Johnson, K. M., Chan, P. W. H., Gherini, S. A., and Chamberlin, C. E. 1985. *Rates, Constants, and Kinetic Formulations in Surface Water Quality Modeling,* 2d ed., EPA/600/3-85/040. Athens, Ga.: Prepared for U.S. EPA, Environmental Research Laboratory.

Bowman, B. T., and Sans, W. W. 1985. Partitioning behavior of insecticides in soil-water systems. I. Adsorption concentration effects. *J. Environ. Qual.* 14:265-273.

Bradley, J.B., and McCutcheon, S. C. 1987. Influence on large suspended-sediment concentrations in rivers. In *Problems of Sediment Transport in Gravel-Bed Rivers,* ed. C. R. Thorne. New York: McGraw-Hill (in press).

Brady, N. C. 1974. *The Nature and Properties of Soil,* 8th ed. New York: Macmillan.

Brown, D. S., Carlton, R. E., and Wool, T. A. 1987. *MINTEQA1: An Equilibrium Metal Speciation Model: Users Manual.* EPA/600/3-87/012. Athens, Ga.: U.S. EPA, Environmental Research Laboratory.

Brownawell, B. J. 1986. *The Role of Colloidal Organic Matter in the Marine Geochemistry of PCB's.* Joint Mass. Inst. Technol./Woods Hole Oceanogr. Inst., Ph.D. Dissertation.

Buchak, E. M., and Edingers, J. E. 1984. *Generalized, Longitudinal-Vertical Hydrodynamics and Transport: Development, Programming and Applications*. Vicksburg, Miss.: Prepared for U.S. Army Corps Engineers, Waterways Experiment Station, June.

Burns, L. A. 1983. Fate of chemicals in aquatic systems: Process models and computer codes. In *Fate of Chemicals in the Environment*, eds. R. L. Swann and A. Eschenroeder, pp. 25–40. Washington, D.C.: American Chemical Society, Symp. Ser. 225.

Callahan, M. A., et al., 1979. *Water-Related Environmental Fate of 129 Priority Pollutants*, vol. I, *Introduction and Technical Background, Metals and Inorganics, Pesticides and PCBs*. Washington, D.C.: U.S. EPA, Office of Water Planning and Standards, Office of Water Waste Management.

Carigan, R., and Tessier, A. 1985. Zinc deposition in acid lakes: The role of diffusion. *Science* 228:1524–1526.

Carigan, R., and Nriagu, J. O. 1985. Trace metal deposition and mobility in the sediments of two lakes near Sudbury, Ontario. *Geochim. Cosmochim. Acta* 49:1753–1764.

Chapman, B. M., et al. 1982. Numerical simulation of the transport and speciation of nonconservative chemical reactants in rivers. *Water Resour. Res.* 18:155–167.

Chapman, B. M., et al. 1983. Processes controlling metal ion attenuation in acid mine drainage streams. *Geochim. Cosmochim. Acta* 47:1957–1973.

Chase, R. R. P. 1979. Settling behavior of natural aquatic particulates. *Limnol. Oceanogr.* 24:417–426.

Chiou, C. L., Malcolm, R. L., Brinton, T. I., and Kile, D. E. 1986. Water solubility enhancement of some organic pollutants and pesticides by dissolved humic and fulvic acids. *Environ. Sci. Technol.* 20:502–508.

Coates, D. 1985. Ph.D. Dissertation. Clemson University, 1985.

Connolly, J. P. 1980. *The Effect of Sediment Suspension on Adsorption and Fate of Kepone*. Ph.D. Dissertation. Austin: University of Texas at Austin.

Crittenden, J. C., Hutzler, N. J., Geyer, D. G., Oravitz, J. L., and Friedman, G. 1986. Transport of organic compounds with saturated groundwater flow: Model development and parameter sensitivity. *Water Resour. Res.* 22:271–284.

Davis, J. A., and Leckie, J. O. 1978. Surface ionization and complexation at the oxide/water interface. *J. Colloid Interface Sci.* 67:90–107.

Davis, J. A., and Leckie, J. O. 1979. Speciation of adsorbed ions at the oxide/water interface. In: *Chemical Modeling in Aqueous Systems*, ed. E. A. Jenne, pp. 299–317. Washington, D.C.: American Chemical Society, Symp. Ser. 93.

Davis, J. A., James, R. O., and Leckie, J. O. 1978. Surface ionization and complexation at the oxide/water interface. 1. Computation of electrical double layer properties in simple electrolytes. *J. Colloid Interface Sci.* 63:480–499.

Davison, W., Hilton, J., Lishman, J. P., and Pennington, W. 1985. Contemporary lake transport processes determined from sedimentary records of copper mining activity. *Environ. Sci. Technol.* 19:356–360.

DeLong, L. L. 1986. Extension of the unsteady one-dimensional open-channel flow equations for flow simulation in meandering channels with flood plains. U.S. Geological Survey, Water Supply Paper 2290, January.

Dhillon, S. K., et al. 1981. Copper adsorption by alkaline soils. *J. Soil Sci.* 32:571–578.

Di Toro, D. M. 1985. A particle interaction model of reversible organic chemical sorption. *Chemosphere* 14:1503–1538.

Di Toro, D. M., and Horzempa, L. M. 1982. Reversible and resistant compounds of PCB adsorption-desorption isotherms. *Environ. Sci. Technol.* 16:594–602.

Donigian, A. S., Lo, T. Y. R., and Shanahan, E. W. 1983. *Rapid Assessment of Potential Ground-Water Contamination under Emergency Response Conditions*. U.S. EPA, unpublished report.

Dzombak, D. A., and Morel, F. M. M. 1987. Adsorption of inorganic pollutants in aquatic systems. *ASCE J. Hydraulic Eng.* 113:430–475.

Einstein, H. A. 1950. The Bed Load Function for Sediment Transportation in Open Channel Flows. Washington, D.C.: U.S. Dept. Agric., Soil Conservation Service, Tech. Bull No. 1026.

Einstein, H. A., and Chien, N. 1955. Effects of heavy sediment concentrations near the bed on velocity and sediment distributions. Berkeley: Report to U.S. Army Corps of Engineers, Missouri River Division, Sediment Series no. 8, Univ. Calif.-Berkeley, Dept. Civil Eng.

Engelund, F., and Fredsoe, J. 1982. Sediment ripples and dunes. *Ann. Rev. Fluid Mech.* 64:1–16.

Felmy, A. R., et al. 1984a. *MINTEQ—A Computer Program for Calculating Aqeuous Geochemical Equilibria.* EPA/600/3-84/032. Athens, Ga.: U.S. EPA, Environmental Research Laboratory.

Felmy, A. R., et al. 1984b. *MEXAMS—The Metals Exposure Analysis Modeling System.* EPA/600/ 3-24/031. Athens, Ga.: U.S. EPA, Environmental Research Laboratory.

Fisher, J. B., Lick, W. J., McCall, P. L., and Robbins, J. A. 1980. Vertical mixing of lake sediments by turbificid oligochaetes. *J. Geophy. Res.* 85:3997–4006.

Fleischhauer, H. L., and Engelder, P. R. 1985. Procedures for reconnaissance stream-sediment sampling. Grand Junction, CO: Prepared for U.S. Dept. of Energy, Nuclear Energy Programs by Bendix Field Engineering Corp., February.

Förstner, U., and Salomons, W. 1983. Trace element speciation in surface waters: Interactions with particulate matter. In *Trace Element Speciation in Surface Waters and Its Ecological Implications,* ed. G. G. Leppard, pp. 245–270. New York: Plenum.

Fread, D. 1974. Numerical Properties of Numerical Four-Point Finite Difference Equations of Unsteady Flow. U.S. National Weather Service Technical Report HYDRO-13.

Gibbs, R. J. 1983. Effect of natural organic coating on the coagulation of particles. *Environ. Sci. Technol.* 17:237–239.

Gschwend, P. M., et al. 1986. *Modeling the Benthos-Water Column Exchange of Hydrophobic Chemicals.* Athens, Ga.: U.S. EPA Report, Environmental Research Laboratory.

Gschwend, P. M., and Wu, S.-C. 1985. On the constancy of sediment–water partition coefficients of hydrophobic organic pollutants. *Environ. Sci. Technol.* 19:90–96.

Guy, H. P., Simons, D. B., and Richardson, E. V. 1966. Summary of alluvial channel data from flume experiments, 1956–1961. *U.S. Geol. Surv. Prof. Pap.* 462-I.

Hamilton-Taylor, J. 1983. Heavy metal enrichments in the recent sediments of six lakes in Northwest England. *Environ. Technol. Lett.* 4:115–122.

Hassett, J. J., Means, J. C., Banwart, W. L., Wood, S. G., Ali, S., and Khan, A. 1980a. Sorption of dibenzothiophene by soils and sediments. *J. Environ. Qual.* 9:184–186.

Hassett, J. J., Means, J. C., Banwart, W. L., and Wood, S. G. 1980b. *Sorption Properties of Sediments and Energy-Related Pollutants.* U.S. EPA, 600/3-80-041.

Hayduk, W., and Laudie, H. 1974. Prediction of diffusion coefficients for non-electrolytes in dilute aqueous solutions. *AIChE J.* 20:611–615.

Hem, J. D. 1972. Chemistry and occurrence of cadmium and zinc in surface water and groundwater. *Water Resour. Res.* 8:661–679.

Hem, J. D. 1977. Reactions of metal ions at surfaces of hydrous iron oxide. *Geochim. Cosmochim. Acta* 41:527–538.

Hendricks, S. W., and Duratti, L. G. 1982. Derivation of an empirical sorption rate equation by analysis of experimental data. *Water Res.* 16:829–837.

Huang, C., and Stumm, W. 1973. Specific adsorption of cations on hydrous gamma alumina. *J. Colloid Interface Sci.* 43:409–420.

Huang, C. P., Elliott, H. A., and Ashmead, R. M. 1977. Interfacial reactions and the fate of heavy metals in soil-water systems. *J. Water Pollut. Control Fed.* 49:745–752.

James, R. O., and Parks, G. A. 1982. Characterization of aqueous colloids by their electrical double-layer and intrinsic surface chemical properties. *Surface Colloid Sci.* 12:119–216.

Jenne, E. A. 1979. Chemical modeling—Goals, problems, approaches, and priorities. In *Chemical Modeling in Aqueous Systems,* ed. E. A. Jenne, pp. 3–24. Washington, D.C.: American Chemical Society, Symp. Ser. 93.

Johanson, R. C., Imhoff, J. C. Kittle, J. L., and Donigian, A. S., Jr. 1984. *Hydrological Simulation Program—FORTRAN (HSPF): User's Manual for Release 8.0.* Athens, Ga.: U.S. EPA, Environmental Research Laboratory, EPA-600/3-84-D66.

Johnson, R. G. 1974. Particulate matter at the sediment-water interface in coastal environments. *J. Marine Res.* 32:313–333.

Karger, B. L., Snyder, L. R., and Hovath, C. 1973. *An Introduction to Separation Science.* New York: Wiley.

Karickhoff, S. W. 1980. Sorption kinetics of hydrophobic pollutants in natural sediments. In *Contaminants and Sediments,* ed. R. A. Baker, pp. 193–206. Ann Arbor, Mich.: Ann Arbor Science.

Karickhoff, S. W. 1984. Organic pollutant sorption in aquatic systems. *ASCE J. Hydraul. Eng.* 110:707–728.

Karickhoff, S. W., and Morris, K. R. 1985. Impact of turbificid oligochaetes on pollutant transport in bottom sediments. *Environ. Sci. Technol.* 19:51–56.

Kay, B. D., and Elrick, D. E. 1967. Adsorption and movement of lindane in soils. *Soil Sci.* 104:314–322.

Kinniburgh, D. G., and Jackson, M. L. 1981. Cation adsorption by hydrous metal oxides and clay. In *Adsorption of Inorganics at Solid-Liquid Interfaces,* eds. M. A. Anderson and A. J. Rubin, pp. 91–160. Ann Arbor, Mich.: Ann Arbor Science.

Knezovich, J. P., Harrison, F. L., and Wilhelm, R. G. 1987. The bioavailability of sediment-sorbed organic chemicals: A review. *Water Air Soil Pollut.* 32:233–245.

Krantzberg, G. 1985. The influence of bioturbation on physical, chemical and biological parameters in aquatic environments: A review. *Environ. Pollut. Ser. A* 39:99–122.

Lambe, T. W. 1951. *Soil Testing for Engineers.* New York: Wiley.

Laxen, D. P. H. 1985. Trace metal adsorption/coprecipitation on hydrous ferric oxide under realistic conditions. *Water Res.* 19:1229–1236.

Lee II, H., and Swartz, R. C. 1980. Biological processes affecting the distribution of pollutants in marine sediments. Part II. Bideposition and bioturbation. In *Contaminants and Sediments,* ed. R. A. Baker, pp. 555–606. Ann Arbor, Mich.: Ann Arbor Science.

Leenheer, J. A., and Ahlrichs, J. L. 1971. A kinetic and equilibrium study of the adsorption of carbaryl and parathion upon soil organic matter surfaces. *Soil Sci. Soc. J.* 35:700–704.

Leo, A. J. 1975. Calculation of partition coefficients useful in the evaluation of the relative hazards of various chemicals in the environment. In *Intern. Joint Commission Symposium on Structure Activity Correlations in Studies of Toxicity and Bioconcentration with Aquatic Organisms,* ed. G. D. Veith. Ontario: IJC.

Leo, A., Hansch, C., and Elkins, D. 1971. Partition coefficients and their uses. *Chem. Rev.* 71:525–616.

Lyman, W. J., Reehl, W. F., and Rosenblatt, K. H., 1982. *Handbook of Chemical Property Estimation Methods.* New York: McGraw-Hill.

Mackay, D. 1979. Finding fugacity feasible. *Environ. Sci. Technol.* 13:1218–1223.

Mackay, D., and Powers, B. 1987. Sorption of hydrophobic chemicals from water: A hypothesis for the mechanism of the particle concentration effect. *Chemosphere* 16:745–757.

Madsen, O. S., and Grant, W. D. 1976. *Sediment Transport in the Coastal Environment.* Ralph M Parsons Laboratory for Water Resour. Hydrodynamics, Report no. 209, Dept. Civil Eng., Mass. Inst. Technol.

Mathews, A. P., and Weber, W. J., Jr. 1976. Effects of external mass transfer and intraparticle diffusion on adsorption rates in slurry reactors. *AIChE Symp. Ser.* 166, 43:91–98.

McIlroy, L. M., DePinto, J. V., Young, T. C., and Martin, S. C. 1986. Partitioning of heavy metals to suspended solids of the Flint River, Michigan. *Environ. Tox. Chem.* 5:609–623.

Medine, A. J., and Bicknell, B. R. 1986. *Case Studies and Model Testing of the Metals Exposure Analysis Modeling System.* EPA/600/3-86/045. Athens, Ga.: U.S. EPA Environmental Research Laboratory.

Medine, A. J., and Porcella, D. B. 1980. Heavy Metal Effects on Photosynthesis/Respiration of Microecosystems Simulating Lake Powell, Utah/Arizona. In *Contaminants and Sediments*, ed. R. A. Baker, pp. 355–390. Ann Arbor, Mich.: Ann Arbor Science.

Medine, A. J., Porcella, D. B., and Adams, V. D. 1980. Heavy metal and nutrient effects on sediment oxygen demand in three-phase aquatic microcosms. In *Microcosms in Ecological Research*, ed. J. P. Giesy, pp. 279–303. Dep. Energy, Symp. Ser. 52, CONF 781101.

Medine, A. J., Lamarra, V. A., and Carter, J. C. 1983. Heavy metal distribution and interactions in the White River ecosystem. In: *Proc. of the 1983 National Conf. on Environmental Engineering.* New York: ASCE.

Mills, W. B., Porcella, D. B., Ungs, M. J., Gherini, S. A., Summers, K. V., Mok, L., Rupp, G. L., Bowie, G. L., and Haith, D. A. 1985. *Water Quality Assessment: A Screening Procedure for Toxic and Conventional Pollutants in Surface and Ground Water.* EPA/600/6-85/002a. Athens, Ga.: Prepared for U.S. EPA, Environmental Research Laboratory.

Morel, F. M. M. 1983. *Principles of Aquatic Chemistry.* New York: Wiley.

Nriagu, J. O., Wong, H. K. T., and Snodgrass, W. J. 1983. Historical records of metal pollution in sediments of Toronto and Hamilton harbours. *J. Great Lakes Res.* 9:365–373.

O'Connor, D. J., and Connolly, J. P. 1980. The effect of concentration of adsorbing solids on the partition coefficient. *Water Res.* 14:1517–1523.

Onishi, Y., and Wise, S. E. 1982. *Mathematical Model, SERATRA, for Sediment-Contaminant Transport in Rivers and its Application to Pesticide Transport in Four Mile and Wolf Creeks in Iowa.* EPA-600/3-82-045. Athens, Ga.: U.S. EPA Environmental Research Laboratory.

Passioura, J. B., Rose, D. A. 1971. Hydrodynamic dispersion in aggregated media. 2. Effects of velocity and aggregate size. *Soil Sci.* 111:345–351.

Plummer, L. N., et al. 1979. Critical review of the kinetics of calcite dissolution and precipitation. In *Chemical Modeling in Aqueous Systems*, ed. E. A. Jenne, pp. 537–573. Washington, D.C.: American Chemical Society, Symp. Ser. 93.

Rai, D., and Zachara, J. M. 1984. *Chemical Attenuation Rates, Coefficients, and Constants in Leachate Migration*, vol. 1, *A Critical Review.* Palo Alto, Calif.: Prepared for Environ. Physics Chem. Program, Energy Analysis and Environ. Div., Electric Power Research Institute, EA-3356.

Rao, P. S., Davidson, J. H., Jessup, R. E., and Selim, H. M. 1979. Evaluation of conceptual models for describing nonequilibrium adsorption-desorption of pesticides during steady flow in soils. *Soil Sci. Soc. Amer. J.* 43:22–28.

Rao, P. S. C., Jessup, R. E., Rolston, D. E., Davidson, J. M., and Dilcrease, D. P. 1980. Experimental and mathematical description of nonadsorbed solute transfer by diffusion in spherical aggregates. *Soil Sci. Soc. Amer.* 44:684–688.

Rao, P. S. C., Davidson, J. M., Berkheiser, V. E., Ou, L. T., Street, J. J., Wheeler, W. B., and Yuan, T. L. 1982. Retention and Transformation of Selected Pesticides and Phosphorus in Soil Water Systems: A Critical Review. EPA-600/3-82-045. Athens, GA.: U.S. EPA, Environmental Research Laboratory.

Reece, D. E., Felkey, J. R., and Wai, C. M. 1978. Heavy metal pollution in the sediments of the Coeur d'Alene River, Idaho. *Environ. Geol.* 2:289–293.

Rhoads, D. 1963. Rates of sediment reworking by *Yoldia limatula* in Buzzards Bay, Mass. and Long Island Sound. *J. Sediment. Petrol.* 33:723–727.

Rodi, W. 1980. *Turbulence Models and Their Application in Hydraulics.* Netherlands: Inter. Assn. Hydraulic Res., State of the Art Paper.

Saar, R. A., and Weber, J. H. 1982. Fulvic acid: Modifier of metal-ion chemistry. *Environ. Sci. Technol.* 16:510A-517A.

Salomons, W. 1985. Sediments and water quality. *Environ. Technol. Lett.* 6:315-326.

Salomons, W., and Förstner, U. 1984. *Metals in the Hydrocycle.* New York: Springer-Verlag.

Schindler, P. W., Furst, B., Dick, P., and Wolf, P. U. 1976. Ligand properties of surface silanol groups: I. Surface complex formation with Fe^{2+}, Cu^{2+}, Cd^{2+}, and Pb^{2+}. *J. Colloid Interface Sci.* 55:469-475.

Schink, D. R., and Guinasso, N. L., Jr. 1978. Redistribution of dissolved and adsorbed materials in abyssal marine sediments undergoing biological stirring. *Am. J. Sci.* 278:687-702.

Schwarzenbach, R. P., and Westall, J. 1981. Transport of nonpolar organic compounds from surface water to groundwater. Laboratory studies. *Environ. Sci. Technol.* 15:1360-1367.

Sheng, Y. P. 1986. Numerical modeling of estuarine hydrodynamics and dispersion of cohesive sediments. In *Sedimentation Control to Reduce Maintenance Dredging in Navigation Channels.* Washington, D.C.: National Academy of Science (in press).

Simons, D. B., and Sentürk, F. 1977. *Sediment Transport Technology.* Fort Collins, Colo.: Water Resour. Pub.

Sly, P. G. 1982. Introduction to second symposium on interactions between sediment and freshwater. *Hydrobiologia (Den.)* 91:1-8.

Steen, W. C., and Karickhoff, S. W. 1981. Biosorption of hydrophobic organic pollutants by mixed microbial populations. *Chemosphere* 10:27-32.

Stevenson, F. J. 1982. *Humus Chemistry, Genesis, Composition Reactions.* New York: Wiley.

Streeter, V. L., and Wylie, E. B. 1975. *Fluid Mechanics,* 6th ed. New York: McGraw-Hill.

Stumm, W., and Morgan, J. J. 1981. *Aquatic Chemistry.* New York: Wiley.

Sudo, Y. D., Misic, D. M., and Suzuki, M. 1978. Concentration dependence of effective surface diffusion coefficients in aqueous phase adsorption on activated carbon. *Chem. Eng. Sci.* 33:1287-1290.

Swakhorne, D., 1985. Ph.D. dissertation presented to Indiana University.

Swallow, K. C., Hume, D. N., and Morel, F. M. M. 1980. Sorption of copper and lead by hydrous ferric oxide. *Environ. Sci. Technol.* 14:1326-1331.

Terzaghi, K., and Peck, R. B. 1967. *Soil Mechanics in Engineering Practice,* 2d ed. New York: Wiley.

Tinsley, I. J. 1979. *Chemical Concepts in Pollutant Behavior.* New York: Wiley-Interscience.

Thibodeaux, L. J., and Boyle, J. D. 1987. Bedform-generated convective transport in bottom sediment. *Nature* 325:341-343.

Thomas, M., Petit, D., and Lamberts, L. 1984. Pond sediments as historical record of heavy metal fallout. *Water Air Soil Pollut.* 23:51-59.

Turk, J. T. 1980. Applications of Hudson River Basin PCB-transport studies. In *Contaminants and Sediments,* ed. R. A. Baker, pp. 171-183. Ann Arbor, Mich.: Ann Arbor Science.

Turk, J.T., and Troutman, B. 1981. U.S. Geological Survey, Water Resources Investigation.

Ullman, W. J., and Aller, R. C. 1982. Diffusion Coefficients in Nearshore Marine Sediments. *Limnol. Oceanogr.* 27:552-556.

van Genuchten, M. T., and Wierenga, P. T. 1976. Mass transfer studies in sorbing porous media. I. Analytical solutions. *Soil Sci. Soc. Amer. J.* 40:278-285.

Weast, R. C., ed. 1986. *CRC Handbook of Chemistry and Physics.* Boca Raton, Fla.: CRC Press.

Weber, W. J., Jr., and Rumer, R. R., Jr. 1965. Intraparticle transport of sulfonated alkylbenzenes in a porous solid: Diffusion with nonlinear adsorption. *Water Resour. Res.* 1:361-373.

Weber, W. J., Jr., and Liu, K. T. 1980. Determination of mass transport parameters for fixed bed adsorbers. *Chem. Eng. Comm.* 6:49-60.

Westall, J. C., et al. 1976. MINEQL: A Computer Program for the Calculation of Chemical

Equilibrium Composition of Aqueous Systems. Mass. Inst. Technol., Dept. of Civil Eng., Water Quality Laboratory, Tech. Note No. 18, 1976.

White, A. F., and Claassen, H. C. 1979. Dissolution kinetics of silicate rocks—Application to solute transport modeling. In *Chemical Modeling in Aqueous Systems,* ed. E. A. Jenne, pp. 447-473. Washington, D.C.: American Chemical Society, Symp. Ser. 93.

Wu, S.-C., and Gschwend, P. M. 1986. Sorption kinetics of hydrophobic organic compounds to natural sediments and soils. *Environ. Sci. Technol.* 20:717-725.

Zabawa, C. F. 1978. Microstructure of agglomerated suspended sediments in northern Chesapeake Bay estuary. *Science* 202:49-51.

Theory and Practice of the Development of a Practical Index of Hazardous Waste Incinerability

Barry Dellinger

Environmental Sciences, University of Dayton Research Institute, Dayton, Ohio

I. INTRODUCTION

The ultimate goal of incineration research is to provide insight into the various processes and technologies that will ensure the environmentally safe continuous operation of hazardous waste thermal disposal facilities. More specifically, the goal of this research should be focused on facility emissions, particularly in regard to a priori, quantitative prediction of partial combustion and pyrolysis effluents and how these effluents are affected by changes in operational and design parameters. However, this goal is not likely to be attained in less than 20 years and is not practical for addressing the immediate environmental problem of safely disposing of large quantities of solid and hazardous materials. Thus, a more realistic set of goals seems necessary.

Emissions of hazardous organic compounds fall into two general categories: those compounds in the waste feed that are not totally destroyed and those compounds formed from the partial degradation of the waste. Designations for

We gratefully acknowledge the contributions of Mr. Duvall, Mr. J. Graham, Mr. D. Hall, Ms. S. Mazer, Ms. J. Mescher, Mr. W. Rubey, Mr. J. Stalter, Dr. P. Taylor, and Mr. J. Torres who developed the instrumentation and generated the majority of the data presented in this review. We also acknowledge the support and guidance of U.S. EPA Project Officers Mr. R. Carnes, Mr. R. Mournighan, and Mr. M. Malanchuk as well as the assistance of numerous graduate and undergraduate students.

We acknowledge the financial support for portions of this work under U.S. EPA cooperative agreements CR-810783 and CR-807815 and contract 68-03-2979.

these classes have been borrowed from the regulatory designations of principal organic hazardous constituents (POHCs) and products of incomplete combustion (PICs). Since regulation of thermal disposal of wastes will always require some type of testing or monitoring of the actual facility, a desirable product of research would be information that can be used to reduce the testing burden and ensure that the proper emissions and operating parameters are being monitored.

The complexity of thermal destruction processes, the differences in combustor and pyrolyzer designs, and the difficulties in monitoring changing operating conditions, make the accurate prediction of *absolute* system performance essentially impossible. A more reasonable goal is the ability to predict the *relative* destruction efficiency of POHCs and the relative emission rate of PICs for a given facility. This is consistent with the goal of reducing the need for full-scale testing, since one could then simply conduct tests focusing on the least incinerable POHCs and the PICs of greatest yield predicted by laboratory testing and research (see the Appendix at the end of this chapter).

In addition, one must have sufficient knowledge of the effect of changes in operating and design parameters on POHC and PIC emissions to correctly define the appropriate conditions for the laboratory and field studies. Based on previous experience, laboratory and field testing under worst case conditions promises to be the best and most prudent mechanisms of evaluating potential environmental impact. These data may then be used to develop a method for continuous monitoring of thermal disposal facilities.

In summary, the broad-based acceptance of thermal disposal techniques as viable disposal methods is contingent upon obtaining satisfactory resolution to a number of issues. These issues, which are addressed in the following paragraphs, include:

- Can guidelines or rules be established for ranking the relative incinerability of POHCs?
- Is this ranking dependent on the composition of the waste or specific operational and design parameters of the disposal system?
- What PICs are formed during the thermal degradation of waste materials?
- Can PIC identity and relative yield be predicted for various waste compositions and system design and operating parameters?
- Can a workable regulatory strategy for the control of PIC emissions be established?
- Can a suitable surrogate for "incineration efficiency" be identified that can be used for continuous compliance monitoring?

II. DEVELOPMENT OF A SIMPLE THERMAL PROCESS MODEL

The first step in determining which chemical and physical parameters significantly affect POHC and PIC emissions was to develop a simple model for use in delineating major effects (Dellinger et al., 1986a). A discussion of the model follows.

When considering the thermal destruction of hazardous organic materials, chemical reactions occurring in condensed phases may effectively be neglected due to mass and heat transfer considerations. The primary concern is with gas-phase chemistry, although the nature of the passage of material from condensed phase into the gas phase by physical processes may have some impact on the overall decomposition process. Once in the gas phase, there exists two modes of destruction of the material. These modes may be designated as direct flame and thermal (nonflame).

Both flame model and thermal destruction studies indicate that any known organic compound can be destroyed in a hazardous waste incinerator to greater than 99.99% destruction efficiency (DE) if the incinerator is operating under theoretically optimum conditions (Dellinger et al., 1984a; Seeker et al., 1984; Wall, 1972). Thermal destruction can be expected at less than 1000°C in flowing air at a mean residence time of 2.0 s. Flame destruction of waste material occurs in flames operating well in excess of 850°C. The fact that these conditions roughly correspond to the mean conditions experienced in an incinerator has caused much confusion. The observation of organic emissions from incinerators is proof that frequent excursions from the optimum (or even the mean) conditions are occurring.

Excursions, or fault modes, are probably the controlling phenomena for incineration efficiency. Four parameters—atomization inefficiency, mixing inefficiency, thermal failure, and quenching—have been identified as failure modes in flames (Seeker et al., 1984). Laboratory studies have shown that relatively small excursions from ideality for these parameters can easily drop measured flame destruction efficiencies from greater than 99.99% to 99% or even less than 90% (three orders of magnitude). Nonflame upset parameters can be related to distributions of oxygen, residence time, and temperature (Dellinger, 1984a; Seeker et al., 1984; Wall, 1972; Graham et al., 1986).

The key to understanding this model is that only a very small fraction of the total volume of the waste needs to experience inadequate conditions to result in significant deviations from the desired destruction efficiencies. To illustrate how laboratory thermal decomposition testing relates to upset modes and can be used to predict observed emissions from full-scale facilities, let us explore a specific example.

Previous research in our laboratories has shown that the destruction kinetics of typical hazardous organic compounds can be described using simple pseudo-first-order kinetics (Dellinger et al., 1984a). Although different or more complex models may be used, the actual model used is not important for the scope of this discussion.

We first examine the case of a simple one-stage combustor where a waste feed mixture is fed directly into a turbulent flame and the hot gases evolving from the flame pass on through a relatively long, high-temperature thermal zone prior to exiting the system. Representative reaction conditions for the flame can be chosen as an average resident time of 0.1 s and a bulk flame temperature of 1700 K. For the postflame zone, we may choose a mean residence time of 2.0 s and a bulk gas phase temperature of 1100 K. Although a range of residence times and temperatures are actually experienced by the individual molecules, the values are chosen are typical effective residence times and temperatures.

As discussed in the previous paragraphs, several destruction failure modes have been identified for the flame. In this model we assume that only 1% of the waste feed escapes the flame undecomposed. This 1% of the waste feed then enters the postflame zone. The overall measured destruction at the stack is the weighted average of the destruction efficiencies of the flame and postflame zones. The results of these calculations for hazardous waste of a range of thermal stabilities are shown in Table 1. From examination of the table it is apparent that each of the compounds is destroyed to essentially the same efficiency in the flame, i.e., $>99.999\%$. In the postflame region significant differences in thermal stability are observed.

As last column of Table 1 shows, the overall destruction efficiency parallels the destruction efficiency in the postflame region. The principal value of the

TABLE 1

Calculated Destruction Efficiencies (DEs) for Representative Hazardous Organics

			Calculated DEs		
Compound	A^a (s^{-1})	E_a (kcal/mole)	(Flame)	(Postflame)	(Overall)
Acetonitrile	$4.7 \ 10^7$	40	99.999+	66.357	99.664
Benzene	$2.8 \ 10^8$	38	99.999+	99.999+	99.999+
Chloroform	$2.9 \ 10^{12}$	49	99.999+	99.999+	99.999+
Tetrachlorobenzene	$1.9 \ 10^6$	30	99.999+	98.556	99.986
Tetrachloroethylene	$2.6 \ 10^6$	33	99.999+	77.127	99.771
Trichlorobenzene	$2.2 \ 10^8$	38	99.999+	99.968	99.999+

$^a A$ is the frequency factor and E_a is the activation energy for a reaction as defined by the Arrhenius equation $k = A \exp(-E_a/RT)$, where, in this case, k is the pseudo-first-order rate constant for the reaction.

overall DE is 99% in all cases, with the variations in DE occurring to the right of the decimal. The destruction achieved in the flame determines the principal value, while the nonflame destruction efficiency determines the approach to complete destruction.

The discussion of the impact of system upsets can be carried a step further to include failure modes in the postflame zone. Poor mixing of waste and oxygen in the afterburner gives rise to a certain fraction of the waste being subjected only to low oxygen conditions. Numerous laboratory studies have shown that destruction of the feed material is much slower under these conditions, and product formation is enhanced. We again have the case where although most of the waste is oxidized and destroyed, the small fraction of the feed experiencing the pyrolytic conditions may be responsible for the emissions.

Although the conclusion that a subfraction of the waste feed is responsible for most hazardous organic emissions may be surprising at first, the same process is generally responsible for emission of most air pollutants. In a power plant for example, one is not really concerned with the major chemistry, i.e., formation of carbon dioxide and water, but instead with the minor reaction pathways which form sulfur dioxide, sulfuric acid, and nitrogen oxides. These pathways are responsible for less than 0.1–1% of the total stack emissions but are the reactions of interest in pollutant formation.

The applicability of this model has recently been confirmed by a more complex model of hazardous waste incineration developed by EER Corp. (Clark et al., 1983, 1985). This model includes considerations of furnace heat transfer, flow, mixing, injection, tracking, and kinetics. Pseudo-first-order thermal decomposition kinetics were used as inputs for the model. Thus far, modeling results for three pilot-scale hazardous waste thermal destruction systems have been reported. These systems are the controlled temperature tower (CTT), the U.S. EPA's Combustion Research Facility's (CRF) rotary kiln system, and the Acurex subscale boiler. The CTT was modeled under several modes of operation and failure modes including standard, cooled, insulated, backheated, fast quench, and various droplet vaporization points. The CRF system was modeled for varying loads, different excess air levels, and kiln or afterburner flameout. The Acurex subscale boiler was modeled for various fuel heating values, heat removal rates, excess air rates, waterwall/nonwaterwall modes, various droplet vaporization points, and temperature profiles.

In each reported case the predicted relative destruction efficiencies correlated almost perfectly with prediction of laboratory thermal decomposition testing as ranked by the temperature required for 99.99% DE in an atmosphere of flowing air at 1.0 s gas-phase residence time (Dellinger et al., 1984a). For the CTT, the agreement was essentially perfect for every case. For the six test compounds modeled for the CRF, only methane exhibited a moderate deviation from the behavior predicted by purely pseudo-first-order postflame kinetics.

For the Acurex boiler, of the eight compounds modeled, only acetonitrile showed significant deviation (Fig. 1).

The excellent agreement between the ranking according to $T_{99.99}(1)$ (i.e., temperature at which 99.99% destruction occurs for 1 s gas phase residence time) and this model are as predicted by the two-zone incineration model, illustrating the importance of postflame reaction kinetics. Although quantitative predictions are available from this model, accurate predictions for complex incineration systems will require additional model development and refinement. However, the significance of postflame chemistry in controlling relative POHC DEs has been clearly demonstrated.

III. COMPARISON OF FLAME AND THERMAL DECOMPOSITION RESULTS

The flow reactor systems at the University of Dayton Research Institute (UDRI) have been used to generate thermal decomposition data on nearly 100 different hazardous organic compounds. The experimental difficulties in generating similar flame data have resulted in a very limited data base.

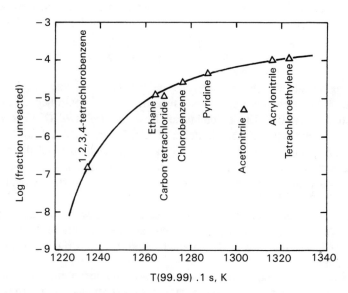

FIG. 1. Comparison of EER Corp. model prediction with predictions of University of Dayton Research Institute (UDRI) two-zone incineration model for the Acurex subscale boiler. Results indicate the control of overall relative destruction efficiencies of test compounds by postflame chemical kinetics.

However, a recently reported study has furnished some data for comparison (Vandell and Shadoff, 1984).

Thirteen compounds of interest from the viewpoint of hazardous waste incineration were combusted in a laboratory diffusion flame. The relative burning rates of these compounds were determined based on their flame front velocities. A list of these compounds and their rankings based on nonflame thermal degradation studies is shown in Table 2. For the six compounds for which thermal decomposition data are available, the nonflame rankings are indicated. Furthermore, the flame mode rankings for the remaining compounds are basically as one would predict for the thermal degradation of untested compounds.

In a second flame experiment, various combinations of dichlorobenzene, benzene, and hydrogen chloride (HCl) were combusted at 40% of stoichiometric air. The identity and yields of the products were found to be essentially invariant as long as the ratios of chlorine, hydrogen, carbon, and oxygen were constant. The observed PICs are listed in Table 3.

We have recently completed a study of PIC formation from the thermal decomposition of a mixture of carbon tetrachloride, toluene, chlorobenzene, trichloroethylene, and Freon 113 (DuPont) (Graham et al., 1986). Those PICs resulting from this mixture that were also found in the flame combustion of chlorobenzene are noted in Table 3.

The agreements between relative POHC stability and PIC production for flame and nonflame studies are striking, particularly for PIC production. Most of the differences in observed PICs are the lack of higher chlorinated compounds from the thermal degradation study. This is probably due to the fact that the chlorine content of the thermal degradation mixtures was only 6 mol % while it was 50 mol % for the flame study, the latter favoring formation of higher chlorinated species. The similarity in results suggests that similar reactions are occurring which are responsible for POHC DE and PIC formation for both the flame and thermal decomposition studies.

It is well documented that hydrocarbon reactions proceed by mechanisms based primarily on attack by molecular species at low temperature (Benson, 1976). At moderate temperatures (250–450 °C) a peroxide-dominated mechanism appears to be active, and at higher temperatures, free-radical mechanisms predominate. The "knee" in the thermal decomposition profiles generated on our laboratory thermal reaction systems denotes the region of transition from a relatively slow to a much faster reaction mechanism, i.e., transition from a peroxide to a free-radical mechanism (Fig. 2). We have performed detailed kinetic calculations for propane which indicate a rapid increase in the concentrations of the free-radical pool, predominately OH, O, and H, in the temperature range of the knee. We have also performed pseudo-equilibrium calculations for other more complex molecules, which also

TABLE 2

Comparison of Flame and Nonflame (Thermal) Stability, Ranking
of Various Test Compounds

Compound	Relative burning rate (flame)	UDRI[a] thermal stability ranking (nonflame)
1,2,4-Trichlorobenzene	10.9	1[b]
m-Dichlorobenzene	13.5	2
o-Dichlorobenzene	12.6	3
1,6-Dichlorohexane	25.6	
Chlorobenzene	28.4	4
1-Chlorohexane	34.7	
Benzene	60.0	5
Dichloroisopropylether	87	
1,2-Dichloropropane	219	
n-Hexane	736	
1,1,-Trichloroethane	844	6
Epichlorohydrin	1142	
1,2-Dichloroethane	1500	

[a]UDRI, University of Dayton Research Institute (Dellinger et al., 1984a).
[b]Ranking of 1 is most stable.

demonstrate a rapid increase in radical concentration in this region. This temperature range of 500–700 °C is also appropriate for unimolecular decomposition reactions to become significant.

The kinetic modeling of propane using an extended, purely gas-phase model, which has been validated against shock tube studies, has been shown to be in good agreement with oxidation results generated on our thermal decomposition unit-gas chromatographic (TDU-GC) flow reactor system (Fig. 3). The agreement between this purely gas-phase kinetic model is especially good in predicting $T_{99.99}(2)$. Some deviation is observed at lower temperatures as peroxide-dominated mechanisms become more important.

One might expect different free-radical pathways to dominate in the very high-temperature regions of the flame. Thus, the demonstrated correlation between flame mode and flow reactor POHC and PIC data further suggests that free-radical reactions occurring in cooler regions of the flame or the postflame zone are responsible for relative POHC destruction efficiencies and PIC formation. The marked agreement in PIC identities, even for dissimilar feed mixtures, further illustrates the importance of the free-radical mechanism. The majority of the products are due to radical-molecule disproportionation reactions or radical addition to aromatic substrates. The lack of oxygen-containing products even under oxidative conditions suggests that

TABLE 3

PICs Found in Diffusion Flame Combustion of Chlorobenzene, Benzene, and HCl Mixture and Thermal Decomposition of a Mixture of Carbon Tetrachloride, Toluene, Chlorobenzene, Trichloroethylene, and Freon 113 (DuPont)

PICs from flame-mode combustion	PICs from thermal decomposition
Anthracene	x
Benzofuran	x
Biphenyl	x
Biphenylene or acenaphthalene	
Chloroacetylene	
Chloroanthracene	
Chlorobenzene	x
Chlorobiphenyl	
Chlorobiphenylene	
Chloronaphthalene	x
Chlorophenylacetylene	x
Chlorostyrene	x
Chlorotoluene	x
Dibenzofuran	x
Dichloroacetylene	x
Dichloroanthracene	
Dichlorobenzene	x
Dichlorobiphenyl	
Dichloromethylstyrene	x
Dichloronaphthalene	x
Dichlorostyrene	x
Fluoranthene	x
Methylnaphthalene	x
Naphthalene (or azulene)	x
Phenol	x
Phenylacetylene	x
Phenylnaphthalene	x
Pyrene	x
Styrene	x
Tetrachlorobiphenyl	
Trichlorobenzene	
Trichlorobiphenyl	
Toluene	x

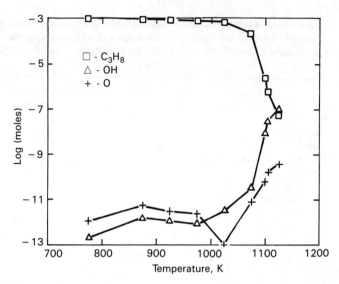

FIG. 2. Concentration vs. temperature for propane oxidation in air at a gas-phase residence time of 2.0 s.

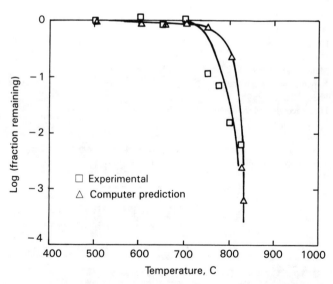

FIG. 3. Comparison of flow reactor generated thermal decomposition profile vs. predicted results from computer modeled gas-phase free radical mechanism for propane oxidation.

abstraction of H by OH and O dominate over addition reactions. Alternatively, addition products such as phenols may be very reactive and rapidly undergo further degradation.

IV. CORRELATION OF LABORATORY PREDICTIONS AND FIELD RESULTS

The ultimate test of the utility of laboratory research is the degree of correlation between experimental or theoretical predictions and actual field results. In an effort to predict the results of full-scale incineration of organic compounds, various researchers have proposed six experimental and theoretical methods (Dellinger et al., 1984a; Seeker et al., 1984; Graham et al., 1986; Crumpler et al., 1981; Cudahy and Troxler, 1983; Tsang and Shaub, 1982; Miller et al., 1984; Lee et al., 1982). These methods for ranking relative compound "incinerability" are:

1. Heat of combustion ($\Delta H_c/g$)
2. Autoignition temperature (AIT)
3. Theoretical flame model kinetics (TFMK)
4. Experimental flame failure modes (EFFM)
5. Ignition delay time (IDT)
6. Gas-phase thermal stability [$T_{99}(2)$, $TSHiO_2$, $TSLoO_2$]

A comparison of predictions which result from application of these methods with actual field results follows (Dellinger et al., 1986b).

The observed incinerability rankings of the test compounds at each source were compared with the prediction of each proposed hierarchy using a rank/order correlation approach (Dickson and Massy, 1957). This method was judged superior to a linear regression analysis since the latter judges the agreement of the data with a best-fit straight line, while the former simply determines if a statistically significant relationship exists between the observed and predicted rankings. The rank/order correlation coefficient (r_s) was used to judge if a correlation existed at the 90% confidence level for a number of test compounds, N.

We have previously proposed the gas-phase thermal stability method based on the results of flow reactor studies. One method of constructing such a ranking is to use laboratory determined thermal stability specified by the temperature required for 99% [$T_{99}(2)$] or 99.99% [$T_{99.99}(2)$] destruction at 2.0 s reactor residence time (Dellinger et al., 1984a; Lee et al., 1982). This scale was originally developed for pure compounds in flowing air. However, recently generated data have shown that relative stability varies as a function of the composition of the waste feed and oxygen concentration (Graham et al., 1986; Taylor and Dellinger, 1986a, 1986b). This has led to modification of the rankings to account for the thermal stability of individual POHCs fed as a mixture in both an oxygen rich ($TSHiO_2$) and an oxygen deficient ($TSLoO_2$) environment. These 3 hierarchies, along with the predictions of the other 5, were applied to predicting results of 10 field studies.

Results of this analysis are summarized in Table 4 for 10 studies judged to meet strict data validation criteria (MRI, 1984; Dellinger et al., 1984b; Wolbach and Garman, 1984; Wyss et al., 1984; RTI/ES, 1984; Kodak Park Division, 1984; R. Bastian, personal communication). Of the eight proposed ranking methods, only $\Delta H_c/g$, AIT, $T_{99}(2)$, $TSHiO_2$, and $TSLoO_2$ had a sufficient data base to make predictions for a significant number of sources. Of these, only the experimentally predicted order under low oxygen conditions, $TSLoO_2$ met with a reasonable success, i.e., 70% (Fig. 4). The other four methods only correlated with field observations 10–20% of the time. More importantly, it was apparent after detailed examination of the individual data plots that certain trends were occurring that could not be explained by the application of single ranking methods. In particular, the compounds that deviated in stability from predictions of the $TSLoO_2$ hierarchy were often the same for the various studies. In many cases this deviation could be explained using other available information.

The degree of success, as indicated by the results reported in Table 4, of predicting the relative thermal stabilities of hazardous organics through laboratory flow reactor studies may appear somewhat surprising considering the complexity of the incineration process. However, the development of the two-zone incineration model illustrates how postflame chemistry controls incinerator emissions and is sufficient to explain general agreement between laboratory-based predictions and field results. However, none of the previously presented incinerability hierarchies directly address the issue of PIC emissions, as they are only concerned with thermal stability of the POHCs in the feed material.

As previously discussed, PICs resulting from the incineration of hazardous wastes are not currently regulated by the U.S. EPA. However, the field data and results of other laboratory, pilot, and full-scale testing programs have shown that toxic products can be formed and are emitted from incinerators (Graham et al., 1986; Vandell and Shadoff, 1984; Taylor and Dellinger, 1986a, 1986b; MRI, 1984; Dellinger et al., 1984b; Wolbach and Garman, 1984; Wyss et al., 1984; RTI/ES, 1984; Kodak Park Division, 1984; Cooke et al., 1983; Hall et al., 1983; R. Bastian, personal communication). Many observed PICs are also potential POHCs; consequently, it is entirely possible that a PIC may also be a POHC selected from the original mixture. Three documented examples are the formation of carbon tetrachloride from chloroform and hexachloro-

TABLE 4

Results of Statistical Analyses of Observed vs. Predicted Thermal Stabilities Rankings

Study	ΔH$_c$/g	AIT	TFMK	EFFM	IDT	T$_{99}$(2)	TSHiO$_2$	TSLoO$_2$
				Hierarchy[a]				
A	-0.300/5[a]	-0.200/4	-	-	-	-	0.000/5	0.900/5*
B	-0.190/8	0.200/4	-	-	-	-0.057/6	0.533/10*	0.529/10*
C	-0.500/5	-	-	-	-	0.500/5	0.400/5	0.600/5
D	-0.100/9	-0.060/	-	-	-	-0.800/4	0.386/9	0.493/9*
E	0.589/7*	0.428/6	-	-	-	-0.300/5	0.425/8	0.429/8
F	0.343/15	0.571/7*	-0.100/5	-	-	-0.425/9	0.041/15	0.073/15
G	0.400/4	-	-	-	-	0.800/4*	0.800/4*	0.900/4*
H	-0.333/7	0.457/6	-	-	-0.300/4	-0.161/7	-0.036	0.655/8*
I	-0.077/10	-0.262/8	0.600/4	0.060/4	-0.100/5	-0.217/9	-0.318/11	0.536/11*
J	-0.291/10	0.147/8	0.800/4*	0.060/4	-0.100/5	-0.202/9	-0.114/11	0.523/11*
No. of successes*	1	1	1	0	0	1	2	7
No. of failures	9	8	2	2	3	8	8	3
Percent success*	10	11	33	0	0	11	20	70

[a] r_s/N, where N is the number of data points (i.e., compounds ranked). Asterisk (*) indicates that correlation was statistically significant at the 90% confidence level.

305

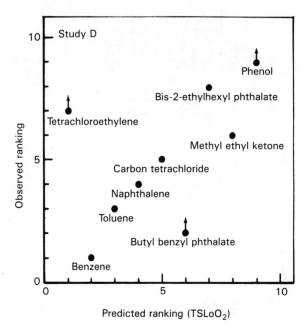

FIG. 4. Correlation between observed and predicted stability for site D. Ranking of 1 is most stable and 10 is least stable. Arrows denote detection limited ranking.

benzene and benzene from chlorobenzene or toluene (Graham et al., 1986; Dellinger et al., 1984b; Hall et al., 1983).

Detailed examination of the previously presented field results identified many such cases. The result is a low apparent destruction and removal efficiency (DRE) for the POHC. Since this effect would be more important when the input concentration of the POHC is low, the result would be a dependence of DRE on input POHC concentration (i.e., the higher the input concentration, the greater the apparent DRE). The true effect, however, is that the emission concentration would be constant if the emissions are due to product formation from other waste components.

The observation of an apparent DRE dependence on concentration has been made for hazardous waste incinerators and attributed to greater than first-order kinetics for individual POHCs (MRI, 1984). While such an effect could be possible for combustion of a pure compound, it is highly improbable when the POHC is only a small portion of a complex waste. The reaction chemistry is determined by the overall waste and fuel composition as opposed to pure compound kinetics. Volatile POHCs in the ambient air as a result of fugitive emissions, volatile POHCs stripped from scrubber waters, and outgassing of phthalate-containing materials would also give rise to apparent concentration dependencies since their emissions levels would be constant while the POHC input rate varies. Specifically, it was observed from the field results that most of the observed deviations from laboratory predicted rankings of incinerability could be attributed to product formation or "contamination" of the stack effluent by volatile POHCs that did not pass through the incinerator (Fig. 5).

While predicting POHC stability is difficult enough, it is obviously desirable to also predict product formation. This is best accomplished by laboratory thermal decomposition testing of the actual waste stream to be incinerated, or a very close simulation. As indicated by the agreement of lab predictions based on LoO_2 conditions, these studies should be conducted under pyrolytic conditions.

An excellent example of this approach is study C (Fig. 6). The incinerability ranking based purely on POHC DRE was successful for four out of the five constituents of the waste, only benzene being apparently more stable than the other components. However, our laboratory has tested a very similar waste stream and (under pyrolytic conditions) significant levels of benzene were observed. Thus, when product formation is included, lab testing of a simulated waste stream would correctly predict the observed field results.

The results of comparison of 10 field studies with thermal stability predictions indicate that no ranking based on *pure* compound properties can provide an appropriate scale of incinerability. However, a ranking based on predicted POHC stability in complex mixtures under low oxygen conditions gave a statistically significant correlation with field results in 7 of 10 cases. More importantly, analysis of results gives strong reason to believe that formation of POHCs in the incineration process may be responsible for their observed DREs.

Pending further confirmatory comparisons with field results, the following comments are offered:

- Relative POHC DREs are determined by postflame thermal reactions.
- Pyrolysis pathways are responsible for most POHC and PIC emissions.
- DREs of very stable or difficult-to-form POHCs are *not* PIC dependent.
- Many POHC DREs are due to PIC formation from other POHCs.
- DREs of many volatile POHCs are controlled by external sources.
- Oxygen-starved atmospheres produce more PICs.
- "Pyrolysis" products dominate even under oxygen-rich conditions.
- Fragile POHCs produce more PICs.
- Most PICs are more complex than the parent POHCs.
- Many PICs persist at higher temperatures than their parent.
- PICs may also be POHCs.
- PICs may form even more stable PICs.

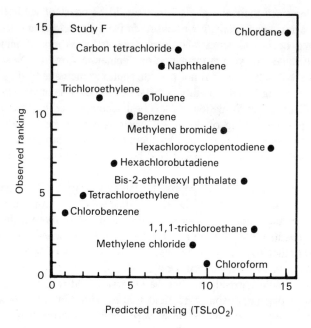

FIG. 5. Correlation between observed and predicted stability for study F. Methylene bromide, bis-2-ethylhexyl phthalate, 1,1,1-trichloroethane, methylene chloride, and chloroform are observed to be more stable than predicted. This is attributed to outgassing of these compounds from contaminated scrubber water or plastics (i.e., phthalates) by hot flue gases.

V. ESTABLISHMENT OF A DATA BASE OF POHC STABILITY

Since 1972 the Environmental Sciences Group at the University of Dayton Research Institute has been deeply involved in research into thermal decomposition of organic materials and the disposal of hazardous organic substances by incineration. Designing state-of-the-art thermal instrumentation systems to accomplish this research has been an integral part of past programs. These systems have been and are currently being used to generate thermal decomposition data on a variety of hazardous organic compounds. A description of these instrumentation systems is presented in the Appendix. The emphasis of past thermal decomposition work has been to determine the relative thermal stability of the parent materials, identify toxic products of incomplete combustion, and identify parameters affecting thermal degradation. Table 5 summarizes some of the thermal decomposition data obtained during the course of previous studies.

High-temperature thermal decomposition kinetic data on 23 pure compounds have been generated in our laboratory. Table 6 is a complete listing of compounds for which pseudo-first-order oxidation kinetic parameters have been determined. The table presents activation energies and frequency factors E_a and A for the compounds ranked by calculated $T_{99}(2)$. The theoretical formalism and experimental design for these studies are available from other reports to which the reader is referred for additional information (Dellinger, 1984a). Kinetic data from these studies for many chemicals (i.e., methane, ethane, benzene, and chloroform) compared favorably to other available high-temperature kinetic data obtained by different techniques (Lee et al., 1982; Dryer and Glassman, 1973; Glassman et al., 1975; Seshadri and Williams, 1975; Fujii and Asaba, 1977; Shilov and Sabirova, 1960).

Although the information presented in Tables 5 and 6 may be used as a guide to determine the thermal stability of the listed materials when combusted as pure compounds, the previously discussed laboratory/field comparison indicates that mixture pyrolysis date is a much better predictor of incinerator emissions. Table 7 summarizes a preliminary ranking of the relative stability of some hazardous organic compounds as defined in 40 Code of Federal Regulations (CFR), Part 261, Appendix VIII. This ranking is limited and more research is clearly necessary to expand this list to make it useful for most applications.

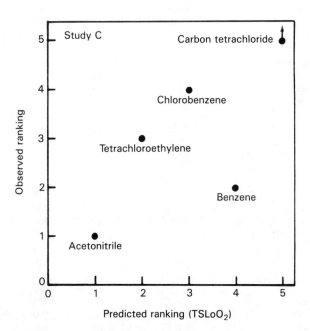

FIG. 6. Correlation between observed and predicted stability for study C. Benzene's apparent stability may be due to its formation as a PIC from combustion of auxiliary fuel.

TABLE 5

Summary of Laboratory-Generated Thermal Decomposition Data[a]

Compound	Empirical formula	Mean residence time (s)	Onset of thermal decomposition (°C)	Temperature for 99% destruction (°C)
Acetonitrile	C_2H_3N	2.0	760	900
Acrylonitrile	C_3H_3N	2.0	650	830
Aniline	C_6H_7N	2.0	620	730
Anthracene[b]	$C_{14}H_{10}$	2.0	400	710
Aroclor 1221	$C_{12}H_{8.8}Cl_{1.2}$	1.0	640	~740
Aroclor 1232	$C_{12}H_{8.1}Cl_{1.9}$	1.0	640	~740
Aroclor 1242	$C_{12}H_{6.9}Cl_{3.1}$	1.0	640	~740
Aroclor 1248	$C_{12}H_{6.1}Cl_{3.9}$	1.0	640	~740
Aroclor 1254	$C_{12}H_5Cl_5$	1.0	640	~740
Aroclor 1260	$C_{12}H_{3.7}Cl_{6.3}$	1.0	640	~740
Aroclor 1262	$C_{12}H_{3.2}Cl_{6.8}$	1.0	640	~740
Azobenzene[c]	$C_{12}H_{10}N_2$	2.0	450	590
Azobenzene[d]	$C_{12}H_{10}N_2$	2.0	475	595
Benzene	C_6H_6	2.0	630	730
Benzo[e]pyrene[b]	$C_{20}H_{10}$	2.0	400	720
Biphenyl	$C_{12}H_{10}$	1.0	590	700
Butyl benzyl phthalate[c]	$C_{19}H_{20}O_4$	2.0	275	395
Butyl benzyl phthalate[d]	$C_{19}H_{20}O_4$	2.0	350	500
Carbon tetrachloride	CCl_4	2.0	600	750
Carbon tetrachloride[e]	CCl_4	2.0	500	670
Carbon tetrachloride[f]	CCl_4	2.0	500	680
Carbon tetrachloride[g]	CCl_4	2.0	500	680
Carbon tetrachloride[h]	CCl_4	2.0	500	790
Carbon tetrachloride[i]	CCl_4	2.0	500	705
Carbon tetrachloride[j]	CCl_4	2.0	450	760
Carbon tetrachloride[k]	CCl_4	2.0	500	880
Chlorobenzene	C_6H_5Cl	2.0	540	700
Chlorobenzene[e]	C_6H_5Cl	2.0	550	730
Chlorobenzene[f]	C_6H_5Cl	2.0	700	>1000
Chlorobenzene[g]	C_6H_5Cl	2.0	700	800
2-Chlorodibenzo-p-dioxin	$C_{12}H_7O_2Cl$	1.0	660	840
2-Chlorodibenzo-p-dioxin[l]	$C_{12}H_7O_2Cl_3$	1.0	730	940
Chloroform	$CHCl_3$	2.0	410	590
Chloroform[h]	$CHCl_3$	2.0	500	680
Chloroform[i]	$CHCl_3$	2.0	540	800
Chloromethane	CH_3Cl	2.0	720	805
Chloromethane[h]	CH_3Cl	2.0	500	900
Chloromethane[i]	CH_3Cl	2.0	500	1100

TABLE 5 *(Continued)*

Compound	Empirical formula	Mean residence time (s)	Onset of thermal decomposition (°C)	Temperature for 99% destruction (°C)
o,p-DDE	$C_{14}H_8Cl_4$	1.0	560	790
p,p-DDE	$C_{14}H_8Cl_4$	1.0	560	790
o,p'-DDT	$C_{14}H_9Cl_5$	1.0	200	480
p,p'-DDT	$C_{14}H_9Cl_5$	1.0	200	480
Decachlorobiphenyl	$C_{12}Cl_{10}$	1.0	590	780
Trans-Decalin[m]	$C_{10}H_{18}$	2.0	710	835
Diazinon	$C_{12}H_{21}N_2O_3PS$	1.0	<280	560
Dibenzofuran	$C_{12}H_8O$	1.0	590	700
Dibenzo-*p*-dioxin	$C_{12}H_8O_2$	1.0	660	840
Dibenzo-*p*-dioxin[l]	$C_{12}H_8O_2$	1.0	730	940
2,2'-Dibromobiphenyl	$C_{12}H_8Br_2$	2.0	670	800
1,2-Dichlorobenzene	$C_6H_4Cl_2$	2.0	630	740
2,6-Dichlorobiphenyl	$C_{12}H_4Cl_2$	2.0	670	800
2,8-Dichlorodibenzofuran	$C_{12}H_6OCl_2$	2.0	670	800
2,7-Dichlorodibenzo-*p*-dioxin	$C_{12}H_6O_2Cl_2$	1.0	660	840
2,7-Dichlorodibenzo-*p*-dioxin	$C_{12}H_6O_2Cl_2$	2.0	670	800
2,7-Dichlorodibenzo-*p*-dioxin[l]	$C_{12}H_6O_2Cl_2$	1.0	730	940
Dichloromethane	CH_2Cl_2	2.0	650	755
Dichloromethane[h]	CH_2Cl_2	2.0	500	760
Dichloromethane[i]	CH_2Cl_2	2.0	500	840
Diphenylamine[c]	$C_{12}H_{11}N$	2.0	400	615
Diphenylamine[d]	$C_{12}H_{11}N$	2.0	600	785
Endrin	$C_{12}H_8Cl_6O$	1.0	<275	<400
Ethane	C_2H_6	2.0	500	735
Freon 113	$C_2F_3Cl_3$	2.0	650	780
Freon 113[e]	$C_2F_3Cl_3$	2.0	600	770
Freon 113[f]	$C_2F_3Cl_3$	2.0	625	780
Freon 113[g]	$C_2F_3Cl_3$	2.0	650	780
Halowax 1001	$C_{10}H_{4.5}Cl_{3.5}$	2.0	650	770
Halowax 1001[m]	$C_{10}H_{4.5}Cl_{3.5}$	2.0	770	1000
Halowax 1001[n]	$C_{10}H_{4.5}Cl_{3.5}$	2.0	520	730
Heptabromobiphenyl	$C_{12}H_3Br_7$	2.0	660	~800
Heptachloronaphthalene	$C_{10}HCl_7$	2.0	510	~760
2,2',4,4',5,5'-Hexabromobiphenyl	$C_{12}H_3Br_6$	2.0	660	800
Hexachlorobenzene	C_6H_6	2.0	650	820
2,2',4,4',5,5'-Hexachlorobiphenyl	$C_{12}H_4Cl_6$	1.0	660	730

See footnotes on p. 313.

TABLE 5 *(Continued)*

Compound	Empirical formula	Mean residence time (s)	Onset of thermal decomposition (°C)	Temperature for 99% destruction (°C)
Hexachlorobutadiene	C_4Cl_6	2.0	−620	750
Hexachlorocyclopentadiene	C_5Cl_6	2.0	340	650
Hexachloroethane	C_2Cl_6	2.0	470	600
Kepone	$C_9Cl_{10}O$	1.0	330	470
Methane	CH_4	2.0	660	830
Methylcyclohexane[m]	C_7H_{14}	2.0	710	850
Mirex	$C_{10}Cl_{12}$	1.0	500	680
Naphthalene[m]	$C_{10}H_8$	2.0	720	830
Nitrobenzene	$C_6H_5NO_2$	2.0	570	670
Nonylphenol[o]	$C_{15}H_{24}O$	2.0	350	480
Nonylphenol[p]	$C_{15}H_{24}O$	2.0	350	565
Octachlorocyclopentene	C_5Cl_8	2.0	290	460
Pentachlorobenzene	C_6H_5Cl	2.0	680	810
Pentabromobiphenyl	$C_{12}H_5Br_5$	2.0	660	~800
2,5,2′,4′,5′-Pentachlorobiphenyl	$C_{12}H_5Cl_5$	2.0	660	740
2,3′4,4′5-Pentachlorobiphenyl[c]	$C_{12}H_5Cl_5$	2.0	575	760
2,3′,4,4′5-Pentachlorobiphenyl[d]	$C_{12}H_5Cl_5$	2.0	650	875
2,5,2′,4′,5′-Pentachlorobiphenyl[m]	$C_{12}H_5Cl_5$	2.0	750	970
2,5,2′,4′,5′-Pentachlorobiphenyl[q]	$C_{12}H_5Cl_5$	2.0	610	730
2,5,2′,4′,5′-Pentachlorobiphenyl[r]	$C_{12}H_5Cl_5$	2.0	700	820
Pentachloronitrobenzene	$C_6Cl_5NO_2$	2.0	490	590
Pentachlorophenol	C_6HOCl_5	2.0	475	635
Pentachlorophenol[m]	C_6HOCl_5	2.0	525	645
Phenol	C_6H_6O	2.0	500	725
Phenol[m]	C_6H_6O	2.0	500	740
Pyrene[d]	$C_{16}H_{10}$	2.0	525	720
Pyrene[r]	$C_{16}H_{10}$	2.0	600	>1000
Pyridine	C_6H_5N	2.0	620	770
Sulfur hexafluoride[j]	SF_6	2.0	~300	>1050
Sulfur hexafluoride[k]	SF_6	2.0	~300	>1050
1,2,3,4-Tetrachlorobenzene	$C_6H_2Cl_4$	2.0	660	800
1,2,3,5-Tetrachlorobenzene	$C_6H_2Cl_4$	2.0	610	760
2,5,2′,5′-Tetrachlorobiphenyl	$C_{12}H_6Cl_4$	2.0	620	740
Tetrachloroethylene	C_2Cl_4	2.0	660	850
Tetrachloroethylene[j]	C_2Cl_4	2.0	500	780
Tetrachloroethylene[k]	C_2Cl_4	2.0	600	920
1,2,3,4-Tetrachloronaphthalene	$C_{10}H_4Cl_4$	2.0	610	760
1,2,3,4-Tetrachloronaphthalene[m]	$C_{10}H_4Cl_4$	2.0	770	970
1,2,3,4-Tetrachloronaphthalene[n]	$C_{10}H_4Cl_4$	2.0	770	970
Tetrahydrodecyclopentadiene[m]	$C_{10}H_{16}$	2.0	655	820

TABLE 5 *(Continued)*

Compound	Empirical formula	Mean residence time (s)	Onset of thermal decomposition (°C)	Temperature for 99% destruction (°C)
Toluene	C_7H_8	2.0	600	680
Toluene[e]	C_7H_8	2.0	525	670
Toluene[f]	C_7H_8	2.0	550	820
Toluene[g]	C_7H_8	2.0	680	820
Toluene[i]	C_7H_8	2.0	500	690
Toluene[k]	C_7H_8	2.0	550	830
Toluene[m]	C_7H_8	2.0	8a40	1040
Triallate	$C_{10}H_{16}NSOCl_3$	2.0	360	470
1,2,4-Trichlorobenzene	$C_6H_3Cl_3$	2.0	640	750
1,2,4-Trichlorobenzene[i]	$C_6H_3Cl_3$	2.0	500	730
1,2,4-Trichlorobenzene[j]	$C_6H_3Cl_3$	2.0	600	890
Trichlorodibenzo-*p*-dioxin	$C_{12}H_5O_2Cl_3$	1.0	660	840
Trichlorodibenzo-*p*-dioxin[k]	$C_{12}H_5O_2Cl_3$	1.0	730	940
1,1,1-Trichloroethane	$C_2H_3Cl_3$	2.0	390	570
Trichloroethylene	C_2HCl_3	2.0	600	780
Trichloroethylene[c]	C_2HCl_3	2.0	550	730
Trichloroethylene[f]	C_2HCl_3	2.0	620	920
Trichloroethylene[g]	C_2HCl_3	2.0	610	790
2,4,5-Trichlorophenol[m]	$C_6H_3OCl_3$	2.0	575	775
Trifluralin	$C_{13}H_{16}N_3O_4F_3$	2.0	360	440
Triphenylene	$C_{18}H_{12}$	2.0	400	720

[a] All data generated as a pure compound in flowing air atmosphere unless otherwise noted.
[b] In polynuclear aromatic hydrocarbon (PNA) mixture.
[c] In mixture in air.
[d] In mixture in N_2.
[e] Five-component mixture #1 in air.
[f] Five-component mixture #1 in N_2.
[g] Five-component mixture, 0% excess air.
[h] Chloromethane mixture in air.
[i] Chloromethane mixture in N^2.
[j] Five-component mixture #2 in air.
[k] Component mixture #2 in N_2.
[l] In helium.
[m] In N_2.
[n] In O_2.
[o] Technical grade mixture in air.
[p] Technical grade mixture in N_2.
[q] O_2 (40%) in N_2.
[r] O_2 (2.5%) in N_2.

VI. FORMATION OF PRODUCTS OF INCOMPLETE COMBUSTION

In a recent laboratory study, the thermal degradation of a mixture of five hazardous organic compounds, Hazardous Waste Mixture #1 (HWM-1), was investigated under a variety of conditions (Graham et al., 1986). The mixture was studied in three reaction atmospheres: oxygen-starved, stoichiometric oxygen, and oxygen-rich. The behavior of the components in the mixture was compared to their behavior when tested as pure compounds and the thermal reaction products were identified. Thermal decomposition behavior was analyzed and related to elementary chemical reaction kinetics. The observed reaction products are presented in Table 8.

Benzaldehyde, phenol, and benzofuran were the only observed *oxidation*

TABLE 6

Summary of First-Order Kinetic Results

Compound	A (s^{-1})	E_a (kcal/mole)	Temperature range (°C)	Calculated $T_{99}(2)$ (°C)
Trichloroethylene	4.2×10^3	18	600–700	913
Acrylonitrile	1.3×10^6	31	750–810	910
Acetonitrile	4.7×10^7	40	800–850	908
Tetrachloroethylene	2.6×10^6	33	725–825	900
Methane	3.5×10^9	48	700–800	874
Hexachlorobenzene	2.5×10^8	41	710–785	845
1,2,3,4-Tetrachlorobenzene	1.9×10^6	30	700–765	834
Ethane	1.3×10^5	24	675–725	830
Carbon tetrachloride	2.8×10^5	26	680–730	824
Monochlorobenzene	8.0×10^4	23	600–670	810
Dichloromethane	3.0×10^{13}	64	700–755	796
1,2,4-Trichlorobenzene	2.2×10^8	39	675–725	789
Pyridine	1.1×10^5	24	700–750	767
1,2-Dichlorobenzene	3.0×10^8	39	685–725	766
Hexachlorobutadiene	6.3×10^{12}	24	700–750	763
Benzene	2.8×10^3	38	685–715	757
Aniline	9.3×10^{15}	71	650–700	726
Nitrobenzene	1.4×10^{15}	64	600–650	672
Hexachloroethane	1.9×10^7	29	500–600	641
Chloroform	2.9×10^{12}	49	520–585	606
1,1,1-Trichloroethane	1.9×10^8	32	475–550	601
Triallate	6.8×10^8	31	360–460	516
Trifluralin	2.7×10^7	25	360–430	483

TABLE 7

**Preliminary Incinerability Ranking of Selected Compounds
of Environmental Significance[a] (U.S. EPA Appendix VIII Compounds)**

Compound	Ranking[b]
Acetonitrile	1
Aniline	9
Benzene	8
Carbon tetrachloride	12
Chlordane	24
Chlorobenzene	4
Chloroform	15
Dichlorobenzene	3
1,2-Dichloroethane	20
Freon 113	19
Hexachlorobutadiene	7
Hexachlorocyclopentadiene	23
Methyl chloride	13
Methyl ethyl ketone	17
Methylene bromide	16
Methylene chloride	14
Naphthalene	11
Phenol	18
Tetrachloroethylene	5
Toluene	10
Trichlorobenzene	2
1,1,1-Trichloroethane	21
1,1,2-Trichloroethane	22
Trichloroethylene	6

[a]Only includes compounds for which lab and field data are available.
[b]Ranking of 1 is most stable or difficult to incinerate.

products under oxidative or pyrolytic conditions while numerous complex pyrolysis-type products were observed. This indicates that most products result from radical-molecule reactions and recombination of radical fragments. OH and O addition products are not significant. The lack of addition products suggests that OH and O may be more likely to participate in abstraction reactions at high temperatures or that the intermediate addition products are not very stable. This is clearly an area for further research.

We have conducted several studies of the formation of PICs from various industrial or U.S. EPA Appendix VIII compounds and mixtures. Table 9 summarizes the results for the majority of these studies.

TABLE 8

Thermal Reaction Products Observed from Thermal Decomposition of HWM-1

Formula	Compound	$\phi = 0.06^a$	$\phi = 1.0$	Pyrolysis
$CHCl_3$	Trichloromethane	x		
$C_2H_3ClF_2$	Chlorodifluoroethane		x	
$C_2H_2Cl_2$	1,1-Dichloroethene	x	x	x
C_2Cl_3F	Trichlorofluoroethene		x	x
C_2Cl_4	Tetrachloroethene		x	
$C_4H_4Cl_2$	Dichlorobutadiene		x	
C_6H_6	Benzene	x	x	x
C_6H_6	1,5-Hexadiyne or 1,5-Hexadien-3-yne	x	x	x
C_6H_6O	Phenol	x	x	x
C_6H_5F	Fluorobenzene		x	x
C_6H_5ClO	Chlorophenol	x	x	
$C_6H_4Cl_2$	Dichlorohexadiyne or Dichlorohexadiene-yne		x	
$C_6H_4Cl_2$	Dichlorobenzene	x	x	
C_7H_8O	Methylphenol	x	x	
C_7H_7Cl	Chloromethylbenzene	x	x	x
$C_7H_6O_2$	Hydroxybenzaldehyde or Benzodioxol	x		
C_7H_6O	Benzaldehyde	x	x	x
C_8H_{10}	Ethylbenzene	x	x	x
C_8H_8	Ethenylbenzene (Styrene)	x	x	x
C_8H_7Cl	Chloroethenylbenzene	x	x	x
C_8H_6	Ethenylbenzene	x	x	x
$C_8H_6Cl_2$	Dichloroethenylbenzene or Chloroethenylchlorobenzene	x	x	x
C_8H_6ClF	Chlorofluoroethenylbenzene		x	x
$C_8H_6F_2$	Difluoroethenylbenzene		x	x
C_8H_6O	Benzofuran	x	x	
C_8H_5Cl	Chloroethenylbenzene			x
$C_8H_5Cl_3$	Trichloroethenylbenzene	x		
C_8H_5ClO	Chlorobenzofuran	x	x	x
$C_8H_5F_3$	?			x
C_9H_8	1H-Indene		x	x
$C_9H_8Cl_2$	Dichloropropenylbenzene Chloropropenylchlorobenzene	x		x
C_9H_8O	Phenylpropenal	x		
$C_9H_7Cl_3$	?		x	x
C_9H_7ClO	Chloromethylbenzofuran		x	
C_9H_6O	Phenylpropynone	x		
$C_{10}H_8$	Azulene or Naphthalene or Methylene-1H-indene		x	x

TABLE 8 *(Continued)*

Formula	Compound	$\phi = 0.06^a$	$\phi = 1.0$	Pyrolysis
$C_{10}H_8ClF$	?			x
$C_{10}H_7Cl$	Chloronaphthalene			x
$C_{10}H_6Cl_2$	Dichloronaphthalene	x	x	x
$C_{10}H_5Cl_3$	Trichloronaphthalene	x		
$C_{11}H_{10}$	1-Methylnaphthalene		x	
$C_{11}H_{10}$	2-Methylnaphthalene		x	
$C_{11}H_{10}$	Methylnaphthalene			x
$C_{12}H_{10}$	Biphenyl		x	x
$C_{12}H_8O$	Dibenzofuran	x	x	
$C_{13}H_{12}$	1,1-Methylenebisbenzene			x
$C_{13}H_{10}$	9H-Fluorene		x	x
$C_{13}H_{10}O$	Diphenylmethanone	x		
$C_{13}H_8O_2$	9H-Xanthen-9-one	x	x	
$C_{14}H_{14}$	1,1'-(1,2-Ethanediyl)bisbenzene	x	x	x
$C_{14}H_{12}$	Methylfluoroene			x
$C_{14}H_{12}$	1,1'-(1,2-Ethenediyl)bis-(E)-benzene		x	
$C_{14}H_{12}$	Dihydrophenanthrene 1,1'-(1,2-Ethynediyl)bis-(Z)-benzene	x		x
$C_{14}H_{10}$	1,1'-(1,2-Ethynediyl) bis-benzene or Phenanthrene or 9-Methylene-9H-fluoroene	x	x	x
$C_{14}H_{10}O$	Anthracenone or Penanthrenol	x	x	
$C_{14}H_9F$	Fluoro-1,1'-(1,2-ethynediyl) bis-benzene or Fluorophenanthrene or Fluoro-methylene-9H-fluoroene			x
$C_{15}H_{12}$	?		x	
$C_{15}H_2$	Methylanthracene or Methylphenanthrene or 2-Phenyl-1H-indene or 9-Ethylidene-9H-fluoroene			x
$C_{16}H_{12}$	1,4-Dihydro-1,4-etheno-anthracene or 1-Phenylnaphthalene or 5-Methylene-5H-dibenzo-[a,d]-cycloheptene		x	x
$C_{16}H_{12}$	2-Phenylnaphthalene		x	x
$C_{16}H_{11}F$	Fluorophenylnaphthalene or			x

See footnotes on p. 318.

TABLE 8 *(Continued)*

Formula	Compound	$\phi = 0.06^a$	$\phi = 1.0$	Pyrolysis
	Fluoro-5-methylene-5*H*-dibenzo [*a,d*]cycloheptene or Fluoro-1,4,-dihydro-1,4- ethenoanthracene			
$C_{16}H_{11}F$	Fluorophenylnaphthalene			
$C_{16}H_{10}$	Pyrene or Fluoroanthene			x
$C_{16}H_{10}ClF$	?			x
$C_{16}H_9F$	Fluoropyrene or Fluorofluoroanthene			x
$C_{17}H_{12}$	11H-Benzo[*a*]fluorene or 11H-Benzo[*b*]fluorene			x

$^a\Phi$ is the fuel/air equivalence ratio.

VII. ESTABLISHING POHC/PIC EMISSIONS LIMITS BASED ON LABORATORY THERMAL DECOMPOSITION TESTING

A key question to all concerned with thermal disposal of hazardous wastes is, how does one ensure that the level of emissions of all toxic organic compounds is sufficiently low to provide protection of human health and the environment? The objective of the following discussion is to develop a method to assist in controlling POHC *and* PIC emissions at an environmentally safe level by establishing relative risk factors based on *laboratory* thermal decomposition research.

Under current regulatory approach, it is assumed that a safe level of organic emissions is achieved by requiring a DRE of 99.99% [99.999% for polychlorinated biphenyls (PCBs), polychlorinated dibenzo-*p*-dioxins (PCDDs), and polychlorinated dibenzoForans (PCDFs)] for a few POHCs in the waste judged to be very stable. Extension of this approach to controlling PIC emissions presents several problems. First of all, one must determine which potential PIC emissions should be tested for during full-scale trial burns. The lack of a comprehensive data base on PIC yield, frequency of formation, and stability makes this especially difficult. Secondly, how does one establish an emissions limit for PICs, since the DRE approach used for POHCs is inapplicable to PICs?

The proposed alternative is based on establishing mass emissions limitations for both POHCs and PICs. These mass emissions limitations would ultimately be set based on allowable ground level concentrations, which in turn are based on available toxicity data. Site specific, stack mass emissions limits could be best set based on back calculations from site-specific dispersion modeling. Alternatively, mass emissions limitations could be set for any source based on

TABLE 9

Summary of Reaction Products Observed from Thermal Decomposition of Various Materials

Parent	Product	Conditions
Kepone	Hexachlorobenzene Hexachlorocyclopentadiene Hexachloroindenone	Air atmosphere
o,p'-DDT	o,'p,-DDT p,p'-DDE o,p'-DDE	Air atmosphere
Mirex	Hexachlorocyclopentadiene Hexachlorobenzene	Air atmosphere
Polyacrylonitrile, polymethacrylonitrile	Acrylonitrile Methacrylonitrile Acetonitrile Hydrogen cyanide Styrene Methyl acrylate	Air and helium atmospheres
Freon 113	Trichlorofluoroethene Tetrachlorodifluoroethane	Air atmosphere, $\bar{t}_r^a = 2.0$ s
Carbon tetrachloride	Tetrachloroethene Hexachloroethane Hexachlorobutadiene	Air atmosphere, $\bar{t}_r = 2.0$ s
Toluene	1,5-Hexadiyne or 1,5-Hexadien-3-yne Benzene Ethylbenzene Ethenylbenzene Benzaldehyde Benzofuran $C_{10}H_8$	Air atmosphere, $\bar{t}_r = 2.0$ s
Chlorobenzene	Benzene Ethynylbenzene Benzaldehyde 1H-Indene $C_{10}H_8$	Air atmosphere, $\bar{t}_r = 2.0$ s
Pentachloronitrobenzene	Hexachlorobenzene	Air atmosphere, $\bar{t}_r = 2.0$ s
Chloroform	CCl_4 1,2-$C_2H_2Cl_2$ C_2HCl_3	$\phi^b = 0.67$, $\bar{t}_r = 2.0$ s

See footnotes on p. 325.

319

TABLE 9 *(Continued)*

Parent	Product	Conditions
	C_2Cl_4	
	C_2HCl_5	
	C_2Cl_2	
	$C_2H_2Cl_4$	
	C_3Cl_4	
	C_4Cl_6	
	C_6Cl_6	
Chloroform	Carbon tetrachloride	ϕ = 0.76 and nitrogen
	Trichloroethene	atmospheres, $\bar{t}_r$ = 2.0 s
	Pentachloroethane	
	Dichloroethyne	
	Tetrachloroethene	
	Tetrachloropropyne	
	1,1,2,4-Tetrachloro-1-buten-3-yne	
	Hexachlorobutadiene	
Pentachloroethane	Chloroform	ϕ = 0.76 and nitrogen
	Carbon tetrachloride	atmospheres, $\bar{t}_r$ = 2.0 s
	Dichloroethane	
	Tetrachloroethane	
	Trichloroethene	
	Dichloroethyne	
	Tetrachloroethene	
	Hexachloroethane	
	Dichloropropane	
	Tetrachloropropyne	
	Hexachlorobutene	
1,2,4-Trichlorobenzene	Chlorobenzene	Air, nitrogen, and ϕ = 1.0
	o-,*m-*, and *p*-Dichlorobenzene	atmospheres, $\bar{t}_r$ = 2.0 s
	Trichlorobenzene isomers	
2,2′,5,5′,-Tetrachlorobiphenyl	Trichlorobenzene	Air atmosphere, $\bar{t}_r$ = 2.0 s
	Biphenyl	
	Tetrachlorobenzene	
	Monochlorobiphenyl	
	Chlorinated compound, MW 204	
	Dichlorobiphenyl	
	Chlorinated compound, MW 230	
	+ (2 isomers)	
	Trichlorobiphenyl (2 isomers)	
	Dichlorodibenzofuran	
	Tetrachlorobiphenyl (2 isomers)	
	Pentachlorobiphenyl (2 isomers)	
	Trichlorodibenzofuran	

TABLE 9 *(Continued)*

Parent	Product	Conditions
2,2′,4,5,5′-Pentachlorobiphenyl	Trichlorobenzene isomer Tetrachlorobenzene (2 isomers) Monochlorobiphenyl Trichlorobiphenyl Pentachlorobenzene Trichlorobiphenyl Tetrachlorobiphenyl Pentachlorobiphenyl Trichlorodibenzofuran Hexachlorobenzene Chlorinated compound, MW 264 + (3 isomers) Tetrachlorodibenzofuran (2 isomers) Hexachlorobiphenyl	Air atmosphere, $\bar{t}_r = 2.0$ s

2,2′,4,4′,5,5′-
  Hexachlorobiphenyl
	Tetrachlorobenzene
	Pentachlorobenzene
	Trichlorobiphenyl
	Hexachlorobenzene
	Hexachlorobiphenyl
	Heptachlorobiphenyl
	Pentachlorodibenzofuran
	Chlorinated compound, MW
	  288 +
		Air atmosphere, $\bar{t}_r = 2.0$ s

2,3′,4,4′,5-Pentachlorobiphenyl	Tetrachlorodibenzofurans Trichlorodibenzofurans Pentachlorodibenzofuran Tetrachlorobiphenyls Trichlorobiphenyl Trichlorobenzene Dichlorobenzene Trichloronaphthalene Tetrachloronaphthalene Trichlorophenylethynes Dichlorophenylethyne Tetrachlorobiphenylenes C_9H_8OCl $C_{10}H_3Cl_3$	$\phi = 0.05, 0.2, 1.0, 3.0$
Pentachlorophenol	Dichlorobutadiyne Tetrachloroethene	Air and nitrogen atmospheres, $\bar{t}_r = 2.0$ s

See footnotes on p. 325.

TABLE 9 *(Continued)*

Parent	Product	Conditions
	Tetrachloropropyne Trichlorofuran 1,1,2,4-Tetrachloro-1-buten-3-yne Tetrachlorofuran Trichlorobenzene Tetrachlorobenzene Pentachlorobenzene Hexachlorobenzene Octachlorostyrene Hexachlorodihydronaphthalene Octachloronaphthalene Octachlorodibenzo-p-dioxin	
Mixture of 2,6-dichlorobiphenyl and 2,2′-dibromobiphenyl	Dibenzofuran Fluorenone Xantheonone	Air atmosphere, $\bar{t}_r = 2.0$ s
Hex wastes (industrial mixture)	Hexachlorobenzene Hexachloroindenone	Air atmosphere, $\bar{t}_r = 2.0$ s
Mixture of methanol, dichloro- methane, acetone, 1,1,1- trichloroethane, acentonitrile, chloroform, methylethyl ke- tone (MEK), 1,2- dichloroethane, toluene, diox- ane, and pyridine	Methyl vinyl ketone Benzaldehyde Cyanobutadiene Acrylonitrile Ethylbenzene Chloroacetone Styrene Acetic acid Benzofuran Benzodioxide Benzylchloride Phenol Dichloroethylene Chlorophenol Cresol	Air atmosphere, $\bar{t}_r = 2.0$ s
Mixture of CCl_4 (53%), $CHCl_3$ (33%), CH_2Cl_2 (7%), and CH_3Cl (7%)	CCl_4 $CHCl_3$ CH_2Cl_2 CH_3Cl $1,1$-$C_2H_2Cl_2$ C_2HCl_3 C_2Cl_4 C_2Cl_6	$\phi = 0.05$, $\bar{t}_r = 2.0$ s

TABLE 9 *(Continued)*

Parent	Product	Conditions
Mixture of CCl_4 (53%), $CHCl_3$ (33%), CH_2Cl_2 (7%), and CH_3Cl (7%)	CCl_4 $CHCl_3$ CH_2Cl_2 CH_3Cl C_2Cl_2 $1,1\text{-}C_2H_2Cl_2$ C_2HCl_3 C_2Cl_4 C_2Cl_6 C_3Cl_4 C_4Cl_4 C_4Cl_6 C_6Cl_6 C_8Cl_8	Pyrolytic, $\bar{t}_r = 2.0$ s
Mixture of CCl_4 (10%), $CHCl_3$ (20%), CH_2Cl_2 (30%), and CH_3Cl (40%)	CCl_4 $CHCl_3$ CH_2Cl_2 CH_3Cl $1,1\text{-}C_2H_2Cl_2$ $1,2\text{-}C_2H_2Cl_2$ $C_2H_4Cl_2$ C_2HCl_3 C_2Cl_4	$\phi = 0.05$, $\bar{t}_r = 2.0$ s
Mixture of CCl_4 (10%), $CHCl_3$ (20%), CH_2Cl_2 (30%), and CH_3Cl (40%)	CCl_4 $CHCl_3$ CH_2Cl_2 CH_3Cl C_2Cl_2 $1,1\text{-}C_2H_2Cl_2$ $1,2\text{-}C_2H_2Cl_2$ $C_2H_4Cl_2$ C_2HCl_3 C_2Cl_4 C_3Cl_4	Pyrolytic, $\bar{t}_r = 2.0$ s
Mixture of heptachlor, pentachlorophenol, diphenylnitrosamine, and pyrene	9H-Carbazole Chlorocyclopentadiene Tetrachlorofuran Trichlorodihydronaphthalene Dichloroacetylene Tetrachloropropyne Pentachloronaphthalene	Air and N_2 atmospheres, $\bar{t}_r$ $= 2.0$ s

See footnotes on p. 325.

TABLE 9 *(Continued)*

Parent	Product	Conditions
	Biphenyldicarbonitrile Diphenylamine or aminobiphenyl Hexachlorodihydronaphthalene Tetrachlorobenzene Pentachlorobenzene Trichloronaphthalene Tetrachloronaphthalene Benzene Acenaphthylene or biphenylene Hexachloromethanocyclohexene Benzonitrile Naphthalene or azulene	
Mixture of butyl benzyl phthalate, 2,3′,4,4′,5-pentachlorobiphenyl, and azobenzene	Toluene Benzaldehyde Methylphenol or benzyl alcohol Benzene Styrene Phenol Unknown	Air and N_2 atmospheres, $\bar{t}_r$ $= 2.0$ s
Nonlyphenol (technical grade)	C_9-Alkene or C_9-cycloalkane Methylbenzofuran Phenol Benzofuran C_3-Alkenylphenol or hydroxyldihydroindene or benzopyran or propenyloxybenzene C_4-Alkenylphenol or hydroxymethyldihydroindene or methylbenzopyran or butenyloxybenzene Naphthalenol C_5-Alkenylphenol or isomer Toluene C_3-Alkenylbenzene Benzene C_3-Benzene 1*H*-Indene Dihydrobenzofuran or hydroxystyrene Phenylacetylene C_6-Alkadienylbenzene	Air and N_2 atmospheres, $\bar{t}_r$ $= 2.0$ s

TABLE 9 *(Continued)*

Parent	Product	Conditions
	C_7-Alkadienylbenzene	
	Ethylbenzene	
	Xylene	
	Styrene	
	Methyl-1H-indene	
	Naphthalene or azulene	
	C_7-Diyne-ene or C_7-cyclodiyne or	
	C_7-cyclodiene-yne	
	C_6-Alkenylphenol or isomer	
	C_7-Alkenylphenol or isomer	
	Phthalic anhydride	
	Cyclohexadiene or C_6-ene-yne	
	Acenaphthylene or biphenylene	
	Biphenyl or acenaphthene	
	Unknown(s)	

^{a} $\bar{t}_r$ is the mean gas-phase residence time.
^{b} ϕ is the fuel/air equivalence ratio.

dispersion modeling for a representative (or worst case) source or source class (e.g., hazardous waste incinerators, municipal incinerators, boilers, kilns).

Regulating mass emissions means that POHCs and PICs can be addressed on a consistent basis. To simplify the permit process, one would like to identify the POHCs and PICs representing the greatest health risk and then conduct stack testing for these compounds only. Factors to consider for POHCs are (1) toxicity, (2) POHC mass feed rate, and (3) POHC stability. For PICs, one must consider (1) toxicity, (2) total waste feed rate, (3) PIC yield, and (4) PIC stability. If a relative risk (*RR*) scale for POHCs and PICs can be developed that addresses all these factors, then one can select with confidence a limited number of species for trial burn testing.

Simple mathematical expressions may be defined to express the relative health risks of POHCs and PICs. Simply stated, Eqs. (1) and (2) define a toxicity-weighted ranking of the mass emission rates of toxic organics.

$$RR_i(\text{POHC}) = M_i \times RT_i \times RS_i \qquad (1)$$

where $RR_i(\text{POHC})$ = relative risk for POHC$_i$
M_i = mass feed rate for POHC$_i$
RT_i = relative toxicity
RS_i = relative stability

$$RR_i(\text{PIC}) = RT_i \times \Sigma_j (M_j \times RY_j^i) \qquad (2)$$

where $RR_i(\text{PIC})$ = relative risk for PIC_i
   $\Sigma_j M_j$ = total mass feed rate of all precursors to PIC_i in the waste
   RY_j^i = is the relative yield of PIC_i from precursor j
The relative toxicity (RT) factor may be based on various testing techniques, a detailed review of which is beyond the scope of this discussion. One possible approach is to use the dose required for 50% mortality for inhalation by rats. This leads to the relation

$$RT_i = \frac{\text{LD}_{50_{MT}}}{\text{LD}_{50_i}} \qquad (3)$$

where LD_{50_i} = lethal dose required for a 50% mortality for rat inhalation studies
   $\text{LD}_{50_{MT}}$ = same parameter for the most toxic U.S. EPA Appendix VIII compound
Note that the LD_{50} is inversely proportional to toxicity. The most toxic U.S. EPA Appendix VIII compound would receive a RT_i of 1.0.

Accurate determinations of RS_i for POHCs and RY_i^j for PICs are clearly crucial factors in the success of this approach. First it should be noted that the RR for POHCs is based on POHC feed rate while the RR for PICs is based on total feed rate of all precursors.

A promising method for determining the RS_i and RY_i for POHCs and PICs is laboratory thermal decomposition testing, discussed previously. Currently available results suggest that testing of simple mixtures (e.g., five components) of U.S. EPA Appendix VIII compounds will achieve the desired results. As an example, let us use the data available from the study involving HWM 1.

Figure 7 presents the thermal decomposition profiles for five U.S. EPA Appendix VIII POHCs originally present in the mixture as well as the formation/destruction profile for PICs listed in the U.S. EPA Appendix VIII. These data were obtained in a nitrogen (pyrolytic) atmosphere at a gas-phase residence time of 2.0 s. The ordinate for the POHC profiles is weight percent remaining (DE = 100 − weight percent remaining) and is a chromatographic response for the PIC profiles. The latter scale could be converted to mass by calibration with appropriate response standards. These curves clearly contain considerable information on the relative stability of POHCs and PICs as well as PIC yield. The question is how to best convert this information to a ranking of POHC stability and PIC yield.

A suitable approach is to select a reference temperature at which to evaluate

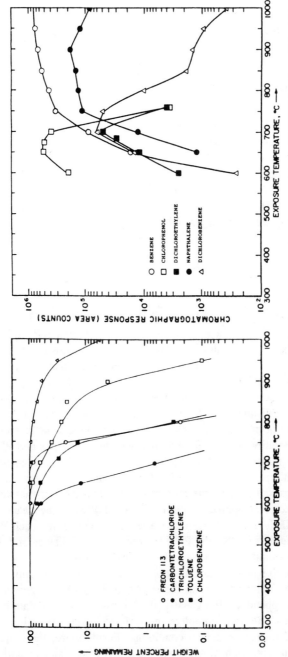

FIG. 7. Thermal decomposition results for HWM-1. As many as 50 such mixtures must be run to complete the POHC/PIC indices. Only stable, high-yield PICs listed in U.S. EPA Appendix VIII are shown.

327

the relative emissions. From this study and other previous studies, 900 °C appears to be a workable choice. One may now define RS_i as:

$$RS_i = \frac{M_i^{900}}{M_i^{300}} \frac{1}{N} \tag{4}$$

where M_i^{900} = mass of POHC$_i$ remaining, following a 900 °C exposure for a duration of 2.0 s

M_i^{300} = mass of POHC$_i$ at 300 °C (i.e., initially in the sample)

N = normalization constant such that the most stable POHC has a RS_i of 1.0. Note that since no degradation occurs at 300 °C, DE_i = 100 $(1 - M_i^{900}/M_i^{300})$

The analogous relationship for RY_i is

$$RY_i = \frac{M_i^{900}}{\Sigma M_j^{300,i}} \frac{1}{N} \tag{5}$$

where $\Sigma M_j^{300,i}$ is the initial total mass of all precursors j, for PIC$_i$.

The normalization constant N is selected to define the RY and RS scale relative to the most "difficult to destroy" compound. If this is a POHC, then N would be M_i^{900}/M_i^{300} for the most stable POHC observed in all the tests. If this is a PIC, then N would be $M_i^{900}/\Sigma M_j^{300,i}$ for the PIC with the highest yield. In practice, the selection of the reference compound would correspond to the POHC or PIC with the largest calculated value of M_i^{900}/M_i^{300} or $M_i^{900}/\Sigma M_j^{300,i}$, respectively. By normalizing the RY and RS scales to the same reference value, one can more readily evaluate the relative risk due to POHCs and PICs. However, one must remember that to determine the relative mass emission rate of a POHC, one multiplies by the POHC concentration, whereas one multiplies by the total feed rate of all possible precursors to obtain the PIC mass emissions.

This approach was applied to the results for HWM-1, the results of which are shown in Table 10. For an actual waste stream containing each POHC at a mass concentration of 10%, the PICs benzene and naphthalene and the POHC chlorobenzene would be expected to be the major emissions.

Extension of this type of ranking approach will require performing extensive thermal decomposition studies on the approximately 400 compounds listed in the U.S. EPA Appendix VIII. Clearly, the currently available data base, especially for PICs, is not complete enough to apply to all incineration systems for all wastes, under all conditions. Furthermore, there are assumptions made in the approach which cannot be tested with available laboratory and field data. However, as our experimental and modeling studies continue and

TABLE 10

Predicted Relative Risk for Most Stable POHCs and PICs for a Hypothetical Waste Stream
Containing Toluene, Chlorobenzene, Trichloroethylene, Carbon Tetrachloride,
and Freon 113

Compound	RS or RY	Relative mass emissions[a]	Relative risk[a]
Benzene (PIC)	3.8×10^{-1}	1.0	1.0
Naphthalene (PIC)	1.3×10^{-1}	0.34	0.94
Chlorobenzene (POHC)	1.0	0.26	0.46
Trichloroethylene (POHC)	8.1×10^{-2}	0.021	0.021
Toluene (POHC)	8.5×10^{-2}	0.022	0.022
Dichloroethylene (PIC)	9.1×10^{-4}	0.0024	0.060

[a]Predicted relative mass emissions and relative risk for full-scale burn of a waste containing each
POHC with mass concentrations of 10%.

more field data become available for comparison, the ranking can be expanded
and refined to serve as a reliable guide for hazardous waste incinerability.

Still, there may be instances where a new technology is being employed or a
unique waste is being incinerated, and the application of this general ranking
index may not be sufficient. Two options are then available to the facility
manager who desires to have a knowledge of potential emissions prior to a full-
scale burn.

One option would be to use a "surrogate" mixture for which laboratory data
are already available. This mixture would have already been determined to
contain POHCs with a range of stabilities, some of which would be extremely
stable and challenge the capabilities of even the most advanced incineration
systems. This surrogate mixture would also produce stable PICs in high yields
to further test the incinerator. Data would be available similar to that presented
for HWMI; however, the initial composition would be selected to fully test the
incinerator's capabilities. This approach has the advantage that the results for a
standardized surrogate mixture could be compared to other incinerators that
also burned the same mixture. If desired, the surrogate mixture could be spiked
into a complex waste matrix.

The last, and most comprehensive, option would be to subject the actual
waste stream(s) selected for full-scale testing, to laboratory thermal decomposition testing. An example of the data to be generated in this procedure is depicted in Fig. 8. The upper tracing is a chromatogram showing the chemical
composition of an actual waste to be incinerated. The lower tracing is a chromatogram of the effluent after this waste has been subjected to pyrolytic thermal degradation using the system for thermal diagnostic studies (STDS) or

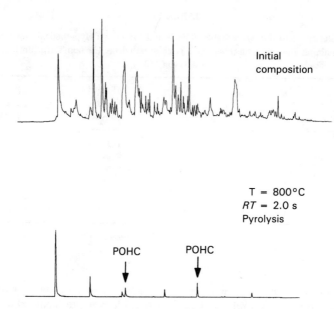

FIG. 8. Chromatogram of a hazardous waste before and after laboratory thermal degradation. Lower GC tracings show composition of effluent from the TDAS following degradation of waste. Peaks in lower trace have been identified using TDAS and would be monitored during trial burn. Arrows denote residual POHCs.

thermal decomposition analytical system (TDAS). As one can see, the waste composition is quite complex, making POHC selection potentially difficult and costly. However, following thermal degradation, the compounds that are likely to be emitted in greatest yields (i.e., thermally stable) are easily identifiable. These emissions include both POHCs and PICs. These POHCs and PICs may now be considered surrogates for the destruction efficiency of the entire waste feed since they have been experimentally shown to be the most stable.

In practice, the incinerator operator would prepare a permit application in which he suggests a trial burn plan to the permit writer and identifies POHCs and/or PICs for monitoring during the trial burn. Thus, the applicant can include laboratory thermal decomposition data with which the permit writer would have considerable confidence. The applicant can then conduct test burns prior to the trial burn to establish the appropriate operating conditions that will destroy the proposed POHCs and PICs. Furthermore, the trial burn plan is greatly simplified since POHCs and PICs have been selected with confidence and fewer compounds must be analyzed and monitored during the trial burn.

Laboratory data can be included in the trial burn plan to show that a very stable waste was selected for the trial burn. It can then be proposed that instead of conducting a new trial burn if different wastes are to be incinerated, the applicant will simply submit laboratory data to show that there are no components in the new waste stream that are more difficult to destroy than the one already tested in the trial burn.

If an additional margin of safety is requested by the applicant or permit writer, continuous monitoring of the incinerator effluent may be utilized. Laboratory data will have already identified the compound likely to have the greatest emission rate. This compound may be monitored on a continuous basis by a suitably modified and tuned process gas chromatograph (GC) unit. This approach assures the continual operation of the incinerator in an optimum manner.

APPENDIX: LABORATORY SYSTEMS FOR DETERMINING HAZARDOUS WASTE INCINERABILITY

The University of Dayton Research Institute has designed and constructed three instrumentation systems that can be used to study the high-temperature thermal decomposition of organic materials under highly controlled conditions. All three systems consist of thermal decomposition units interfaced with an in-line GC or gas chromatograph-mass spectrometer (GCMS). The thermal units are comprised of a control console, a sample introduction region, a high-temperature reactor, and heated transfer lines. Effluents from the thermal reactor (both remaining starting materials and thermal reaction production) are cryogenically trapped at the head of the in-line gas chromatographs, and then are analyzed by GCMS (for both qualitative and quantitative analysis) or GC (for quantitative analysis only).

The thermal decomposition analytical system (TDAS) consists of a controlled, high-temperature, quartz reactor with a dedicated in-line LKB 2091 GCMS to determine the thermal stability of organic materials in addition to identification of thermal reaction products (Rubey, 1980). Fig. A.1 shows a block diagram of the TDAS and an artist's rendering of this assembled instrumentation system. Basically, the TDAS consists of (1) a special thermal decomposition unit with its own dedicated instrumentation assembled to a control console, (2) an in-line high-resolution gas chromatograph (HRGC) which is coupled to (3) a mass spectrometer which, in turn, is served by (4) a dedicated minicomputer system.

The thermal decomposition unit-gas chromatographic (TDU-GC) system uses essentially the same thermal decomposition unit design but incorporates a HRGC with flame ionization and other detectors (Rubey et al., 1984). The

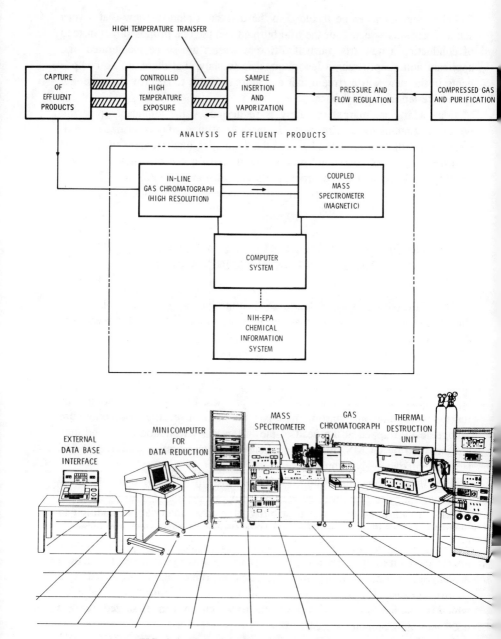

FIG. A.1. Thermal decomposition analytical system (TDAS).

TDU-GC has a Varian 4600 GC and a dedicated CDS 401 data terminal. A block diagram and artist's rendering of the TDU-GC are given in Fig. A.2.

A key component of both the TDAS and TDU-GC is the precisely controlled high-temperature reactor. The major portion of the reactor consists of a narrow-bore (nominal 1 mm ID) quart tube flowpath in a race-track configuration (3.5 cycles, 1 m in length). The quartz tube reactor assembly fits into a high-temperature, three-zone Lindberg furnace designed for continuous operation at temperatures up to 1200°C with control to ±1°C. Fine-bore entrance and exit tubes transport the sample rapidly into and out of the central portion of the reactor. This flow design provides a narrow residence time distribution and square wave exposure temperature profile. The inert surface provided by the all-quartz construction minimizes wall reactions.

The most advanced instrumentation assembly, the system for thermal

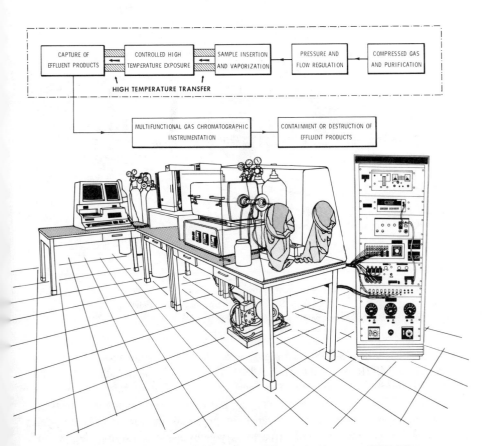

FIG. A.2. Thermal decomposition unit-gas chromatographic system (TDU-GC).

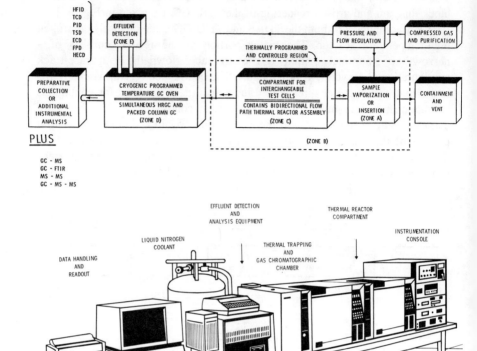

FIG. A.3. System for thermal diagnostic studies

diagnostic studies (STDS), is a highly versatile thermal simulation unit with an in-line GCMS (Rubey and Carnes, 1984). The STDS has a modular design with interchangeable components. The most salient feature of this system is its ability to accommodate a variety of intricate, custom-designed, test-cell assemblies to conduct a broad range of physical simulations. The test-cell assemblies include test cells (reactors) which produce varying degrees of gas-phase mixing (such as may be found in incinerators), test cells wherein two-zone thermal decomposition occurs (such as may be found in incinerators), test cells wherein two-zone thermal decomposition occurs (such as in a rotary kiln incinerator with afterburner), a flame-mode test cell, and many others. The STDS has an added advantage in that the MS may be used in either the conventional full-scan mode for identification of bulk thermal reaction products, or in the selected ion monitoring mode for detection of trace products, such as chlorinated dibenzo-*p*-dioxins and dibenzofurans.

A block diagram and photo of the STDS are shown in Fig. A.3. The foundation equipment for the STDS consists of two Hewlett Packard 5890A GCs and a Hewlett Packard 5970B mass selective detector (MSD), which functions as a quadrupole MS. One of the GCs is used in the standard mode as a capillary GC. The other contains a specially designed and fabricated high-temperature furnace which houses a quartz reactor assembly and can be operated at temperatures up to approximately 1050 °C. The primary advantage in housing the thermal reactor inside the GC oven lies in the fact that the precise temperature control possible with a GC permits highly controlled and quantitative transport of materials into and out of the reactor.

All of these systems can be used for conducting precise laboratory-scale experiments and evaluating the effects of major decomposition variables, such as exposure temperature, residence time, reaction atmosphere, and waste feed composition.

There are several advantages in using these thermal decomposition systems for identifying the most probable emissions from hazardous waste incinerators. Specifically: (1) These systems have already been constructed, tested, and used successfully. (2) The variables affecting thermal decomposition can be well controlled and measured. (3) To facilitate valid comparisons, materials can be tested under well-defined, reproducible, standard conditions. (4) Both quantitative and qualitative data can be obtained relatively quickly and easily. (5) The capability of the in-line analytical instrumentation to trap thermal reaction products obviates the need for time-consuming sample extractions and other manipulations.

REFERENCES

Benson, S. W. 1976. *Thermochemical Kinetics*. New York: Wiley.

Clark, W. D., et al. 1983. The prediction of liquid injection hazardous waste incinerator performance, paper 84-HT-13, August 1984, presented at the ASME/AIChE National Heat Transfer Conference.

Clark, W. D., et al. 1985. Engineering analysis of hazardous waste incineration: Failure mode analysis for two pilot-scale incinerators, *Proceedings of the 11th Annual Symposium on Incineration and Treatment of Hazardous Waste*, Cincinnati, Ohio, April.

Cooke, M., et al. 1983. PNA emissions in industrial coal-fired stoker boilers, presented at the

185th National American Chemical Society Meeting in Seattle, Washington, March. Division of Environmental Chemistry paper no. 3, vol. 23, no. 1, pp. 6–7.

Crumpler, E. P., et al. 1981. Best engineering judgment for permitting hazardous waste incinerators, presented at the ASME/EPA Hazardous Waste Incineration Conference, Williamsburg, Virginia, May.

Cudahy, J. J., and Troxler, W. L. 1983. Autoignition temperature as an indicator of thermal oxidation stability. J. Hazard. Mater. 8:59–68.

Dellinger, B., et al. 1984a. Determination of the thermal stability of selected hazardous organic compounds. Hazard. Waste 1(2):137–157.

Dellinger, B., et al. 1984b. Predicting emissions from the thermal processing of hazardous wastes. Destruction efficiency testing of selected compounds and wastes, Final Report to Eastman Kodak Company, September.

Dellinger B., et al. 1986a. Incinerability of hazardous wastes. Hazard. Waste Hazard. Mater. 3(2):139–150.

Dellinger, B., et al. 1986b. Hazard. Waste Hazard. Mater. 3(3):293–307.

Dickson, W. J., and Massy, Jr., F. J. 1957. Introduction to Statistical Analysis, 2nd ed. New York: McGraw-Hill.

Dryer, F. L., and Glassman, I. 1973. High-temperature oxidation of CO and CH_4, 14th Symposium (International) on Combustion, p. 987. Pittsburgh: The Combustion Institute.

Fujii, N., and Asaba, T. 1977. High temperature reaction of benzene. J. Faculty Eng. Univ. Tokyo B 34(1):189–224.

Glassman, I., et al. 1975. Combustion of hydrocarbons in an adiabatic flow reactor: Some considerations and overall correlations of reaction rate, paper presented at Joint Meeting of the Central and Western States Sections of the Combustion Institute, San Antonio, Texas, April.

Graham, J. L., et al. 1986. Laboratory investigation of thermal degradation of hazardous organic compounds. Environ. Sci. Technol. 20(7):703–710.

Hall, D. L., Dellinger, B., Rubey, W. A., and Carnes, R. A. 1983. Considerations for the thermal degradation of hazardous waste, presented at the International Symposium on Environmental Pollution, Miami Beach, Florida, October.

Hall, D. L., Dellinger, B., Fraken, J. L., and Rubey, W. A. 1986. Thermal deposition of a twelve component organic mixture. Hazard. Waste Hazard. Mater. 34(4):441–449.

Kozlov, G. I. 1959. On High-Temperature Oxidation of Methane, Seventh Symposium (International) on Combustion, p. 142. Pittsburgh: The Combustion Institute.

Lee, K. C., et al. 1982. Revised model for the prediction of the time-temperature requirements for thermal destruction of dilute organic vapors and its usage for predicting compound destructibility, presented at the 75th Annual Meeting of the Air Pollution Control Association, New Orleans, June.

Miller, D. L., Cundy, V. A., and Matula, R. A. 1984. Incinerability characteristics of selected chlorinated hydrocarbons, Proceedings of the Ninth Annual Research Symposium on Incineration and Treatment of Hazardous Waste, EPA-600/94-015, 113–128. Springfield, Va.: National Technical Information Service.

MRI. 1984. Performance evaluation of full-scale hazardous waste incinerators. Final Report, MRI Report submitted to the U.S. Environmental Protection Agency, Contract 68-02-3177, August.

Nemeth, A., and Sawyer, R. F. 1969. The overall kinetics of high temperature method oxidation in a plow reactor. J. Phys. Chem. 73:2421.

RTI/ES (Research Triangle Institute and Engineering Science). 1984. Evaluation of waste combustion in cement kilns at General Paulding, Inc., Paulding, Ohio. Draft Final Report to U.S. Environmental Protection Agency, March.

Rubey, W. A. 1980. Design considerations for a thermal decomposition analytical system. EPA-600/2-80-098. Springfield, Va.: National Technical Information Service.

Rubey, W. A., and Carnes, R. A. 1984. Review of laboratory systems developed for studying gas-phase thermal decomposition, paper presented at the 11th Biennial National Waste Processing Conference in Orlando, Florida, June.

Rubey, W. A., et al. 1984. Description and operation of a thermal decomposition unit-gas chromatographic system. EPA-600/a52-84-149. Springfield, Va.: National Technical Information Service.

Seeker, R. L., et al. 1984. Laboratory-scale flame mode study of hazardous waste incineration. *Proceedings of the Ninth Annual Research Symposium on Incineration and Treatment of Hazardous Waste*, U.S. Environmental Protection Agency, IERL, EPA-600/9-84-015, PB 84-234525, pp. 79–94.

Shilov, A. E., and Sabirova, R. D. 1960. The mechanism of the first stage in the thermal decomposition of chloromethanes: The decomposition of chloroform. *Russian J. Phys. Chem.* 34(4):408–411.

Taylor, P. H., and Dellinger, B. 1986a. Thermal degradation characteristics of chloromethane mixtures. Submitted to Environ. Sci. Technol.

Trial Burn Report for Kodak Park Division Chemical Waste Incinerator. 1985. U.S. Environmental Protection Agency Id. No. NYD980592497. Rochester, NY: Eastman Kodak Co.

Tsang, W., and Shaub, W. 1982. Chemical processes in the incineration of hazardous materials. In Detoxification of Hazardous Waste, ed. J. H. Exner, Chap. 2, pp. 41–60. Ann Arbor, Mich.: Ann Arbor Science.

Vandell, R. D., and Shadoff, L. A. 1984. Relative rates and partial combustion products from the burning of chlorobenzene and chlorobenzene mixtures. Chemosphere 13(11):1177–1192.

Wall, L. A., ed. 1972. The mechanisms of pyrolysis, oxidation, and burning of organic materials, Proceedings of the Fourth Materials Research Symposium, held by National Bureau of Standards, Gaithersburg, Maryland, NBS Special Publication 357, CODEN:XNBSAV.

Williams, G. C., et al. 1969. The Combustion of Methane in a Jet-Mixed Reactor, 12th Symposium (International) on Combustion, p. 913. Pittsburgh: The Combustion Institute.

Wolbach, C. D., and Garman, A. R. 1984. Destruction of hazardous wastes cofired in industrial boilers: Pilot scale parametrics testing, Acurex Technical Report FR-84-46/EE, February.

Wyss, A. W., Castaldini, C., and Murray, M. M. 1984. Field evaluation of resource recovery of hazardous wastes, Acurex Technical Report TR-84-160/EE, August.

Estimating Emissions from the Synthetic Organic Chemical Manufacturing Industry: An Overview

Clay E. Carpenter, Cindy R. Lewis, and Walter C. Crenshaw

Versar, Inc., Springfield, Virginia

I. TYPES OF ENVIRONMENTAL EMISSIONS DATA

The problems associated with assessing emissions from facilities of the synthetic organic chemical manufacturing industry (SOCMI) are complex because emissions are affected by many diverse factors (e.g., process type and operating parameters) and emanate from numerous sources. For example, approximately 16,000 industries manufacture or use organic chemicals (DMI, 1986). The majority of these companies are broadly classified under standard industrial classification (SIC code) 2800. However, this is a broad classification that includes first tier industries that produce organic chemicals as well as second tier industries that use organic chemicals during processing. The general characteristics of these industries, four-digit SIC codes, and the number of manufacturers in each industry are listed in Table 1.

These diverse industries produce a vast array of products. Combined production of all synthetic organic chemicals, tar, and primary products from petroleum and natural gas was 338,025 million pounds in 1984 (U.S. ITC, 1985). General categories of chemicals and production and sales data are listed in Table 2. This diversity of industries and chemicals leads to difficulty in estimating emissions from these facilities.

This chapter provides an overview of the process of estimating emissions from SOCMI. It is intended to provide general guidance for preparing exposure assessments, emissions inventories, or any other types of documents designed to characterize releases from SOCMI facilities. It is not designed to provide a "how to" or methods approach to this subject; rather, it focuses on identifying the types and locations of releases and the factors required to quantify them. Simple estimation methods, however, have been included to provide a clearer understanding of the interrelationships of the required factors.

Before considering environmental releases from SOCMI facilities, it is im-

TABLE 1

Industries That Produce or Use Synthetic Organic Chemicals in the United States

Industry	SIC code	Number of manufacturers
Plastics	2821	1313
Synthetic rubber	2822	290
Cellulosic manmade fibers	2823	64
Synthetic organic fibers except cellulosic	2824	148
Biological products	2831	383
Medicinal chemicals and botanical products	2833	416
Pharmaceutical preparations	2834	1342
Soap and other detergents except specialty cleaners	2841	590
Specialty cleaning, polishing, and sanitation preparations	2842	1462
Surface active agents, finishing agents, and sulfonated oils and assistants	2843	167
Perfumes, cosmetics, and other toilet preparations	2844	1138
Paints, varnishes, lacquers, enamels, and allied products	2851	1664
Gum and wood chemicals	2861	164
Cyclic crudes and cyclic intermediates, dyes, and organic pigments	2865	296
Industrial organic chemicals not classified elsewhere	2869	1053
Nitrogenous fertilizer	2873	513
Phosphatic fertilizers	2874	238
Fertilizers, mixing only	2875	543
Pesticides and agricultural chemicals not classified elsewhere	2879	734
Adhesives and sealants	2891	926
Explosives	2892	159
Printing ink	2893	565
Chemicals and chemical preparations not classified elsewhere	2899	1805

Source: DMI (1986).

TABLE 2

Synthetic Organic Chemicals and Their Raw Materials: U.S. Production and Sales, 1984

Chemical classes	Production (million pounds)	Sales (million pounds)	Sales (million dollars)
Cyclic intermediates	47,052	19,957	6,930
Dyes	233	221	691
Organic pigments	86	76	493
Medicinal chemicals	279	152	1,369
Flavor and perfume materials	179	115	637
Plastics and resin materials	48,255	40,751	20,923
Rubber-processing chemicals	288	176	287
Elastomer (synthetic rubber)	4,609	2,686	2,266
Plasticizers	1,788	1,685	849
Surface-active agents	5,519	3,433	1,874
Pesticides and related products	1,189	1,108	4,730
Miscellaneous end-use chemicals and chemical products	23,731	14,931	3,834
Miscellaneous cyclic and acyclic chemicals	92,009	40,386	12,043
Synthetic organic	225,215	125,659	56,968
Tarhemicals, total[a]	4,144	2,223	311
Primary products from petroleum and natural gas	108,666	51,178	8,256
Grand total[a]	338,025	179,061	65,535

[a]Because of rounding, figures may not add to the totals shown.
Source: U.S. ITC (1985).

portant to understand the different techniques available to characterize these releases. A release or an environmental loading is a measure of the emission rate to air, land, or water expressed in units of mass per time (e.g., g/s, kkg/ yr). Numerous types of environmental loadings data exist. Typically, they can be organized into three categories, in descending order of preference: (1) site-specific monitoring data, (2) environmental materials balance data, and (3) emission factors.

A. Site-Specific Monitoring Data

Plant- or site-specific monitoring data for the chemical(s) of concern provide the most reliable and accurate means of estimating emissions. Unfortunately, relevant site-specific monitoring data are rarely available. Some data are obtainable in the literature and through several U.S. Environmental Protection Agency (U.S. EPA) data bases; however, they are limited and of inconsistent quality. For example, most of the available data bases have a wide array of monitoring data from different sources, which vary in the methods of sampling and analysis and in the detection limits. In addition, monitoring data are frequently mixed with emission factors or engineering estimates. A discussion of the data bases that contain monitoring data is included in Section III.

B. Materials Balance

A materials balance is a tool for comparing inputs to and outputs from a system in order to account for all the material passing through the system. Materials balance data provide a means of estimating environmental emissions based on the theory that mass can neither be created nor destroyed. According to this concept, all the material entering a system will either leave, be transformed, or remain in the system. A system can be a unit process, a series of unit processes, a complete facility, or an entire industry. This concept is extremely useful because it helps identify and quantify environmental releases. It also enables an environmental analyst to identify all the possible release points as a chemical passes from "cradle to grave."

A major problem with materials balances, however, is finding sufficient data to completely balance all of the inputs and outputs for a system. This is especially true for environmental materials balances where releases to the environment are generally insignificant compared to the total throughput of the system (e.g., production volume).

C. Emission Factors

The least desirable method to use when estimating releases from production or processing is the application of emission factors to volume throughputs. This method expresses the approximate amount of a substance released to a specific medium as a percentage of production or process volume. Note that emission factors are usually crude approximations of releases and should be used only as a last resort when all other techniques have been exhausted.

Emission factors are developed from many sources including (in order of preference) monitoring and modeling data and engineering judgment. Besides the method used to develop the factors, a major problem with this type of estimating technique is the frequent misuse of these factors as surrogates for operations for which they were not derived; for example, an emission factor developed for benzene releases from storage tanks should not be applied automatically to other types of organic chemical storage. Consequently, emission factors should only be used for the processes for which they were developed.

II. OVERVIEW OF SOURCES AND TYPES OF ENVIRONMENTAL RELEASES

The potential for environmental release exists through every phase of a chemical's life—from the time it is synthesized until it no longer exists (either through degradation or consumptive processing). In order to approach the problem of estimating environmental releases from the diverse industries in SOCMI, the assessor must devise a method of systematically dividing the life cycle of the chemical into categories, which can be investigated individually. Judicious categorization at this stage of the assessment can facilitate the analysis and enhance the quality of the results. Criteria for category selection include the chemical(s) of concern, the level of detail desired, and the focus of the study. A general approach is depicted in Fig. 1. This figure shows the different stages in the life cycle of a chemical and the types of emissions that can be expected from each stage. The emissions shown in this figure can lead to contamination of air, groundwater, surface water, or soil, depending on the physical characteristics of the emission source and the physicochemical characteristics of the chemical itself.

The following overview ties together the typical life cycle of chemicals produced, processed, transported, stored, and disposed by SOCMI. In later sections the individual life cycle stages are discussed in more detail.

A. Production and Processing

As shown in Fig. 1, the production (first tier) and processing (second and subsequent tiers) stages provide the opportunity for the same types of releases. This is not surprising considering that the distinction between production and processing is relative, rather than absolute. For example, the production stage for a pesticide would be pesticide manufacturing, rather than organic chemicals manufacturing, which makes the reactants used to manufacture the pesticide. This is because the first time that the pesticide actually exists is during its manufacture and not during the production of individual reactants.

Production industries are those that modify raw materials to produce either an intermediate (to be processed further) or a finished product. Processing industries, on the other hand, are those industries that further process, modify, or fabricate the material from the production stage to produce either an interme-

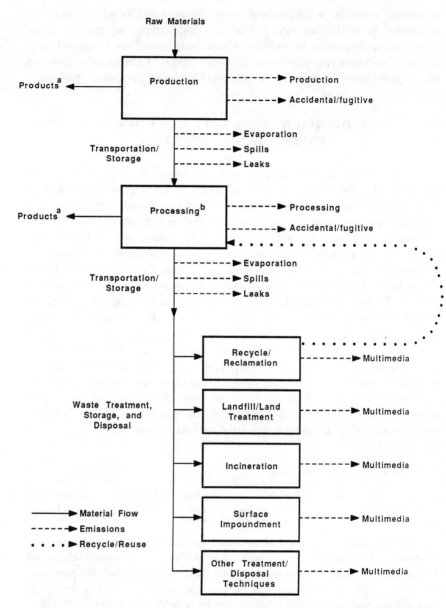

FIG. 1. Life cycle of a chemical. [a]Additional emissions will also occur from products, including commercial and consumer products; [b]processing may involve many steps, including the formation of consumer products.

diate product (bound for further processing) or a finished product. The processing stages thus continue until the chemical of concern degrades, has been consumed by the process, or shows up in a final product.

It is common for the product from a production step to be used as an intermediate in the manufacture of several different products. Therefore, multiple processing steps can follow a single production stage to produce many different products. Thus, the distinction between production and processing depends on the specific chemical under study. This principle is illustrated in Fig. 2.

The estimation of inherent and accidental or fugitive releases from production/processing industries is discussed further in Section III.

B. Transportation and Storage

As shown in Fig. 1, transportation and storage of chemicals may result in releases from evaporation, spills, or leaks. Transportation and storage are discussed together because of similarities in environmental release pathways. For example, evaporative emission can occur from a sealed tank whenever head space over a volatile compound is vented, whether the tank is mounted on a truck, train, ship, or stationary pad.

The potential for spills exists whenever a vessel is filled or emptied, regardless of the purpose of that vessel. The potential for catastrophic spills is, however, greater during bulk transport of a chemical. Although the initial releases may be comparable whether the spill occurs within the confines of a tank farm or a busy interstate highway, emissions and their impact would vary considerably because the handling and containment procedures that are available at the tank farm, obviously, do not exist on a highway.

Leaks can result from a variety of factors including degradation, cracks, or punctures. Leaking usually occurs first in seals and gaskets causing air, soil, surface water, or groundwater releases, depending on the location of the vessel and the physiochemical properties of the chemical of concern. Leakage from the vessel itself, caused by corrosion, cracks, or punctures, can be significant, particularly if it goes undetected for a long period of time. This often happens with underground storage tanks resulting in significant soil and groundwater contamination. Transportation and storage releases are discussed more fully in Section IV.

C. Waste Treatment, Storage, and Disposal

Most production/processing procedures generate some type of waste stream along with a finished product. This waste must then be treated, stored, and disposed of; environmental emissions can result from all of these steps. Because of highly publicized incidents like that which occurred at Love Canal, New

Production/Processing Tree

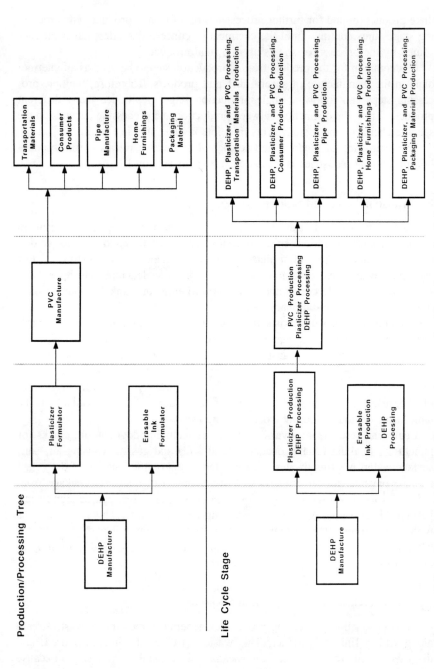

Life Cycle Stage

FIG. 2. Partial production/processing tree and corresponding life cycle stage for diethyl hexyl phthalate (DEHP).

346

York, the responsible treatment, storage, and disposal (TSD) of industrial wastes has become a national issue. Federal laws, such as the Resource Conservation and Recovery Act (RCRA), the Clean Air and Clean Water Acts, and various state laws have been enacted that severely restrict allowable waste management practices. Because these regulations change quickly, any estimation of environmental emissions from treatment, storage, and disposal of wastes must be based on current allowable practices. The federal regulations governing TSD are summarized briefly in Section V.

In addition to this changing regulatory climate, environmental releases from waste TSD are difficult to assess because both the emission type and TSD options are waste specific. Two facilities even within the same company rarely generate identical waste streams. The relevant factors associated with estimating environmental loadings from the treatment, storage, and disposal of wastes are discussed in Section V.

III. PRODUCTION AND PROCESSING

Production and processing of chemicals result in chemical releases to the air, surface water, groundwater, or soil through production and processing and accidental/fugitive releases. These releases can be quantified through monitoring data, materials balance information, or emission factors. A complete set of monitoring data for a chemical would include time-averaged concentrations of the chemical of concern in the air, groundwater, surface water, and soil, at enough different distances and directions from the facility to produce a complete, three-dimensional picture. These data would need to be available for every facility producing or processing this chemical. Because such complete data are rarely, if ever, available, what monitoring data are obtainable are typically used to confirm release estimates that have been based either on a materials balance or an emission factors approach.

Any attempt to quantify releases via these approaches must begin with an investigation of the chemical itself, production/processing synthesis trains, the geographic location of the facilities employing each type of synthesis, and the production volume at each location. Some references that may provide these data are listed in Table 3. Most of the references noted in this table are encyclopedias, which provide a basic overview of the chemical in question. Information that should be retrieved from these references includes the type of chemical, its physical state at ambient conditions, its method of manufacture, and its major uses.

Detailed production/processing synthesis routes can be determined from a number of sources; the major sources are noted in Table 4. In general, the more commercially important the chemical of concern, the more information that will

TABLE 3

References Used to Obtain a General Overview of Chemical Manufacture

Title	Reference
Chemical Process Industries	Shreve, 1977
Chemical and Process Technology Encyclopedia	Considine, 1974
Faith, Keyes, and Clark Industrial Chemicals	Lowenheim and Moran, 1975
Encyclopedia of Polymer Science and Technology	Mark et al., 1985–1986
Encyclopedia of Chemical Technology	Kirk and Othmer, 1978–1984
Modern Plastics Encyclopedia	Agranoff, 1986
Pesticides Manufacturing and Toxic Materials Control Encyclopedia	Sittig, 1981
Reigel's Handbook of Industrial Chemistry	Kent, 1983
Mineral Commodity Profiles	Bureau of Mines, 1986a

be available in published literature concerning its synthesis routes. If no data can be found, an analogy can often be drawn between the chemical of concern and a structurally similar chemical for which data are available. Patent applications for the chemical or a structural analog can also help to pinpoint synthesis routes. If no well-known analog is known, the synthesis routes may have to be assumed, based on the presence of structural groups and other physicochemical properties of the chemical.

Once this information is known, processing conditions can be combined with the physicochemical properties of the chemical and the geographic location of the facility to estimate the location of releases. The probability of accidental/fugitive releases can also be determined, either on a facility-by-facility basis or in general, depending on the level of detail desired.

At this stage the level of detail desired in the analysis becomes crucial. The availability of resources—time, manpower, and money—determines the depth to which the analysis can be conducted. Each of the factors discussed in this section should be addressed, even for a cursory level of detail, but the depth of the investigation at every stage varies in response to the resources available. For instance, if a chemical is produced by a number of different processes, but one process accounts for 85% of production, a detailed assessment would investigate all of the processes, whereas one with limited resources may investigate only the process accounting for the highest production volume. This concept extends throughout the evaluation procedure.

Sources and methods discussed in this chapter are commonly applicable to any level of detail. In general, however, those offering the broadest base of information are given the most attention.

TABLE 4

Information Resources for Synthesis Routes and Their Flow Diagrams

Sources of information	Data provided		Comments
	Lists different synthesis routes	Provides process flow diagrams	
I. References for major chemicals			
Chemical Economics Handbook (SRI, 1987)[a]	X	X	Very good source
Chemical and Process Technology Encyclopedia (Considine, 1974)[a]	X	X	Very good source for process descriptions
Encyclopedia of Chemical Technology (Kirk and Othmer, 1978–1984)[a]	X	X	Good source for general overview of chemical
Chemical Process Industries (Shreve, 1977)[a]	X	X	Good source for process descriptions
Encyclopedia of Polymer Science and Technology—Plastics, Resins, Rubbers, Fibers (Mark et al., 1985–1986)	X	X	Very good source for polymer process descriptions
Modern Plastics Encyclopedia (Agranoff, 1986)	X	X	Very good source for polymer process descriptions
Organic Chemicals Manufacturing (IT Enviroscience, 1980)	X	X	Excellent source for the organic chemicals reviewed
Kline Industrial Marketing Guides (Kline, 1979–1982)	X	X	Good references for chemical, paper, plastics, paint, and packaging industries
Faith, Keyes and Clark Industrial Chemicals (Lowenheim and Moran, 1975)[a]	X	X	Major synthesis routes discussed; lists other routes
Unit Processes Guide to Organic Chemical Industries (Herrick et al., 1979)	X		Only lists reactions for major organic chemicals
Organic Chemical Producers Data Base (Radian Corp., 1979)	X		Lists synthesis routes

See footnote on p. 351.

349

TABLE 4 *(Continued)*

Sources of information	Data provided		Comments
	Lists different synthesis routes	Provides process flow diagrams	
MEDLARS/TDB Toxicology Data Bank	X		Lists synthesis routes
Reigel's Handbook of Industrial Chemistry (Kent, 1983)	X		Some process descriptions. Discusses wastewater treatment and air pollution control technology
Chemical Marketing Journals (magazines)[a]	X	X	Must search for appropriate issue
Industrial Minerals and Rocks (American Institute of Mining, Metallurgical and Petroleum Engineers, Inc., 1983)	X	X	Very good source for mining industry
Industrial Minerals (journal)	X	X	Articles will provide process descriptions
Mineral Commodity Profiles (Bureau of Mines, 1986a)	X	X	Describes mining techniques used for chemicals
Minerals Yearbook (Bureau of Mines, 1986b)	X	X	Describes mining techniques used for chemicals
Industrial Process Profiles for Environmental Use	X	X	30-volume set providing a general source of information on various industries, including mining and metals
II. Government agencies, general publications U.S. EPA Office of Toxic Substances[a] Office of Air Quality Planning and Standards Office of Water and Waste Management Office of Solid Waste	X	X	If a document has been written on the chemical or its industry, it will generally have process descriptions. The development documents for air and water standards are good references
Locating and Estimating Air Emissions from Sources of Ethylene Dichloride (U.S. EPA, 1984)	X	X	Good source for chemical-specific data
Dept. of Health and Human Services; National Institute for Occupational Safety and Health (NIOSH)[a]	X	X	Control technology assessments for industries are useful. Describes industrial exposure

350

Source			Description
Dept. of Labor; Occupational Safety and Health Administration (OSHA)[a]	X		Documents on industry have descriptions of occupational exposure
Dept. of the Interior; Bureau of Mines[a]	X		Documents on minerals
III. Specialty literature search on DIALOG			
Patent files[a]		X	Discusses process and chemical
Chemical abstracts[a]		X	Discusses process and chemical
IV. Industry public relations material[a]		X	Company brochures on chemicals and/or processes are useful in confirming information from other sources

[a]These sources address a variety of industries.

A. Production and Processing Releases

1. Background Data Requirements and Sources

Background information, such as the geographic location and production volume of the chemical, must also be gathered. These data are important in determining the receiving media and in quantifying the release. Geographic locations of the producers are important because they identify the points of release from production and thus the likely areas where the environmental loadings will be the greatest. In addition, geographic location determines geologic patterns, such as aquifer depth and location and soil porosity, which, in turn, affect the media to which the chemical partitions.

Annual production volumes indicate the amount of chemical that can potentially be released. Because companies view their production volumes as confidential, most sources list plant capacity. If capacities are given, a general rule of thumb is to estimate production volume as 80% of capacity.

Table 5 presents references, including data bases, that can be used to determine producers, estimate production volumes, and locate manufacturing facilities. Comments on the variety of chemicals covered, the frequency or publication, and the ease of use are provided in the table. Generally, the information is stored under the chemical name. Two corroborating sources are needed to provide medium to high confidence in production volumes and locations.

Once these background data are gathered, inherent releases from production/ processing can be estimated. These estimations are often based on engineering judgment combined with background physicochemical properties of the chemical, process operating conditions, or unit operations used. To increase the accuracy of the results, any available monitoring data, materials balance data, or emission factors are also used. These data have been compiled in several computerized national data bases.

2. Computerized Data Sources

Several computerized national data bases, taken together, constitute the most comprehensive source of data from which emission data can be derived. Computerized data bases containing information on emissions to air are most common, although several data bases pertain to water discharges. Emission factors concerning releases to land are not practical because of the strong dependence on waste management practices (e.g., solid waste may be incinerated, landfilled, or stored in waste piles).

Data bases containing point source discharges to air include the National Emission Data System (NEDS), Hazardous Trace Emissions (HATREMS), the Compliance Data System (CDS), and the National Air Toxics Information Clearinghouse (NATICH). These data bases are maintained by the U.S. EPA's Office of Air Quality Planning and Standards located in Research Triangle

TABLE 5

Sources of Information of Production Volumes and Geographic Location of Production/Processing Facilities

Sources of information	Production volume	Geographic location	Comments
I. Reference journals, government agency publications			
Directory of Chemical Producers (Stanford Research Institute, 1986)[a]	X	X	Good source, lists plant capacities instead of production for 10,000 chemicals with production > 5000 lb or sales > $5000
Chemical Economics Handbook (Stanford Research Institute, 1987)[a]	X	X	Very good source; contains materials balances
Synthetic Organic Chemicals—U.S. Production and Sales 1984 (U.S. ITC, 1985)[a]	X		Good source; lists producers that produce > 5000 lb/yr or > $5000 of sales
Minerals Yearbook (Bureau of Mines, 1986b)[a]		X	Good source for production and use patterns of minerals
Mineral Inventory Surveys (Bureau of Mines, 1986c)[a]		X	Good source for production and use patterns of minerals
Mineral Commodity Profiles (Bureau of Mines, 1986a)[a]		X	Good source for production and use patterns of minerals
Chemical Marketing Reporter (chemical marketing journal)[a]	X	X	Must search for appropriate issue and journals (use DIALOG)
Thomas Register of American Manufacturers (Duchaine, 1987)[a]		X	Lists producers and locations of a wide variety of chemicals
Kline Industrial Marketing Guide (Kline, 1979–1982)		X	Industries covered are: chemical, paper, plastics, paint, and packaging; additional data on use patterns and process descriptions are provided
Chemical Products Synopsis (Mannsville Chemical Products, 1976)	X	X	Lists capacities, production, imports, exports, and uses. Estimates provided are based on 1975 data

See footnote on p. 354.

353

TABLE 5 *(Continued)*

Sources of information	Production volume	Geographic location	Comments
II. Buyers guides			
Chem Sources—USA (Chem Sources, 1987)[a]		X	No production values; covers 90,500 chemicals
Aldrich Catalog Handbook of Organic and Biochemicals (Aldrich, 1987)[a]		X	No production values; covers 8,500 chemicals
Chemical Week—Buyers Guide Issue (Chemical Week, 1986)[a]		X	No production values
1988 CPD Chemical Buyer's Directory (Chemical Marketing Reporter, 1987)[a]		X	No production values; covers 15,000 chemicals
III. Data bases			
PTS PROMPT Predicasts Overview of Market and Technology[a]	X	X	Access by DIALOG on line; good source for end uses, capacities, new products, and foreign trade
PTS Funk and Scott Indexes[a]	X	X	Subset of PTS PROMPT, access by DIALOG, forecasts of sales
Radian—U.S. EPA Organic Chemicals Producers' Data Base[a]		X	Good source; additional data on use patterns and release during production
IPC Chemical Production Data[a]	X	X	Production, imports, and exports
MEDLARS/TDB Toxicology Data Bank[a]	X	X	Production, imports, and exports for 3600 chemicals; easy access
CIS Chemicals in Commerce Information Service	X	X	Production values and imports based on 1977 data and growth rates; easy access; covers 55,000 chemicals
N Chemical Industry Notes	X	X	Access by DIALOG on line, contains production, sales, products, and processes

[a]These sources address a variety of industries.

Park, North Carolina, and contain the majority of computerized point source discharge data available.

NEDS describes the annual emissions characteristics of all point source emitters in the United States. Although it contains data on such pollutants as specific hydrocarbons and volatile organic compounds, most of the data are on criteria pollutants—lead, particulates, sulfur dioxide, nitrogen oxides, carbon monoxide, total hydrocarbons, and ozone. All states are required to submit this information to the U.S. EPA on an annual basis.

HATREMS, which is operated in parallel with NEDS, stores data on emissions of air pollutants that are not included in NEDS (i.e., pollutants other than the criteria pollutants). It contains data on the emissions of 27 noncriteria pollutants including metals and other toxics, although quantity data are limited for most pollutants.

CDS is a management information system that contains data on more than 30,000 plants, factories, and other stationary point sources of air pollution. Pollutants include hydrocarbons, vinyl chloride, and heavy metals. Nonpollutant parameters include compliance data, discharge points, and production levels (U.S. EPA, 1985a).

NATICH computerized data base contains information on toxic and potentially toxic air pollutants. It is intended to support state and local programs in the control of noncriteria air pollutants. This data base contains two major categories of data: (1) data from state and local air pollution control agencies on their toxic chemical-related activities and (2) citations and abstracts to published documents and ongoing research related to toxic air pollutants. Specifically, this data base contains source testing information (end-of-pipe monitoring data) by pollutant and SIC code and ambient monitoring data by pollutant (Gates and Smith, 1985).

Data bases concerned with point source discharge to water include Best Available Technology (BAT) Review Studies (industry specific); Data Collection Portfolios (industry specific); National Pollutant Discharge Elimination System (NPDES); and Permit Compliance Systems (PCS). These data bases contain much of the point source discharge data available.

The BAT Review Studies are industry-specific programs sponsored by the Effluent Guidelines Division (EGD) of EPA. Pollutants include the 129 Consent Decree Priority Pollutants, and nonpollutant data include discharge points, flow rates, production levels, and treatment. Most of the specific plant data in these studies are confidential; however, industry-wide averages and ranges may be obtained from the development documents published from these studies.

The Data Collection Portfolios contain point source effluent data on an industry-specific basis. Pollutants include conventional and priority pollutants. Nonpollutant data include discharge points, flow rates, production levels, treatment, and water usage. Measurement of waste volume and pollutant load per ton of finished product and applied waste disposal scheme and capability are also included in this EGD data base.

NPDES Compliance Files contain information relating to the compliance status of both industrial and municipal permit holders. Point source data collection dates back to 1974, and the resulting discharge monitoring reports are maintained on a state and regional basis. Which pollutants are monitored depends on permit reporting requirements. Nonpollutant data include flow rates, production levels, discharge points, treatment, and volume/mass measures.

PCS is similar to the NPDES file and is structured to compare effluent data and NPDES limits in order to identify violations. This data base contains information on more than 65,000 active water discharge permits issued to facilities throughout the nation. The Office of Water Enforcement and Permits (OWEP) at the U.S. EPA is responsible for operation and maintenance of PCS (U.S. EPA, 1985a).

3. Estimating Emissions

After all available data have been gathered and reviewed, the emissions estimation process can begin. In general, this process consists of sorting through the available data, assessing the accuracy of each piece of data, and combining them in such a way as to derive the most meaningful estimate of emissions possible. The best method for estimating emissions will vary for each study depending on the chemical of concern, availability of data, and level of detail.

Monitoring and material balance emission data provide the most accurate emission estimates and require the most resources. Emissions factors, on the other hand, can provide reasonable estimates of emissions with only moderate expenditures of resources. The accuracy of industrywide estimates based on emission factors can be improved by comparing monitored emissions of the chemical of concern, or a structural analog, for a few locations with those predicted by the emission factors approach. Emissions factors can be found in the data sources described previously.

On a less detailed level, emissions can be more quickly estimated based on the unit processes used in the synthesis of the chemical. Within SOCMI, the synthesis of the large-volume chemicals can be described by 23 unit processes. Table 6 lists these unit processes along with the number of compounds produced commercially using these processes. Emission factors characterizing releases from these unit processes have been derived for the major high-volume chemicals. These factors are summarized in Table 7. More detail on the derivation of this table is available in *Development of the Environmental Release Data Base for the Synthetic Organic Chemicals Manufacturing Industry* (U.S. EPA, 1982). Although it is unlikely that the chemical of concern will be found in this table, the major unit process in the synthesis of the chemical of concern is likely to be

TABLE 6

Summary of the 23 Major Unit Processes Commonly Used in Synthetic Organic Chemical Manufacturing Industry

Process	Compounds
Alkylation	15
Amination by ammonolysis	13
Ammoxidation	10
Carbonylation (oxo)	10
Condensation	55
Cracking (catalytic)	3
Dehydration	6
Dehydrogenation	15
Dehydrohalogenation	6
Esterification	24
Halogenation	54
Hydrodealkylation	3
Hydrogenation	27
Hydrohalogenation	7
Hydrolysis (hydration)	28
Nitration	12
Oxidation	47
Oxyhalogenation	5
Phosgenation	3
Polymerization	34
Pyrolysis	20
Reforming (steam)-Water Gas Reaction	1
Sulfonation	11
TOTAL	409

Source: U.S. EPA (1982).

included. Table 7 can then provide an order of magnitude estimate of the emissions that may be expected from that unit process.

B. Accidental/Fugitive Releases

The category of accidental/fugitive releases from production/processing operations encompasses those environmental releases that cannot be accurately predicted. Accidental emissions, by definition, fall into this category. Fugitive emissions are air emissions from leaking pieces of ancillary process equipment.

Accidental emissions can release solids, liquids, or gases to any receiving medium, at any time. However, accidental emissions generally account for a small percentage of the total emissions. This apparent contradiction results from the periodic and infrequent nature of their occurrence, which is, in turn, a

TABLE 7

Summary of Emission Factors for Major SOCMI Unit Processes[a]

Unit process	Product	Type of release[b]	Receiving medium[c]	Emissions (kg/kkg prod.)	Form[d]	Typical pollutants released[e]
Alkylation	Cumene	p	A	0.00117–0.0235	g	C₂, C₃, C₄ hydrocarbons; benzene; alkylbenzenes
		f	A	0.24	g	Total VOCs
	Ethylbenzene	p	A	0.29–1.7	g	Benzene, other VOCs
		p	W	1.9–21.5	l	Hydrogen chloride, aluminum salts, sodium hydroxide, sodium chloride
	Lead alkyls	p	W	7–723	l	Ethylchloride, sodium chloride, organic by-products
Amination by ammonolysis	Ethanolamine	p	L	22–40	l, s	Triethanolamine, tars, bottom products
		p	W	8	l	Ethanolamine
	Hexamethylenetetramine	p	A, W	1.5–2.5	g, l	Formaldehyde, ammonia, hexamethylenetetramine
Ammoxidation	Acrylonitrile	f	A (to flare)	0.806–0.844	g	Acrylonitrile, other VOCs
		p	A (to flare)	Trace to 200	g	Carbon monoxide, propane, ammonia, acetonitrile
		p	A	0.017–200	g	Sulfur dioxide, butane, nitrogen oxides, carbon monoxide, acrylonitrile, other VOCs
		p	W	0.0002–34.1	l	Ammonium compounds, chromium, nitriles, sulfate, other VOCs

Process	Product					Constituents
Carbonylation	Hydrogen cyanide	p	A	6.2–829	g	Hydrogen cyanide, hydrogen, carbon monoxide, methane
	Acetic acid	p	A	2–204	g	Hydrogen, carbon monoxide, methane, methanol, light ends, halogens, VOCs
	Ethyl acrylate	p	L	20	s, sl	Propionic acid, higher organics
		p	A	6.1	g	Acetylene, carbon monoxide
		p	W	4.8–43.2	l	Ethanol, sodium carbonate, sodium acrylate
		p	L	0.8–63	s	Polymers, hydroquinone, nickel chloride, sodium salts
Condensation	Acetic anhydride	p	A	0.217–31	g	Carbon monoxide, benzene, other VOCs
	Methyl parathion	p	A	58–260	g	Hydrogen sulfide, organophosphorus compounds, sulfur dioxide
		p	W	5–188	l	Hydrogen chloride, parathion, p-nitrophenol, organophosphates
	Polyethylene terephthalate	p	L	0.085–23.1	s, sl	Ethylene glycol, polyethylene glycol, sodium hydroxide, sodium bisulfite
	Pyridine	p	W, L	0.0003–44	l, sl	Phenol, paraldehyde, picolines, acetates, soluble amides, chlorides, sulfuric acid
Catalytic cracking	Vinyl chloride	p	A	0.10–50	g	Hydrogen chloride, vinyl chloride, 1,2-dichloroethylene, 1,1-dichloroethylene, acetylene
		p	L	0.05–37.0	s	Ethylene chloride, tars, solids (ash), heavy ends

See footnotes on p. 366.

TABLE 7 (*Continued*)

Unit process[f]	Product	Type of release[b]	Receiving medium[c]	Emissions (kg/kg prod.)	Form[d]	Typical pollutants released[e]
Dehydration[f]						
Dehydrogenation	Styrene (from ethylbenzene)	p	W	0.393–5.9	l	Heavy ends, tarry matter, aromatics, phenol
	Isoprene (from tertiary amines from C_5 hydrocarbons)	p	W	0.5–19	l	Isoprene, amylene, *n*-hexane, acetonitrile, sodium hydroxide, sodium sulfate
Dehydrohalogenation	Acetaldehyde	p	A	0.013–0.228	g	Acetaldehyde, ethylene, ethane
	Hydrogen chloride	p	A	0.4–15.0	g	Vinyl chloride, 1,2-dichloroethylene, 1,1-dichloroethylene, acetylene
		p	L	0.05–0.8	s	Ethylene dichloride, tars (as C_2HCl), solids (as C)
	Vinylidene chloride	p	A	0.7–7.1	g	Vinylidene chloride, monochloroacetylene, VOCs
		p	W	45.5–605.5	l	Sodium hydroxide, sodium chloride
Esterification	Dimethyl terephthalate	p	W	Trace to 54	l, s	Acetic acid, methanol, terephthalic acid, zinc dust, antimony oxide, dimethyl terephthalate, *p*-formylbenzoic acid
		p	A	0.02–1.0	g	Total VOC
		f	A	0.66	g	Total VOC
	Ethyl acrylate	p	A	0.265–265	g	Ethanol, ethylacrylate, ethylacetate, sodium carbonate
	Methyl methacrylate	p	A	3.15–29.5	g	Acetone, hydrogen cyanide

360

	p	W	0.25–26.45	l	Acetone, hydrogen cyanide, methyl methacrylate, methanol, polymers, ammonium compounds
Polyethylene terephthalate	p	L	0.095–23.1	s, l	Ethylene glycol, polyethylene glycols, sodium hydroxide, sodium bisulfite
Diethyl phthalate	p	W	0.0001–8.04	l	Cadmium, ethanol, iron, oil, phenol, sulfates, zinc, ethyl phthalate monoester
Chlorobenzene	f	A	0.58–1.39	g	Benzene, other VOCs
	p	A	<0.001–1.4	g	Hydrogen chloride, mono- and dichlorobenzene, benzene, other VOCs
Halogenation					
Vinyl chloride monomer	p	W	0.33–4.35	l	Ethylene, sodium hydroxide, sodium chloride, vinyl chloride, ethyl chloride
	p	L	0.005–4.0	l, s	Tars, tetrachloroethane, ethylene, carbon solids, mercuric chloride
Trichloroethylene	p	A	4.9–12	g	Ethylene, methane
	p	A	0.05–1.25	g	Ethane, methane, tetrachloroethane, hydrogen chloride, trichloroethylene, perchloroethylene
Perchloroethylene	p	A	0.001–2.8	g	Ethane, methane, antimony trichloride, tetrachloroethane, trichloroethane, tri- and tetrachloroethylene, dichloroethane
	p	W	Trace to 344	l, s	Chlorinated solvents (trace), calcium, calcium chloride

See footnotes on p. 366.

361

TABLE 7 (Continued)

Unit process	Product	Type of release[b]	Receiving medium[c]	Emissions (kg/kkg prod.)	Form[d]	Typical pollutants released[e]
	Ethylene dichloride	p	L	10–230	l, s	Hexachlorobutadiene, chloroethanes, chlorobutadienes, tars, residue
		f	A, W	0.265–0.40	g, l	Chlorine, hydrogen chloride, VOCs
		p	A	0.0005–3.0	g	Ethane, ethylene dichloride, methane, chlorine, other VOCs
		p	W	0.05–3.8	l	Ethylene dichloride, hydrochloric acid, vinyl chloride, methyl chloride, ethyl chloride, caustic
		p	L	0.0035–2.45	s, l	Mercuric hydroxide, tars, carbon solids, 1,1,2-trichloroethane, tetrachloroethane
	Chloromethanes	p	A	0.052–13	g	Methane, methyl chloride, methylene chloride, chloroform, carbon, other VOCs
		p	W	0.02–0.03	l	Methyl chloride, methylene chloride, chloroform, carbon, perchloroethylene
Hydrodealkylation[f] Hydrogenation	Aniline	p	A	0.0012–0.095	g	Benzene, aniline, nitrobenzene, VOC

Process	Product					Components
	Cyclohexane	p	A	0.0032–1.419	g	Benzene, total VOC
	Cyclohexanol	p	A	0.0315–1.52	g	Benzene, total VOC
	Hexamethylenediamine	p	W	3.0–9.35	l	Hexamethylenediamine, 1,2-cyclohexanediamine, acrylonitrile
Hydrohalogenation	Ethyl chloride	p	L	1.2–30.8	l	Trichloroethylene, dichloroethane, hexachlorobutadiene, chlorobenzene, chloroethane, chlorobutadiene
	1,1,1-trichloroethane	p	A	0.001–9.0	g	Vinyl chloride, ethylene dichloride, dichloroethane, trichloroethane, VOCs
	Vinyl chloride	p	A	0.75–5.3	g	Acetylene, hydrogen chloride, vinyl chloride
Hydrolysis and hydration	Glycerin (from epichlorohydrin)	p	W	0.8–61.0	l	Glycerin, toluene, light impurities
	Glycerin (from allyl alcohol)	p	A	<0.1–2.8	g	Total VOC
		p	W	3.5–23.5	l	Allyl alcohol, glycerin, light impurities
		p	A	0.15–22	g	Allyl alcohol, butanol, acrolein, light impurities
	Propylene oxide (chlorohydrin process)	p	A	0.005–8.5	g	Benzaldehyde, ethane, toluene
	Chlorobenzene	p	W	0.70–12.2	l	Phenol, benzene, chlorotoluene
		p	L	0.60–19.3	s	Diphenyl ether, ortho-dichlorobenzene, chlorotoluene

See footnotes on p. 366.

363

TABLE 7 *(Continued)*

Unit process	Product	Type of release[b]	Receiving medium[c]	Emissions (kg/kkg prod.)	Form[d]	Typical pollutants released[e]
Nitration	Nitrobenzene	p	A	0.0011–7.5	g	Benzene, nitrogen oxide, nitrobenzene, total VOC
		p	W	2–20	1	Nitrobenzene, nitrated phenols, sodium sulfate, sodium carbonate, sulfuric acid
	Toluene diisocyanate	p	A	0.025–12	g	VOCs, nitro aromatics, sulfur dioxide, nitrogen oxides, sulfuric acid vapor
Oxidation	Acetaldehyde	p	W	0.6–13.9	1	Acetic acid, acetyl chloride, chloral, aldehydes
	Acrylic acid, acrylic ester	p	A	0.1–118	g	Acetaldehyde, acetic acid, acetone, acrolein, propane, carbon monoxide, propylene
	Adipic acid cyclohexane	p	A	1.1–58	g	Hydrocarbons, carbon monoxide, nitrogen oxides
	Cyclohexane/cyclohexyl alcohol	p	A	0.091–42	g	VOCs, carbon monoxide
	Ethylene oxide	p	A	0.25–47.5	g	Ethylene, ethylene oxide
	Maleic anhydride		p	A	2.8–86	g
Maleic acid, formaldehyde, formic acid, benzene						
Oxohalogenation	Ethylene dichloride	p	A	<0.003–18	g	Ethylene dichloride, ethylene, ethane, vinyl chloride, dichloroethylene, methane, carbon tetrachloride, methylene chloride

364

Process	Product					By-products / materials
Trichloroethylene and perchloroethylene		p	A	0.012–21.3	g	Vinyl chloride, vinylidene chloride, perchloroethylene, trichloroethylene, dichloroethylene
Phosgenation	Toluene diisocyanate	p	A	4×10^{-7} to 4.6	g	Phosgene, chlorinated hydrocarbons, organic amines, dichlorobenzene
		p	L	1.4–19.1	s	Polymers, tarry matter, ferric chloride
Polymerization	Epoxy resins	p	L	14.8–45	s	Sodium monohydrogen phosphate, epichlorohydrin, methyl isobutyl ketone
	Polypropylene	p	A	0.2–40	g	Ethane, propylene, propane, isopropyl alcohol, heptane
	Acrylonitrile-butadiene-styrene resins	p	W	0.25–8.8	1	Styrene, acrylonitrile, nacconol, sulfuric acid, calcium chloride
	Polyethylene	p	A	1–22.5	g	Ethylene, methanol, n-hexane, ethane
	Polyethylene (autoclave reactor)	p	A	1.05–42.6	g	Ethylene, ethane, butane
	Polyethylene (tubular reactor)	p	A	0.9–30.8	g	Ethylene, ethane, butane
	Polyvinyl acetate		p.	A	0.5–20	g
Vinyl acetate, methanol, hydroquinone	Polyvinyl alcohol	p	W	0.5–250	1	Methane, ethyl acetate, sodium sulfite
	Amino resin	p	W	20–35	1	Urea, melamine, formaldehyde
	Nylon 66	p	W	1.55–4.0	1	Methanol, gluteric acid, succinic acid, 1,2-cyclohexadiamine

See footnotes on p. 366.

TABLE 7 (Continued)

Unit process	Product	Type of release[b]	Receiving medium[c]	Emissions (kg/kkg prod.)	Form[d]	Typical pollutants released[e]
	Polyethylene terephthalate	p	L	0.1–23.1	s, sl	Ethylene glycol, polyethylene glycol, sodium hydroxide, sodium bisulfite
Pyrolysis	Ethylene	p	A	0.001–80.0	g	Benzene, total VOC
		f	A	8.3–4935	g	Benzene, total VOC
Reforming (steam)–water gas reaction	Natural gas	p	A	0.049–1.1	g	VOCs
		f	A	0.578	g	VOCs
Sulfonation and sulfation	Methyl methacrylate	p	A	0.004–29	g	Acetone, hydrogen cyanide, sulfur dioxide, carbon monoxide, VOCs
		p	L	29–100	s, sl	Sulfate salt crystal, polymers, hydroquinone, acetone

[a]Emissions factors noted in this table are of uncertain accuracy but serve to give an order of magnitude estimate of releases. It was beyond the scope of this table to include all the possible pollutants and their corresponding emissions factors; additional information is presented in U.S. EPA (1982). The emission factor given must be multiplied by the throughput volume to obtain an emission estimate. For production volumes see SRI (1986) and U.S. ITC (1985).

[b]Key: f, fugitive; p, production/processing.

[c]Key: A, air; water; L, land. Note that many of the pollutants slated for discharge to water and land may be incinerated or undergo some other form of waste treatment instead of or before being discharged to water or land.

[d]Key: g, gas; l, liquid; s, solid; sl, sludge.

[e]VOCs, volatile organic compounds.

[f]The source does not contain emission factors for these unit processes.

Source: U.S. EPA (1982).

consequence of SOCMI's focus on accident prevention. Unless the emission assessment is very detailed, these emissions are usually neglected because of their low volumes. If they are to be considered, the most accurate method of quantification includes statistical analysis of previously quantified accidental releases from similar units that process similar compounds.

On the other hand, fugitive emissions account for a significant portion of the total emissions. IT Enviroscience (1980) estimated that 32% of the 1.4 billion pounds of volatile organic compounds (VOCs) released to the air by SOCMI in 1982 were attributable to fugitive releases.

Fugitive releases occur in different quantities from both equipment known to be leaking and equipment thought to be secure. Ancillary pieces of equipment from which fugitive emissions occur are valves, pumps, compressors, pressure relief devices, open-ended valves or lines, sampling connections, and flanges and other connectors.

One study (Blacksmith et al., 1980) detected leakage from as many as 26% of the pumps used in light liquid service. The U.S. EPA recently completed a study of fugitive emission factors for seven types of equipment as shown in Table 8. These emission factors were derived from data published in several references and account for releases from both leaking and nonleaking equipment. Note that the factors are on a per-source basis; therefore, in order for them to be useful, the number of sources (e.g., flanges) within a facility must be known. In the absence of hard data, this parameter can be estimated based on the process flow sheet and engineering judgment.

TABLE 8

Average Emission Factors for Fugitive Emissions in SOCMI

Equipment component	Average SOCMI factors (kg/h/source)
Pump seals	
Light liquid	0.0494
Heavy liquid	0.0214
Valves	
Gas	0.0056
Light liquid	0.0071
Heavy liquid	0.00023
Compressor seals	0.228
Safety relief valves: gas	0.104
Flanges	0.00083
Open-ended lines	0.0017
Sampling connections	0.0150

Source: U.S. EPA (1986a).

IV. STORAGE AND TRANSPORTATION

All chemicals must be stored and transported at some stage of their life cycle. Therefore, emissions from storage and transportation must be investigated.

A. Storage Tanks

Storage tanks are located at chemical manufacturing and production facilities as well as bulk liquid transfer terminals and waste treatment facilities. Tanks used by organic chemical manufacturers are predominately constructed above ground with the axis perpendicular to the foundation. Horizontal and underground tanks are also used, but much less frequently than vertical aboveground tanks. Occasionally, but not frequently, pressure vessels may be used (U.S. EPA, 1983).

There are three main types of vertical aboveground storage tanks: fixed-roof, external floating roof, and internal floating roof. The fixed-roof tank is the least expensive to build. It is constructed as a cylindrical steel shell with a permanently affixed cone or dome-shaped roof. A breather valve in the roof prevents the buildup of excessive internal pressure. These valves operate at either a slight internal pressure or vacuum depending on the contents of the tank. External floating roof tanks are constructed of a cylindrical steel shell with a deck or roof that floats on or slightly above the surface of the stored liquid. The floating roof rises or falls with the changing liquid levels. Internal floating roof tanks combine the attributes of the two previously mentioned vertical tanks. Each is not only constructed with a permanent roof but also has an internal floating roof. As with external floating roofs, these internal roofs may float on pontoons a few inches above the surface of the liquid or may rest on the liquid itself.

Emissions can occur because of accidents, structural failure, defective hardware, and breathing/working losses. These factors result in spills, leaks, and evaporation. Spills that result in chemical emissions may be relatively small, as in the case of a small leak that occurs over a short period of time. On the other hand, overfilling a tank or failure to adequately close the proper valves can cause significant releases. Structural failure of a tank can result in substantial releases of organic chemicals. Corrosion can lead to liquid release and is more likely to occur at welded seams, rivets, and joints. If these leaks occur aboveground, they can generally be seen and remedied before significant amounts of a chemical have escaped. If a leak occurs in a tank that is underground, or in any part of a tank that is underground, then a seemingly tiny flow can result in a sizable release over a long period of time.

Trying to quantitatively estimate the release of organic feedstocks, intermediates, products, or wastes resulting from spills or leaks is a risky undertaking. These incidents occur randomly depending on the chemical characteristics of the stored material, inspection and maintenance of individual facilities, and chance. Thus, it is extremely difficult to accurately estimate emissions from these occurrences.

On the other hand, breathing and working losses from fixed roof, external floating roof, and internal floating roof tanks are easily estimated. Because the majority of organic liquid storage occurs in these types of tanks and the major emissions from them are breathing and working losses, any attempt to estimate emissions from storage should focus on these types of tanks (U.S. EPA, 1983).

For fixed-roof tanks, breathing and working losses can be significant. Breathing losses occur without any change in the liquid level of the tank. Daily changes in barometric pressure and temperature cause the vapor space in the tank to expand or contract. Contraction causes air from outside the tank to be drawn in through the breather valve. Expansion, however, causes the expulsion of vapor through the breather valve into the atmosphere. Working losses occur when the tank is filled or emptied. As the tank is filled, the pressure in the vapor space increases so that organic vapor is expelled. When the tank is emptied, air is drawn into the tank and becomes saturated with organic vapor. This causes the mixture to expand, increasing the pressure in the vapor space so that organic vapor is expelled into the atmosphere (U.S. EPA, 1983).

The following are equations for estimating breathing and working losses from fixed roof tanks (U.S. EPA, 1983, 1985b):

$$L_T = L_B + L_W \tag{1}$$

$$L_B = 2.26 \times 10^{-2} M_v \left(\frac{P}{P_A - P}\right)^{0.68} D^{1.73} H^{0.51} \Delta T^{0.5} F_P C K_c \tag{2}$$

$$L_W = 2.40 \times 10^{-5} M_v P V N K_n K_c \tag{3}$$

where L_T = total loss (lb/yr)
  L_B = breathing loss (lb/yr)
  L_W = working loss (lb/yr)
  M_v = molecular weight of product vapor (lb/lb mol); 80 assumed as a typical value for volatile organic liquids
  P = true vapor pressure of product (psia)
  P_A = average atmospheric pressure at tank location (psia)
  D = tank diameter (ft)
  H = average vapor space height (ft); use tank-specific values or an assumed value of one-half the tank height

ΔT = average daily temperature change in °F; 20 °F assumed as a typical value

F_P = paint factor (dimensionless); 1.0 for clean white paint

C = tank diameter factor (dimensionless): for diameter ≥ 30 feet, C = 1; for diameter <30 feet, $C = 0.0771D - 0.0013D^2 - 0.1334$

K_c = product factor (dimensionless) = 1.0 for volatile organic liquids

V = tank capacity (gal)

N = number of turnovers per year (dimensionless)

K_n = turnover factor (dimensionless): for turnovers >36, $K_n = (180 + N)/6N$; for turnovers ≤ 36, $K_n = 1$

Equations (4)–(8) can be used for estimating breathing and working losses from floating roof tanks (U.S. EPA, 1983, 1985b).

$$L_T = L_w + L_r + L_f + L_d$$
$$= \text{the total loss (lb/yr)} \tag{4}$$

$$L_w = \text{withdrawal factor} = \frac{(0.943)QCW_L}{D}\left(1 + \frac{N_cF_c}{D}\right) \tag{5}$$

where D = tank diameter (ft)

Q = product average throughput (bbl/year); tank capacity (bbl/turnover) × turnovers/yr

C = product withdrawal shell clingage factor (bbl/10^3 ft^2); use 0.0015 bbl/10^3 ft^2 for volatile organic liquids in a welded steel tank with light rust (0.0075 for dense rust)

W_L = density of the product (lb/gal)

N_c = number of columns (dimensionless); obtain from Table 10

F_c = effective column diameter (ft); 1.0 assumed

$$L_r = \text{the rim seal loss (lb/yr)} = K_sV^nP^*DM_vK_c \tag{6}$$

$$L_f = \text{the fitting loss (lb/yr)} = F_fP^*M_vK_c \tag{7}$$

$$L_d = \text{the seam loss (lb/yr)} = S_dK_dD^2P^*M_vK_c \tag{8}$$

where K_s = seal factor (Table 9)

N = seal windspeed exponent (Table 9)

V = average windspeed at tank site (miles/h)

$P^* = (P/P_A)/[1 + (1 - P/P_A)^{0.5}]^2$

K_c = product factor (dimensionless) = 1.0 for volatile organic liquids

F_f = total fitting loss factor (lb mol/yr) $[(N_{F_1}K_{F_1}) + (N_{F_2}K_{F_2}) + \cdots + (N_{F_n}K_{F_n})]$

N_{Fi} = typical number of fittings (Table 11)

K_{Fi} = fitting loss factor (Table 11)

S_d = the seam length factor (ft/ft^2)

    = 0.14, for continuous metal sheets with a 7-ft spacing between seams

    = 0.33, for rectangular panels 5 ft by 7.5 ft

    = 0.20, an approximate value for use when no construction details are known

K_d = the seam loss factor (lb mol/ft yr)

    = 0.34 for internal floating roof with nonwelded seams

    = 0 for external floating roof and internal floating roof in which the floating roof has welded seams

Several factors affect emissions from these tanks including: (1) seal factors and seal windspeed exponents, (2) typical number of columns, and (3) fitting loss factors. An extensive discussion of these equations and their applications is presented in U.S. EPA Compilation of Air Pollution Emission Factors Volume (U.S. EPA, 1985b).

Emissions from floating-roof tanks occur mainly at the seal between the roof and the interior wall. Emission levels may vary with different seal types and combinations. Other important factors are the age and the condition of the seal(s). Obviously, higher emission rates can be expected around seals that are old, excessively worn, or damaged, resulting in a poor fit between the roof and tank wall. If there is an improper fit, then air flowing across the seal forms a pressure differential that accelerates vapor loss.

When estimating emissions from floating roof tanks, losses through the seal material should be considered. This is especially important when dealing with volatile organic liquids rather than petroleum. In these cases, the permeability of the stored product must be considered. For example, benzene is suspected of having a permeability loss that equals or exceeds loss from the seal because of convection and diffusion (U.S. EPA, 1983). Seal types, seal factors, and seal windspeed exponents all affect the emissions rates from floating roof tanks; these are shown in Table 9. Additional losses occur when interior columns support the floating roof. Table 10 presents the typical number of columns as a function of tank diameter. The various types and number of fittings in the floating roof itself need to be considered when estimating emission losses. Emissions can vary depending on fitting loss factors, such as whether or not the floating roof has welded or nonwelded seams. Factors relating losses to fitting type are presented in Table 11.

This is by no means the only approach for estimating emissions from storage

TABLE 9

Seal-Related Factors for Floating Roof Tanks[a]

Seal type	$(K_s)^b$	$(n)^c$
External floating roof tanks		
Metallic shoe seal		
Primary seal only	1.2	1.5
With shoe mounted secondary seal	0.8	1.2
With rim mounted secondary seal	0.2	1.0
Liquid mounted resilient seal		
Primary seal only	1.1	1.0
With weather shield	0.8	0.9
With rim mounted secondary seal	0.7	0.4
Vapor mounted resilient seal		
Primary seal only	1.2	2.3
With weather shield	0.9	2.2
With rim mounted secondary seal	0.2	2.6
Internal floating roof tanks		
Liquid mounted resilient seal		
Primary seal only	3.0	0
With rim mounted secondary seal	1.6	0
Vapor mounted resilient seal		
Primary seal only	6.7	0
With rim mounted secondary seal	2.5	0

[a]Based on emissions from a tank seal system with emissions control devices (roof, seals, etc.) in reasonably good working condition with no visible holes, tears, or unusually large gaps between the seals and the tank wall.
[b]K_s, seal factor in Eq. (6).
[c]n, seal windspeed exponent (dimensionless) in Eq. (6).
Source: U.S. EPA (1985b).

tanks. A good deal of research has been done in this area, and the following sources may provide the assessor with useful information: Erickson, 1980 (information on SOCMI storage and handling practices on an industrywide basis); IT Enviroscience, 1980 (provides useful data on the number of storage vessels used for selected "model" processes); and U.S. EPA, 1985b (information on petroleum refining on an industrywide basis).

B. Transportation

Transportation of feedstocks, intermediates, final products, by-products, and wastes provides an additional source of emissions. When trying to estimate the quantities of transportation related chemical releases, one must consider several issues: (1) the complexity and magnitude of the chemical transportation sector,

(2) factors that affect or cause releases, and (3) factors that affect release/ emission estimates.

Characterizing the sector of the transportation industry that transports organic products, their feedstocks, and associated waste is by no means a trivial undertaking. According to the U.S. ITC (1985), the combined production of all synthetic organic chemicals, tar, and primary products in 1984 was 3.38×10^{11} pounds (this does not include the quantity of waste produced). Even though not all of this amount is shipped, the chemical industry is one of the largest users of commercial transportation. Moving such vast quantities of materials requires the services of rail, motor, water, air, and pipeline carriers. The large number of different organic chemicals manufactured annually adds to the complexity of the transportation sector. Additionally, the multiplicity of chemical and physical characteristics of substances, locations of manufacturing facilities in relation to final destinations, and volume of movement tend to further complicate the transportation issue. Another factor to consider is that packaging can vary significantly (Hoffman, 1983). For example, formaldehyde may be transported in

TABLE 10

Typical Number of Interior Support Columns in a Floating Roof of a Tank as a Function of Tank Diameter[a]

Tank diameter (ft)	Typical number of columns N_c
$0 < D \leq 85$	1
$85 < D \leq 100$	6
$100 < D \leq 120$	7
$120 < D \leq 135$	8
$135 < D \leq 150$	9
$150 < D \leq 170$	16
$170 < D \leq 190$	19
$190 < D \leq 220$	22
$220 < D \leq 235$	31
$235 < D \leq 270$	37
$270 < D \leq 275$	43
$275 < D \leq 290$	49
$290 < D \leq 330$	61
$330 < D \leq 360$	71
$360 < D \leq 400$	81

[a]For use in Eq. (5). This table was derived from a survey of users and manufacturers. The actual number of columns in a particular tank may vary greatly depending on age, roof style, loading specifications, and manufacturing prerogatives. This table should not supersede information based on actual tank data.
Source: U.S. EPA (1983).

TABLE 11

Summary of Fitting Loss Factors (K_f) and Typical Number of Fittings (N_f) in Floating Roof Tanks

Fitting type	Fitting loss factor, K_f (lb-mol/yr)	Typical number of fittings N_f
Access hatch		1
Bolted cover, gasketed	1.6	
Unbolted cover, gasketed	11	
Unbolted cover, ungasketed	25	
Automatic gauge float well		1
Bolted cover, gasketed	5.1	
Unbolted cover, gasketed	15	
Unbolted cover, ungasketed	28	
Column well		(see Table 10)
Built-up column—sliding cover, gasketed	33	
Built-up column—sliding cover, ungasketed	47	
Pipe column—flexible fabric sleeve seal	10	
Pipe column—sliding cover, gasketed	19	
Pipe column—sliding cover, ungasketed	32	
Ladder well		1
Sliding cover, gasketed	56	
Sliding cover, ungasketed	76	
Roof leg or hanger well		$\left(5 + \dfrac{D}{10} + \dfrac{D^2}{600}\right)^a$
Adjustable	7.9	
Fixed	0	
Sample pipe or well		1
Slotted pipe—sliding cover, gasketed	44	
Slotted pipe—sliding cover, ungasketed	57	
Sample well—slit fabric seal, 10% open area	12	
Stub drain,[b] 1-in. diameter	1.2	$\left(\dfrac{D^2}{125}\right)^a$
Vacuum breaker		1
Weighted mechanical actuation, gasketed	0.7	
Weighted mechanical actuation, ungasketed	0.9	

[a]D, tank diameter (ft).
[b]Not used on welded, contact internal floating decks.
Source: U.S. EPA (1985b).

gallon-sized containers, 55-gallon drums, 6000-gallon tank trucks, or 20,000-gallon rail tank cars.

Emissions emanating from chemical transport may vary considerably. They may simply be the result of vapor loss from loading or unloading. As in the case of storage tanks, an incompletely closed valve, a defective hose, or a corroded seam can yield a relatively small loss of material. On the other hand, a ruptured pipeline, overturned tank truck, or derailed tank car can emit large amounts of chemicals, resulting in a catastrophic event.

Data that can be used to determine releases of organic chemicals from the various modes of transportation are sparse. The complexity of the transportation industry makes it difficult, if not impossible, to obtain accurate monitoring data or emission factors. The different modes of transportation coupled with the categories within each mode (e.g., rail transport of a chemical may range from bulk transport in tank cars to a few drums carried in freight cars) make it extremely difficult to quantify releases that occur because of normal working losses. These can include tank breathing losses, spills during loading and unloading, and leaks resulting from defective valves, worn seals, or corroded vessels.

Concern over the transportation of hazardous materials and transportation spills, however, has led to the development of methods for estimating the amount of chemical released in transit because of accidents. Factors that should be considered when estimating the quantity of a release include quantity of chemical shipped, distance shipped, accident rate, and probability of a release given an accident.

An analysis of releases of chemicals distributed in commerce requires information on commodity shipping patterns, i.e., the quantities of chemicals and distances they must be shipped by various modes of transportation.

At this time commodity shipping data and historical accident data are not compiled in any one system, nor are the available records stored by any one classification scheme. Rather, these data are compiled by various federal agencies in distinct and separate data bases. The assessor will need to consult a number of references in order to calculate the quantity of chemical released as a result of transportation accidents.

If the investigator does not have firsthand knowledge of the actual amounts of chemical shipped, its shipping distance, or its mode of transportation, information concerning commodity shipping patterns may be obtained from the commodity flow data bases listed in Table 12. Two particularly helpful sources are the Census of Transportation and the Interstate Commerce Commission's Waybill File. Information obtained from the data bases may be quite general. However, this information may allow the investigator to estimate the amount of a particular chemical shipped by the various modes of transportation and the corresponding shipping distances.

TABLE 12

Commodity Flow Data Bases[a]

Data bases	Kept by	Years	Modes	Strengths	Weaknesses or drawbacks
Commodity Transportation Survey (CTS)	U.S. Bureau of the Census	1977	All	• Multimodal • Consistent selection procedure for all sample data for all modes of transportation • Cross-checked against the Census of Manufacturers	• Only 5-digit level of commodities • No hazardous materials flags • Only shipments from manufacturing sites to first destinations • Only principal mode is reported
Truck Inventory and Use Survey	U.S. Bureau of the Census	1977, 1982	Highway	• Covers all trucks used in the United States • Contains hazardous materials-related data items • Sample emphasis on heavy trucks	• No flow data • Only rudimentary commodity information • Tractor data base, not a trailer data base, reflects tractor use not trailer use
Motor Carrier Census	Bureau of Motor Carrier Safety, FHWA	Most recent 5 years	Highway	• Comprehensive listing of carriers and truck fleet operators	• No flow data • Mileage and fleet size data are sparse
TRANSEARCH, FREIGHTSCAN, etc.	Consulting firms	Varies	All	• Cross-checked against other production/consumption data • Melding of the best data available for each mode	• Truck flows predominantly based on the CTS data (see above) • Not in the public domain

376

Data source	Collected by	Time period	Mode	Advantages	Limitations
National Motor Truck Database	Consulting firms	1977 to present	Highway	• Focuses on long-distance highway flows • True flow data • Describes the vehicle used to carry the commodity	• Purposely excludes short-haul truck movements, especially in the Northeast • Not in the public domain
Waybill File	Interstate Commerce Commission	At least past 12 years	Rail, TOFC/COFC	• Well-organized sample (1%) of all rail flows • Data base is consistent enough to allow trend analyses • Contains some routing information	• Not all hazardous material flows use the special Hazardous Materials STCC
Waterborne Commodity Statistics	Army Corps of Engineers	At least last 12 years	Water, domestic and international	• 100% sample of all vessel movements • Complete routing information	• Only 163 commodity codes in all, so level of detail is weak • Conversion table has some incorrect cross-references
TRAIN II	Association of American Railroads	Current	Rail, TOFC/COFC	• 100% data on all movements for participating railroads • Routing information	• Not specifically designed to record car movement histories • Not in the public domain
Hazardous Waste Shipment Data	States for EPA	Varies	Primarily highway	• 100% sample of all hazardous waste shipments • Actual flow data	• Many states do not computerize the data • No consistency to commodity code usage • No routing information

*a*Acronyms: EPA, U.S. Environmental Protection Agency; FHWA, Federal Highway Administration; OHMT, Office of Hazardous Materials Transportation, Research and Special Programs Administration; STCC, Standard Transportation Commodity Code; TOFC/COFC, trailer on flatcar (piggyback)/container on flatcar; WCSC, Waterborne Commerce Statistics Center (U.S. Army Corps of Engineers).
Source: Office of Technology Assessment (1986).

Once the shipping patterns have been established, it is necessary to determine the accident rate for the different modes of transportation. To consider releases resulting from transportation accidents such as collisions, derailments, or groundings, historical accident data should be obtained on the chemical in question. Currently, the accident rate data on waterborne transport of chemicals is being updated by the U.S. Coast Guard's Office of Marine Safety. It is expected that these data will be available in 1988. Accident rate data for truck and rail transport can be obtained from the sources cited in Table 13. Note, however, that obtaining up-to-date accident rates from these sources may require extensive statistical analysis of the data at a considerable cost. An alternative approach is to use published accident rates. Some of these rates are presented in Table 14. Even though some of the data are relatively old, the average accident rate is not expected to change significantly over time.

Once an accident rate has been estimated or selected, the investigator must establish the probability that the accident will result in a release. As with accident rates, extensive statistical analysis of the previously mentioned accident/incident data bases can yield accurate probability-of-release factors. On the other hand, Clarke et al. (1976) have reported probability-of-accident-severity factors that include release factors. They assume that any accident more severe than a minor collision will result in a release. This approach results in an estimate of 0.021 release per truck accident. Additionally, the probability of a rail accident resulting in a release has been estimated as 0.130 release per accident (U.S. DOT, 1986). Probability-of-release factors have also been included in the accident rates reported by ICF, Inc. (1984) in Table 15. Please note that these probability factors are presented to serve as examples only and are not intended as the sole factors to be considered.

The investigator can estimate the number of expected spills through use of a simple approach and the information previously obtained. This information includes: shipping distances; number of shipments; accident rate; and probability of a release.

The annual number of releases can be estimated by the linear equation

$$N = D \times S \times R \times P$$

where N = number of releases per year
D = distance per shipment (miles)
S = average number of shipments
R = average accident rate (accidents per mile)
P = probability of a release given an accident

This is by no means the only method for estimating the annual number of spills resulting from transportation accidents. Menzie (1979) presents an interesting approach for predicting the expected number of spills, and ICF, Inc.

TABLE 13

Accident/Incident Data Bases

Database	Kept by	Years	Modes	Accidents	Incidents
Hazardous Materials Information System	U.S. DOT, Office of Hazardous Materials Transportation, Research and Special Programs Administration	1977 to present	All	Yes	Yes
Commercial Vessel Casualty File	U.S. Coast Guard	1963 to present	Marine	Yes	No
Pollution Incident Reporting System	U.S. Coast Guard	1971–1985	All	Yes	Yes
Truck Accident File	DOT, Bureau of Motor Carrier Safety, Federal Highway Administration	1973 to present	Highway	Yes	No
Railroad Accident File	Association of American Railroads	1973 to present	Rail	Yes	Yes
Air Accident File	Federal Aviation Administration	NA[a]	Air	Yes	Yes
National Accident Sampling System	National Highway Traffic Safety Administration	1983 (hazardous materials flags added in 1983)	Highway	Yes	No
Fatal Accident Reporting System	National Highway Traffic Safety Administration	1983 (hazardous materials flags added in 1983)	Highway	Yes	No
National Response Center	U.S. Coast Guard	NA	All	Yes	Yes
National Transportation Safety Board File	National Transportation Safety Board	NA	All	Yes	No
U.S. Department of Energy Data Base	Sandia National Laboratories	1979 to present	All	Yes	Yes
Washington State Accident File	Washington State Utility and Transportation Commission	1978	Highway	Yes	No

[a]NA, Not available.
Source: Office of Technology Assessment.

TABLE 14

Example Accident Rates for Truck and Rail Transportation

Mode of transportation	Accident rate, $\times\ 10^{-6}$ (accidents/mile)	Comment	Source
Truck	2.5	Average for 1966–1970	Menzie (1979)
Freight car	1.8	Average for 1972–1974	Menzie (1979)
Rail (general)	6.01		U.S. DOT (1986)

(1984) has completed an extensive statistical analysis of releases from trucks. Nevertheless, the method presented is applicable to the majority of cases and provides an intuitive basis for alternate analysis.

V. WASTE TREATMENT, STORAGE, AND DISPOSAL

Most SOCMI processes generate waste that must be handled in some manner. All methods of waste handling can potentially release chemicals to the environment. In order to quantify these releases, the waste streams and handling methods must be identified. Waste streams are identified from knowledge of the process flow diagram, equipment operating parameters, and physico-chemical properties of the chemical of concern. Methods for determining these parameters have been discussed in previous sections. Handling methods depend on the waste stream characteristics and current state and federal governmental regulations. One federal act that has had enormous impact on waste handling is the Resource Conservation and Recovery Act (RCRA).

RCRA was developed in 1976 in an effort to responsibly manage the large volumes of waste generated annually. For instance, in 1981 the U.S. EPA estimated that the United States generated 264 million metric tons of hazardous waste. The goals of RCRA are to protect human health and the environment, to

TABLE 15

Release-per-Mile Rates for Truck Transportation

Release rate, $\times\ 10^{-6}$ (release/mile)	Site of measurement
0.13	Interstate highways
0.45	U.S. and state highways
0.73	Urban areas
0.28	Composite

Source: Adapted from ICF, Inc. (1984).

reduce waste and conserve energy and natural resources, and to reduce or eliminate the generation of hazardous waste as expeditiously as possible (U.S. EPA, 1986b).

Three distinct programs were outlined to achieve these goals. In Subtitle C of the act, a system is devised to manage hazardous waste from the time it is generated until the time it is disposed. This established the well-known "cradle-to-grave" management requirements. Subtitle D encourages states to develop plans for managing municipal solid waste. Subtitle I deals with the management of underground storage tanks. Because SOCMI hazardous wastes are regulated under Subtitle C, only that portion of RCRA will be discussed further.

Under the Hazardous and Solid Waste Amendments of 1984 (HSWA), EPA is required to develop a comprehensive set of regulations to implement the programs outlined by RCRA. These regulations concern the generation, transportation, treatment, storage, and disposal of hazardous wastes. According to Section 1004(5) of RCRA, a hazardous waste is a solid waste, or combination of solid wastes, which because of its quantity, concentration, or physical, chemical, or infectious characteristics, may:

(A) cause, or significantly contribute to an increase in mortality or an increase in serious irreversible, or incapacitating reversible, illness; or

(B) pose a substantial present or potential hazard to human health or the environment when improperly treated, stored, transported, or disposed of, or otherwise managed.

Notice that these definitions require that the waste be a solid, as defined by RCRA. As it pertains to industrial waste, a solid may be a sludge or other discarded material, including solid, semisolid, liquid, or contained gas resulting from industrial, commercial, mining, agricultural, and community activities (U.S. EPA, 1986b). Therefore, the term solid in the legal definition of hazardous wastes does not exclude from regulation substances that are physically semisolids, liquids, or gases.

In order to make RCRA's definition workable, however, the U.S. EPA had to develop a specific set of attributes that would identify a waste as hazardous. These attributes are (40 CFR Part 261) that the waste:

1. Exhibits, on analysis, any of the characteristics of a hazardous waste
2. Has been named as a hazardous waste and listed
3. Is a mixture containing a listed hazardous waste and a nonhazardous solid waste (unless the mixture is specifically excluded or no longer exhibits any of the characteristics of hazardous waste)

4. Is not excluded from regulation as a hazardous waste

Furthermore, the by-products of the treatment of any hazardous waste are also considered hazardous unless specifically excluded.

The characteristics referred to in item 1 above are: (a) ignitability, (b) corrosivity, (c) reactivity, and (d) extraction procedure toxicity. The requirement that a waste be listed refers to a comprehensive, ongoing effort undertaken by the U.S. EPA to list all known hazardous wastes. Some of the listed wastes are source specific (i.e., industry specific), and some are product specific. These lists can be helpful in determining the handling practices for a waste stream generated by the process of concern, even if the chemical of concern is relatively innocuous. If, for instance, a chemical is used in formulating veterinary pharmaceuticals that also contain organo-arsenic compounds, it is likely to be present in the wastewater treatment sludges generated during the formulation. Because these sludges are listed as hazardous by the U.S. EPA, the allowable handling practices are limited.

In addition, the U.S. EPA is in the process of establishing the Best Demonstrated Available Technology (BDAT) for treating listed wastes. This is being done in support of an RCRA mandate to ban land disposal of these wastes, in a graduated program to be completed in 1990. Although the technologies for the majority of the listed wastes have not yet been identified, the program, when completed, is expected to be useful in estimating releases from the processes.

Until that resource is available, the assessor must use traditional methods. The most accurate estimates may be derived on a facility-by-facility basis, from the plant itself. Although direct contact with plant personnel is rarely an option, information is often available in the form of operating and discharge permit applications.

In many cases, however, the level of detail desired in the investigation does not support facility-specific estimates. In these cases, releases from the treatment, storage, and disposal of wastes must be estimated on an industrywide basis. For these investigations, physicochemical properties of the chemical of concern and the characteristics of the waste stream itself may be combined with the types of waste treatment, storage, and disposal operations employed to estimate releases.

To estimate the characteristics of the waste stream, the process flow diagram should be analyzed to determine likely sources of wastes. For instance, a polymerization reaction that releases water is likely to generate an aqueous waste at some point in the process. These characteristics are further clarified with data concerning the type of waste treatment, storage, and disposal options likely to be employed. In many cases, waste treatment devices, such as distillation columns and filters, are part of the flow diagram. In other cases, the waste characteristics will imply a treatment process and/or disposal option. For example, if the product is liquid and an analysis of the flow diagram indicates that the process effluent is a mixed liquid and solid stream, it can probably be assumed in the absence of further information that the stream is separated and the solid disposed. Additional engineering judgments can often indicate the type of separation process likely to be used.

Once these factors are known the types of releases that are expected can be qualitatively determined. Table 16 identifies the types of releases that can be expected from several forms of waste treatment and disposal methods. Waste storage releases are similar to product storage releases, which were discussed in Section IV.

Available data on process flow rates, chemical kinetics, and waste management practices can then be used to quantify the releases. These data can either be facility-specific values or generic values. An example of the generic data that can be used to estimate environmental loadings involves estimating volatiliza-

TABLE 16

Summary of Emissions Potential for Treatment and Disposal of Wastes

Treatment or disposal method	Waste stream character[a]		
	Liquid organic	Aqueous	Solid, sludge
Steam stripping	X	A	L
Conventional distillation	L	L	X
Liquid-liquid phase separation	A	A	X
Solid-liquid phase separation	L	A	A
Neutralization	X	A	X
Precipitation/coagulation	X	A	X
Chemical oxidation	X	L	X
Thermal destruction[b]	A	L	A
Biological treatment	X	A	X
Landfill[c]	A, W, S	X	A, W, S
Surface impoundment	X[d]	A, W, S	A

[a]Key: X, method is generally not applicable to this type of waste; L, low potential for any emissions; A, potential for significant air emissions; W, potential for significant water emissions; S, potential for significant soil emissions.

[b]Although current regulations require the constituents of hazardous waste to be destroyed to a 99.99% efficiency, significant emission can result from incomplete combustion, formation of new compounds, and from varying incinerabilities among constituents of a complex waste (Dellinger et al., 1986).

[c]Additional information on releases from landfills can be found in Adkins et al. (1985).

[d]This practice is not acceptable; therefore, no information is available concerning its impact on national emissions.

Source: IT Enviroscience (1980).

tion from a surface impoundment. The surface area of the surface impoundment must be known to determine this quantity. If the size of the specific impoundment is not known, a generic estimate that 95% of the industrial surface impoundments cover 2.5 acres and the remaining 5% cover 20 acres (Adkins et al., 1985) can be used. These generic data can be applied to one surface impoundment, or the distribution can be used to estimate the releases from a population of surface impoundments.

As is the case with all stages of a chemical's life cycle, more specific data provide more accurate results but also require more resources.

VI. SUMMARY

The factors involved in estimating emissions from the synthetic organic chemical manufacturing industry (SOCMI) are useful to industry representatives, special interest groups, state agencies, and the federal government. This chapter has identified the types of locations of SOCMI releases and other factors that must be considered to characterize emissions from processes used in this industry. Emissions may be characterized by analyzing the life cycle of the chemical from cradle to grave (production and processing; transportation and storage; and waste treatment, storage, and disposal). Generic emissions data are available for each step of a chemical's life cycle; however, it was recommended that these be used only in the absence of more site-specific information. It is hoped that this broad overview of this complicated subject matter will provide the reader with a better foundation and approach to future endeavors in this area.

REFERENCES

Adkins, et al. 1985. *Methods for Assessing Exposure to Chemical Substances*, vol. 3, *Methods for Assessing Exposure from Disposal of Chemical Substances*. EPA 560/5-85-003. Washington, D.C.: U.S. Environmental Protection Agency, Office of Toxic Substances.

Agranoff, T. (ed.). 1986. *Modern Plastics Encyclopedia*. New York: McGraw-Hill.

Aldrich Chemical Company, Inc. 1987. *The Aldrich Catalog Handbook of Organic and Biochemicals*. Milwaukee: Aldrich Chemical Company, Inc.

American Institute of Mining, Metallurgical and Petroleum Engineers, Inc. 1983. *Industrial Minerals and Rocks*. New York: author.

Blacksmith, J. R., Harris, G. E., and Langley, G. L. 1980. *Frequency of Leak Occurrence for Fittings in Synthetic Organic Chemical Plant Process Units*. EPA 600/2-81-003. Washington, D.C.: Office of Research and Development, U.S. Environmental Protection Agency.

Bureau of Mines. 1986a. *Mineral Commodity Profiles*. Washington, D.C.: Bureau of Mines, U.S. Department of the Interior.

Bureau of Mines. 1986b. *Minerals Yearbook*. Washington, D.C.: U.S. Government Printing Office.

Bureau of Mines. 1986c. *Mineral Inventory Surveys*. Washington, D.C.: U.S. Government Printing Office.

Chemical Marketing Reporter. 1987. *1988 CPD Chemical Buyer's Directory*. New York: Schnell Publishing.

Chem Sources. 1987. *Chem Sources—USA*. Ormond Beach, Fla.: Directories Publishing Co.

Chemical Week. 1986. *Buyer's Guide Issue*. New York: McGraw-Hill.

Clarke, R. K., Foley, J. T., Hartman, W. F., and Larson, D. W. 1976. *Severities of Transportation Accidents* [microfiche]. NTIS SLA 74 0001. Albuquerque, N.M.: Sandia National Laboratories, for Energy Research Development Administration and Department of Transportation.

Considine, D. M. 1974. *Chemical and Process Technology Encyclopedia*. New York: McGraw-Hill.

Dellinger, B., Graham, M., and Tirey, D. 1986. Predicting emissions from the thermal processing of hazardous wastes. *Hazard. Waste Hazard. Mater.* 3(3):293–307.

DMI (Dun's Market Identifiers). 1986. Computer printout: Listing of Manufacturers and Users of Organic Chemicals by SIC code. Retrieved December 16.

Duchaine, R. U. (ed.). 1987. *Thomas Register of American Manufacturers*. New York: Thomas Co.

Erickson. 1980. *Storage and Handling*. Research Triangle Park, N.C.: U.S. Environmental Protection Agency, Office of Air Quality Planning and Standards.

Gates, N., and Smith, S. 1985. *National Air Toxics Information Clearinghouse (NATICH) Data Base Users Guide for Data Viewing*. Research Triangle Park, N.C.: U.S. Environmental Protection Agency, Office of Air Quality Planning and Standards.

Herrick, E. C., King, J. A., Ouellette, R. P., and Chermisinoff, P. N. 1979. *Unit Processes Guide to Organic Chemical Industries*. Ann Arbor, Mich.: Ann Arbor Science.

Hoffman, S. 1983. Transportation. In *Encyclopedia of Chemical Technology*, vol. 23, ed. R. E. Kirk and D. F. Othmer, pp. 375–398. New York: Wiley-Interscience.

ICF Inc. 1984. *Assessing the Releases and Costs Associated with Truck Transport of Hazardous Wastes*. PB84-224468. Washington, D.C.: U.S. Environmental Protection Agency.

IT Enviroscience. 1980. *Organic Chemicals Manufacturing*, vols. 1–10. EPA-450/3-80-023. Research Triangle Park, N.C.: U.S. Environmental Protection Agency, Office of Air Quality Planning and Standards.

Kent, J. A. (ed.). 1983. *Reigel's Handbook of Industrial Chemistry*, 7th ed. New York: van Nostrand-Reinhold.

Kirk, R. E., and Othmer, D. F. (eds.). 1978–1984. *Encyclopedia of Chemical Technology*, 3rd ed. New York: Wiley-Interscience.

Kline, C. H. 1979–1982. *Kline Industrial Marketing Guides*. Fairfield, N. J.: Charles H. Kline.

Lowenheim, F. A., and Moran, M. K. 1975. *Faith, Keyes and Clark Industrial Chemicals*, 4th ed. New York: Wiley-Interscience.

Mannsville Chemical Products. 1976. *Chemical Products Synopsis*. Mannsville, N.Y.

Mark, H. F., et al. (eds.). 1985–1986. *Encyclopedia of Polymer Science and Technology—Plastics, Resins, Rubbers, Fibers*. New York: Wiley-Interscience.

Menzie, C. A. 1979. An approach to estimating probabilities of transportation-related spills of hazardous materials. *Environ. Sci. Technol.* 13(2).

Office of Technology Assessment. 1986. *Transportation of Hazardous Materials*. OTA-SET-304. Washington, D.C.: Office of Technology Assessment.

Radian Corp. 1979. Organic chemical producers data base.

Shreve N. 1977. *Chemical Process Industries*, 4th ed. New York: McGraw-Hill.

Sittig M. 1986. *Pesticides Manufacturing and Toxic Materials Control Encyclopedia*. Oarkridge, N.J.: Noyes Data.

Stanford Research Institute. 1987. *Chemical Economics Handbook*. (updated yearly) Menlo Park, Calif. SRI International.

Stanford Research Institute. 1986. *Directory of Chemical Producers.* Annual publication. Menlo Park, Calif.: SRI International.

Thomas. 1987. *Thomas Register of American Manufacturers.* New York: Thomas Co.

U.S. DOT (U.S. Department of Transportation). 1986. *Accident/Incident Bulletin No. 154.* Washington, D.C.: Federal Railroad Administration Office of Safety.

U.S. EPA. 1979. *Measurements of Benzene Emissions for a Floating Roof Tank Test.* EPA-450/3-79-020. Research Triangle Park, N.C.: U.S. Environmental Protection Agency.

U.S. EPA. 1986. Compilation of Air Pollution Emission Factors. Report No. AP-42; Suppl. 12. Research Triangle Park, N.C.: U.S. Environmental Protection Agency.

U.S. EPA. 1982. *Development of the Environmental Release Data Base for Synthetic Organic Chemicals Manufacturing Industry.* Washington, D.C.: U.S. Environmental Protection Agency, Office of Research and Development. Draft Report. Springfield, Va.: Versar, Inc. (EPA Contract No. 68-03-3061).

U.S. EPA. 1983. *VOC Emissions for Volatile Organic Liquid Storage Tanks—Background Information and Proposed Standards.* Preliminary Draft. EPA 450/3-81-003s. Research Triangle Park, N.C.: U.S. Environmental Protection Agency, Office of Air Quality Planning and Standards.

U.S. EPA. 1984. *Locating and Estimating Air Emissions from Sources of Ethylene Dichloride.* EPA-450/4-84-007D. Research Triangle Park, N.C.: U.S. Environmental Protection Agency, Office of Air Quality Planning and Standards.

U.S. EPA. 1985a. *Information Systems Inventory,* NTIS PB87-131017. Washington, D.C.: U.S. Environmental Protection Agency.

U.S. EPA. 1985b. *Compilation of Air Pollutant Emission Factors.* Research Triangle Park, N.C.: U.S. Environmental Protection Agency, Office of Air Quality Planning and Standards.

U.S. EPA. 1986a. *Emission Factors for Equipment Leaks of Volatile Organic Compounds and Hazardous Air Pollutants.* EPA-450/3-86-002. Research Triangle Park, N.C.: U.S. Environmental Protection Agency, Office of Air Quality Planning and Standards.

U.S. EPA. 1986b. *RCRA Orientation Manual.* EPA/530-SW-86-001. Washington, D.C.: U.S. Environmental Protection Agency, Office of Solid Waste.

U.S. ITC. 1985. *Synthetic Organic Chemicals—United States Production and Sales 1984.* U.S. International Trade Commission publ. 1745. Washington, D.C.: U.S. Government Printing Office.

Chemical Substance Index

387

Subject Index

Acceptable daily intake (ADI), quantitative estimate of, 2
Acetylcholinesterase (AChE), toxic effect by inhibition of, 7
N-acetyltransferase:
  phenotype, 5
  variations in activity of, 3
Acid neutralizing capacity (ANC), 188–214
Acidic deposition, loading scenarios of, 187
Additive index, 152
Adenylate cyclase activity, 17
Adsorption, role in contaminant fate and transport, 226–228, 232, 234–262
Advective exchange, with groundwater systems, 228, 273
Advective exchange/transport, 228, 230, 232, 249
Alanine aminotransferase (ALT) activity, 133, 171
American Society for Testing and Materials (ASTM), 135–136
Ames *Salmonella* assay, 88–89, 106–115
Antagonism, 43–44
Army Corps of Engineers, U.S.:
  commodity flow data base report by, 377
  use of Sheng's model for contaminant fate, 271
Aryl hydrocarbon hydroxylase (AHH), 5–6
Association of American Railroads, railroad accident files, 377, 379
Asthmatics, immunologically sensitive individuals, 14
ATP levels, measurement by chemiluminescence, 60
Autoimmune disease, 53, 75

Bacterial lipopolysaccharide (LPS), 57
Benthic organisms, bioaccumulation in, 225

Bioaccumulation, 225
Biogeochemical processes, within watershed-lake systems, 188
Biological degradation, role in contaminant fate and transport, 226–228, 270
Bioturbation (biological mixing), 228, 237–238, 273–274
Birkenes model, simulation of water quality in streams by, 207–209, 215
Bureau of Census, U.S., 376
Bureau of Motor Carrier Safety, 376, 379

California's Proposition 65, 85
Carcinogens:
  activation of, 5
  altered susceptibility with age, 13
  DNA binding of, 8
  identification methods, 87
  pulmonary exposure to, 35
Cation exchange capacity (CEC), 189
Chemical emissions:
  computerized data sources, 352–356
  estimation of, 356–357
Chemical manufacture, references for, 348
Chemical precipitation, 226, 266–269
Cholinesterase (CHE), serum, 7, 9–10
Cirrhosis, pathological condition, 15
Clean Air Act, 347
Clean Water Act, 347
Coast Guard, U.S., Office of Marine Safety, 378–379
COBAS Bact apparatus, automate Ames test by, 114
Complexation, role in contaminant fate and transport, 226–228, 230, 232, 245, 265–267
Creatinine clearance test, 15

QH
545
.A1
H38

QH
545
.A1
H38

35.88